2nd International Conference on Innovative Engineering Materials (ICIEM 2017)

IOP Conference Series: Materials Science and Engineering
Volume 284

Philadelphia, Pennsylvania, USA
21 – 23 October 2017

ISBN: 978-1-5108-5746-9
ISSN: 1757-8981

Printed from e-media with permission by:

Curran Associates, Inc.
57 Morehouse Lane
Red Hook, NY 12571

Some format issues inherent in the e-media version may also appear in this print version.

Printed by Curran Associates, Inc. (2018)

Institute of Physics
Dirac House, Temple Back
Bristol BS1 6BE UK

Phone: 44 1 17 929 7481
Fax: 44 1 17 920 0979

techtracking@iop.org

Additional copies of this publication are available from:

Curran Associates, Inc.
57 Morehouse Lane
Red Hook, NY 12571 USA
Phone: 845-758-0400
Fax: 845-758-2633
Email: curran@proceedings.com
Web: www.proceedings.com

Table of contents

Volume 284

2017 2nd International Conference on Innovative Engineering Materials (ICIEM 2017)

21–23 October 2017, Philadelphia, USA

Accepted papers received: 14 December 2017
Published online: 30 January 2018

Preface

Papers

Design and modeling of functional materials

Material processing and preparation

Materials Physics

IOP Conf. Series: Materials Science and Engineering **284** (2017) 011001 doi:10.1088/1757-899X/284/1/011001

PREFACE

It was a great pleasure to welcome all the participants from many countries to the 2017 2nd International Conference on Innovative Engineering Materials (ICIEM 2017) in Philadelphia, USA, 21-23 October, 2017. ICIEM 2017 is dedicated to the dissemination of the state of the art knowledge and technical issues related to Innovative Engineering Materials.

The major goal of the conference was to bring academic scientists, engineers, and industry researchers together to exchange and share their research and discuss the practical challenges encountered in their research and explore opportunities for collaboration. Professors from various prestigious universities in USA and abroad were invited to deliver keynote and plenary talks describing the latest information in their respective areas of expertise. The conference offered a golden opportunity for the students, researchers and engineers to interact with the experts to get their advice and consult on technical matters, and collaboration opportunities and strategies.

The conference proceeding present a selection of peer reviewed papers submitted to the conference from universities, research institutes and industry. All the papers were subjected to rigorous peer-review by conference committee members and international reviewers. The papers were selected based on their quality and relevancy to the conference. The proceedings present to the readers the recent advances in the field of Materials Science and Engineering and their Applications, and in Materials Manufacturing and Processing etc.

On behalf of the conference committee and organizers, I would like to thank all the authors who contributed to this volume as well as to the reviewers, speakers, chairpersons, sponsors and all the conference participants for their support to ICIEM 2017.

Prof. Ramesh K. Agarwal
Washington University, USA
November 14, 2017

IOP Conf. Series: Materials Science and Engineering **284** (2017) 011002 doi:10.1088/1757-899X/284/1/011002

Organizing Committee

Conference Chairs

Prof. Karen Winey, University of Pennsylvania, USA

Prof. Ramesh K. Agarwal, Washington University, USA

Technical Program Committee Chairs

Assoc. Prof. Jing Wang, University of South Florida, USA

Assoc. Prof. Ouyang Jianyong, National University of Singapore, Singapore/ Stanford University, USA

Organizing Chair

Ms. Yang Dan, Hong Kong Chemical, Biological & Environmental Engineering Society, Hong Kong

Publication Chair

Ms. Tammy Luo, Chemistry and Materials Society, Hong Kong Chemical, Biological & Environmental Engineering Society, Hong Kong (CMS-HKCBEES)

Technical Program Committee

Prof. Pankaj Vadgama, Queen Mary University of London, UK

Prof. Roberto Capata, University of Roma Sapienza, Italy

Prof. Filippo Berto, Norwegian University of Science and Technology, Norway

Prof. Carlo Montemagno, University of Alberta, Canada

Prof. Dongfang Yang, National Research Council Canada, Canada

Prof. Shuming Chen, Jilin University, China

Prof. Luqman Chuah Abdullah, Universiti Putra Malaysia, Malaysia

Assoc. Prof. Vincent Teng, Swansea University, UK

Assoc. Prof. Kazem Abhary, University of South Australia, Australia

Assoc. Prof. Waail Mahmod Lafta, Griffith University, Australia

Assoc. Prof. Sen Qian, Chinese Academy of Sciences, China

Assoc. Prof. Krzysztof Górski, Kazimierz Pulaski University of Technology and Humanities in Radom, Poland

Assoc. Prof. A. Nirmala Grace, VIT University, India

Dr. Xianfeng Fan, University of Edinburgh, UK

Dr. Abbas Amini, Western Sydney University, Australia

Dr. Meiyong Liao, National Institute for Materials Science, Japan

Dr. Vimal S. Joshi, Gujarat University, India

Dr. Manojkumar N. Parmar, Hemchadrachyara North Gujarat University, India

Peer review statement

All papers published in this volume of *IOP Conference Series: Materials Science and Engineering* have been peer reviewed through processes administered by the proceedings Editors. Reviews were conducted by expert referees to the professional and scientific standards expected of a proceedings journal published by IOP Publishing.

IOP Conf. Series: Materials Science and Engineering **284** (2017) 012001 doi:10.1088/1757-899X/284/1/012001

A concentrated electrolyte for zinc hexacyanoferrate electrodes in aqueous rechargeable zinc-ion batteries

D Kim, C Lee and S Jeong*

Department of Chemical Engineering, Soonchunhyang University, Asan, Chungnam 336-745, Republic of Korea
*E-mail: hamin611@sch.ac.kr

Abstract. In this study, a concentrated electrolyte was applied in an aqueous rechargeable zinc-ion battery system with a zinc hexacyanoferrate (ZnHCF) electrode to improve the electrochemical performance by changing the hydration number of the zinc ions. To optimize the active material, ZnHCF was synthesized using aqueous solutions of zinc nitrate with three different concentrations. The synthesized materials exhibited some differences in structure, crystallinity, and particle size, as observed by X-ray diffraction and scanning electron microscopy. Subsequently, these well-structured materials were applied in electrochemical tests. A more than two-fold improvement in the charge/discharge capacities was observed when the concentrated electrolyte was used instead of the dilute electrolyte. Additionally, the cycling performance observed in the concentrated electrolyte was superior to that in the dilute electrolyte. This improvement in the electrochemical performance may result from a decrease in the hydration number of the zinc ions in the concentrated electrolyte.

1. Introduction

In the area of energy storage research, aqueous rechargeable metal-ion batteries have attracted interest as possible substitutes for commercial aqueous batteries, such as lead–acid, Ni–Cd, and Ni–MH, because of their safety, low cost, high ionic conductivity, and environmental friendliness [1–8]. Among the various available systems, aqueous rechargeable zinc-ion batteries (ARZIBs) based on electrochemical insertion/extraction reactions of zinc ions have been widely studied and developed because of their advantages, including nontoxicity and safety [9–12]. These characteristics have been demonstrated by the wide application of ARZIBs to devices used in the body, such as hearing aids. In addition, ARZIBs are expected to have relatively high capacities because of the high theoretical value for zinc metal anodes (820 mAh g^{-1}), as well as low costs owing to their relatively simple manufacturing processes and the abundance of zinc resources [13–15].

In recent years, various materials, such as manganese dioxide and Prussian blue analogues (PBA), have been reported as cathode materials for ARZIBs [16, 17]. In particular, PBAs are notable potential cathode materials for application in ARZIBs owing to their superior characteristics, such as structural stability, large channels, and ease of synthesis [9, 18]. Additionally, these materials have large open sites (4.6 Å in diameter) and <100> channels (3.2 Å in diameter), which enable the rapid diffusion of various hydrated metal ions, including zinc ions [19–21]. However, the electrochemical performance of these materials has to be improved, as they exhibit relatively low capacities and capacity reduction during the initial cycles.

In this work, we applied a concentrated electrolyte in ARZIBs with PBA electrodes to improve the electrochemical performance by changing the hydration number of zinc ions. Our previous study demonstrated that the electrochemical performance of aqueous rechargeable calcium-ion batteries could be improved by applying a superconcentrated electrolyte, and the improved performance was attributed to a decrease of the hydration number and radius of calcium ions [22]. Therefore, it was expected that the hydration number and radius of zinc ions could be changed by applying a concentrated electrolyte in ARZIB systems, which, in turn, would positively influence the electrochemical performance.

2. Experimental

Zinc hexacyanoferrate ($Zn_3[Fe(CN)_6]_2$, ZnHCF) powder, which is a PBA, was used as the active material. The synthesis process follows the method reported by Wang et al. [15]. However, as we wanted to further explore the ZnHCF synthesis method to optimize the cathode material for ARZIBs, in this study, three materials were synthesized by using aqueous solutions with three different concentrations of $Zn(NO_3)_2$ (Alfa Aesar, 99%). In brief, the materials were synthesized by simultaneously adding 120 mL of 0.05, 0.1, or 0.3 mol dm^{-3} $Zn(NO_3)_2$ and 120 mL of 0.05 mol dm^{-3} $K_3Fe(CN)_6$ (Sigma Aldrich, 99%) in a dropwise manner to 60 mL of deionized water under constant stirring at 80 °C. After sonicating for 30 min at 80 °C, the suspensions were aged for 6 h at 80 °C. Then, the precipitates were filtered and dried under vacuum at 30 °C for 24 h.

The structures and morphologies of the ZnHCF powders were analyzed by X-ray diffraction (XRD) and scanning electron microscopy (SEM), respectively. The XRD patterns were collected using two different devices. The ZnHCF powders synthesized using 0.1 and 0.3 mol dm^{-3} $Zn(NO_3)_2$ were analyzed using a D8 ADVANCE diffractometer (Bruker AXS, Germany), whereas the powder synthesized using 0.05 mol dm^{-3} $Zn(NO_3)_2$ was analyzed using a D/Max-2200PC diffractometer (Rigaku, Japan) equipped with a Cu Kα source (40 kV, 40 mA). The SEM images were collected using a SNE-3000M microscope (SEC, South Korea).

Charge/discharge and cycling performance tests were performed using a three-electrode cell with a ZnHCF electrode as the working electrode. A saturated calomel electrode (SCE) and activated charcoal were used as the reference and counter electrodes, respectively. The ZnHCF electrodes were prepared by casting a slurry including the ZnHCF active material powder (80 wt%), super P (9 wt%), graphite powder (2 wt%), and polyvinylidene fluoride (PVDF) as a binder (9 wt%, Sigma Aldrich) onto a carbon cloth (Fuel Cell Earth, USA) as a current collector. The coated electrode was dried under vacuum at 80 °C for 12 h. Prior to the electrochemical tests, the ZnHCF electrode was discharged from an open circuit voltage to 1 V in deionized water over 12 h to extract potassium ions. Then, the electrode was rinsed with deionized water and dried at 60 °C for 12 h. The dilute and concentrated electrolytes consisted of 1 and 3 mol dm^{-3} $Zn(NO_3)_2$ dissolved in water, respectively. The electrochemical tests were performed on a WBCS3000 battery test system (WonATech, South Korea).

3. Results and discussion

First, to analyze the structure of the three synthesized materials, XRD patterns were collected. As shown in Fig. 1a and 1b, the peaks corresponding to the materials synthesized using 0.05 and 0.1 mol dm^{-3} $Zn(NO_3)_2$, respectively, are well-indexed to $Zn_3[Fe(CN)_6]_2$ (JCPDS 38-0688). However, as shown in Fig. 1c, different peaks are observed for the material synthesized using 0.3 mol dm^{-3} $Zn(NO_3)_2$, which means that this material does not have a structure that is consistent with $Zn_3[Fe(CN)_6]_2$. Therefore, only the materials synthesized using 0.05 and 0.1 mol dm^{-3} $Zn(NO_3)_2$ were used in the subsequent analyses and tests. Notably, the peaks in Fig. 1a are broader than those in Fig. 1b, which indicates that the material synthesized using 0.1 mol dm^{-3} $Zn(NO_3)_2$ has superior crystallinity to that synthesized using 0.05 mol dm^{-3} $Zn(NO_3)_2$.

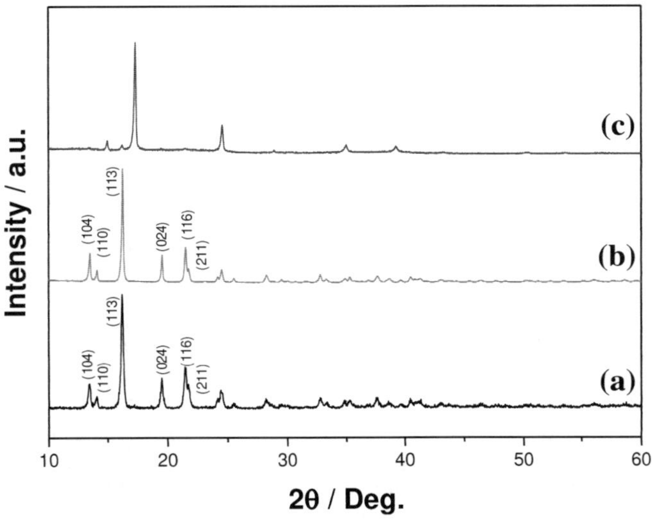

Figure 1. XRD patterns of the materials synthesized using (a) 0.05, (b) 0.1, and (c) 0.3 mol dm^{-3} $Zn(NO_3)_2$ dissolved in water.

Figure 2 shows SEM images of the two materials synthesized using 0.05 and 0.1 mol dm^{-3} $Zn(NO_3)_2$ dissolved in water. A comparison of Fig. 2a and 2b shows evenly distributed particles with relatively uniform sizes in the material synthesized using 0.1 mol dm^{-3} $Zn(NO_3)_2$, whereas particles with nonuniform sizes were observed in the material synthesized using 0.05 mol dm^{-3} $Zn(NO_3)_2$. Thus, the materials synthesized using 0.05 and 0.1 mol dm^{-3} $Zn(NO_3)_2$ dissolved in water showed differences in crystallinity and the degree of particle size uniformity.

To confirm differences in electrochemical performance between dilute and concentrated electrolytes, charge/discharge tests were performed using two different electrodes. Figures 3 and 4 show the charge/discharge curves of the ZnHCF electrodes synthesized using 0.05 and 0.1 mol dm^{-3} $Zn(NO_3)_2$ dissolved in water, respectively, when the dilute and concentrated electrolytes were used. First, a comparison of Figs. 3 and 4 show that better electrochemical performance was exhibited by the ZnHCF electrode obtained from 0.1 mol dm^{-3} $Zn(NO_3)_2$ than by that obtained from 0.05 mol dm^{-3} $Zn(NO_3)_2$. In our previous study, it was demonstrated that a PBA electrode with a collapsed crystalline structure realizes relatively poor electrochemical performances in an aqueous electrolyte [22]. As indicated by the XRD results, the crystallinity of ZnHCF synthesized using 0.1 mol dm^{-3} $Zn(NO_3)_2$ is to that of ZnHCF synthesized using 0.05 mol dm^{-3} $Zn(NO_3)_2$, which likely accounts for the differences in the charge/discharge properties of the two ZnHCF electrodes.

Figure 2. SEM images of the active materials synthesized using (a) 0.05 and (b) 0.1 mol dm^{-3} $Zn(NO_3)_2$ dissolved in water.

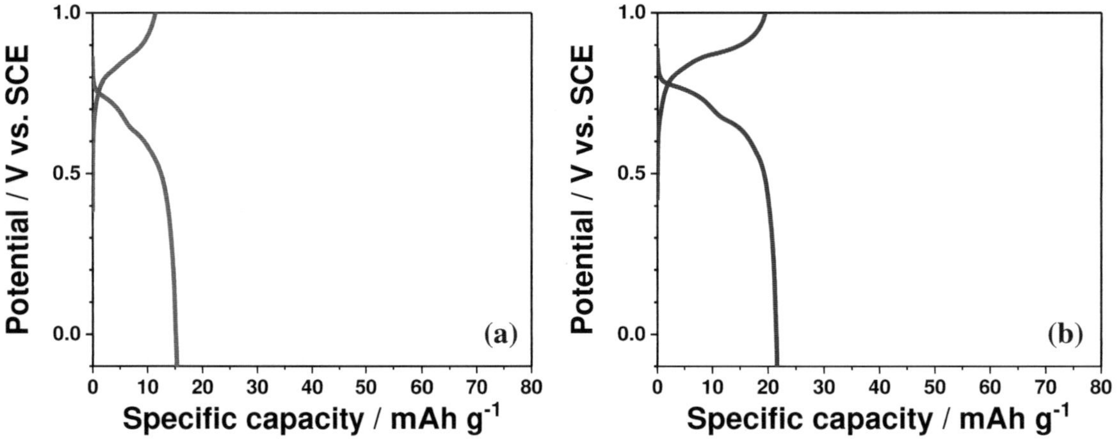

Figure 3. Charge/discharge curves at 0.1 C of the ZnHCF electrode synthesized using 0.05 mol dm^{-3} $Zn(NO_3)_2$ in (a) 1 and (b) 3 mol dm^{-3} $Zn(NO_3)_2$.

On the other hand, in the charge/discharge curves, the electrochemical performance is notable improved in the concentrated electrolyte. As shown in Figs. 3 and 4, improved charge/discharge capacities are achieved in the concentrated electrolyte relative to those in the dilute electrolyte. In particular, as shown in Fig. 4 for the electrode obtained from 0.1 mol dm^{-3} $Zn(NO_3)_2$, the reversible charge/discharge capacity in the concentrated electrolyte is more than twice that in the dilute electrolyte. This increased capacity may result from the zinc ions having a decreased hydration number in the concentrated electrolyte. In the 1 and 3 mol dm^{-3} $Zn(NO_3)_2$ electrolytes, the molar ratios of zinc ions to water molecules is 1:50 and 1:13, respectively. It has been reported that a zinc ion can have a hydration number of up to six in the first shell. If the secondary shell is included, the hydration number of a zinc ion increases to 18 [23, 24]. Additionally, the hydration number of a nitrate anion is known to be 3.7 [25]. Therefore, zinc ions in the concentrated electrolyte may have a smaller hydration number because there are insufficient water molecules in the concentrated electrolyte, whereas zinc ions in the dilute electrolyte can have the maximum hydration number. Zinc ions with a smaller hydration number can be inserted into and extracted from the ZnHCF electrode to a greater degree, which leads to increased charge/discharge capacities in the concentrated electrolyte.

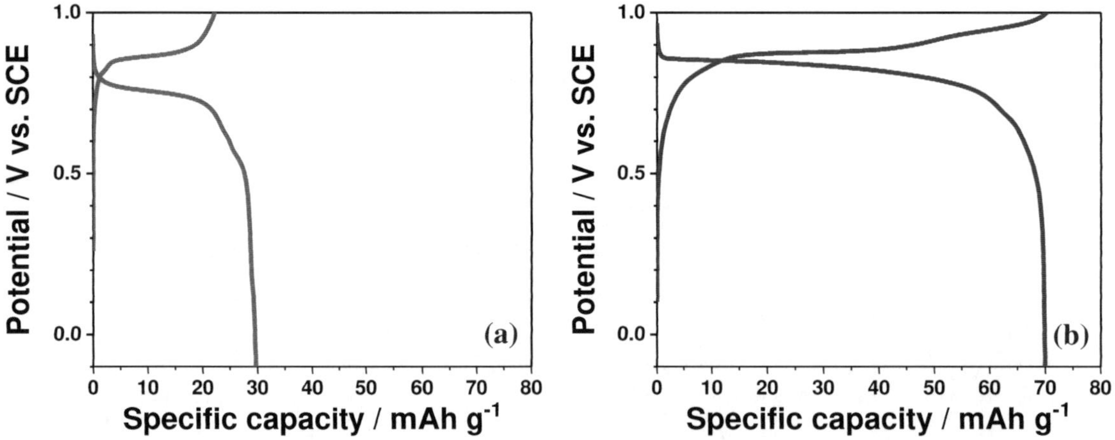

Figure 4. Charge/discharge curves at 0.1 C of the ZnHCF electrode synthesized using 0.1 mol dm^{-3} $Zn(NO_3)_2$ in (a) 1 and (b) 3 mol dm^{-3} $Zn(NO_3)_2$.

Improved electrochemical performance was observed not only in terms of charge/discharge capacities, but also in cycling performance. Cycling tests were carried out for the ZnHCF electrode synthesized using 0.1 mol dm^{-3} $Zn(NO_3)_2$ dissolved in water owing to the better charge/discharge capacities of this electrode. Figure 5 shows the cycling performance over 150 cycles at 5 C in both dilute and concentrated electrolytes. The capacity gradually increased during the initial cycles in both electrolytes. The reason for this phenomenon is unclear at present and a detailed interpretation is beyond the scope of this work, but it may be related to activation or conditioning effects [20, 21, 26].

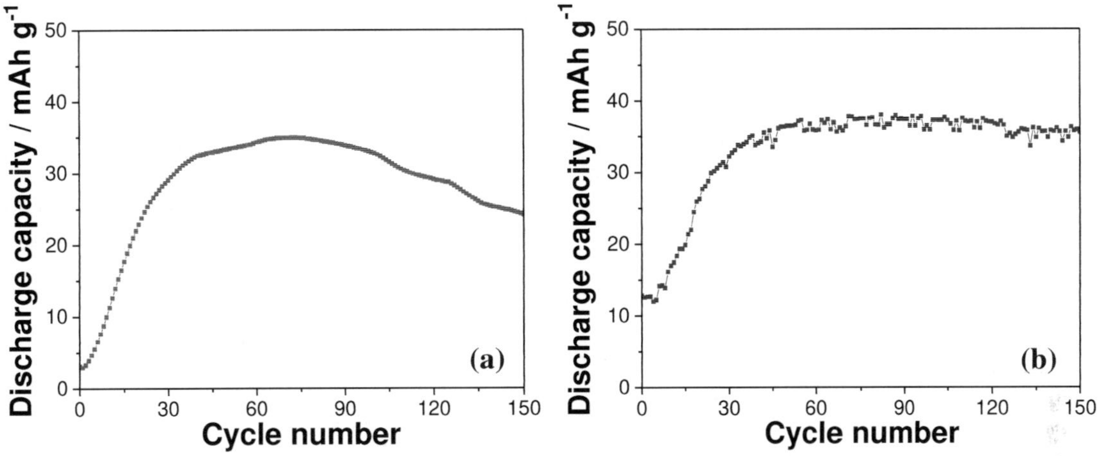

Figure 5. Cycling performance curves for ZnHCF electrodes synthesized using 0.1 mol dm^{-3} $Zn(NO_3)_2$ in (a) 1 and (b) 3 mol dm^{-3} $Zn(NO_3)_2$ over 150 cycles at 5 C.

As shown in Figs. 5a and 5b, improved cycling performance is observed in the concentrated electrolyte. In our previous study, it was demonstrated that the crystalline lattice structure of an electrode could be maintained after cycling if the hydration number of the cations were decreased, leading to improved cycling performance [22]. Therefore, in this study, the reason for the improved cycling performance in the concentrated electrolyte may be the decreased hydration number of the zinc ions. The ZnHCF electrode may experience structural stress owing to repeated insertion/extraction of hydrated zinc ions with a larger hydration number and radius during cycling in the dilute electrolyte, resulting in capacity reduction. However, this phenomenon might be suppressed in the concentrated electrolyte because the zinc ions have a relatively smaller hydration number and radius, resulting in improved cycling performance.

4. Conclusions

We applied a concentrated electrolyte to improve the electrochemical performance in an ARZIB system. In charge/discharge tests, improved charge/discharge capacities were observed in the concentrated electrolyte, which may be attributed to a small hydration number and radius of zinc ions. Additionally, the cycling performance was also improved in the concentrated electrolyte, probably because the structural stress experience by the ZnHCF electrode was suppressed in the concentrated electrolyte during cycling owing to the decrease in the hydration number of the zinc ions. Based on these data, our future work will focus on investigating various electrodes and electrolytes in ARZIB systems.

5. References

[1] Kohler J, Makihara H, Uegaito H, Inoue H and Toki M 2000 *Electrochim. Acta* **46** 59
[2] Wang G, Fu L, Zhao N, Yang L, Wu Y and Wu H 2007 *Angew. Chem.* **119** 299
[3] Luo J Y, Cui W J, He P and Xia Y Y 2010 *Nature Chemistry* **2** 760
[4] Whitacre J F, Tevar A and Sharma S 2010 *Electrochem. Comm.* **12** 463
[5] Wu X, Cao Y, Ai X, Qian J and Yang H 2013 *Electrochem. Comm.* **31** 145

[6] Soloveichik G L 2011 *Annu. Rev. Chem. Biomol. Eng.* **2** 503
[7] Wang H, Huang K, Zeng Y, Yang S and Chen L 2007 *Electrochim. Acta* **52** 3280
[8] Wu X Y, Sun M Y, Shen Y F, Qian J F, Cao Y L, Ai X P and Yang H X 2014 *ChemSusChem* **7** 407
[9] Zhang B, Liu Y, Wu X, Yang Y, Chang Z, Wen Z and Wu Y 2014 *Chem. Commun.* **50** 1209
[10] Liu Z, Pulletikurthi G and Endres F 2016 *ACS APPI. Mater. Interfaces* **8** 12158
[11] Yan J, Wang J, Liu H, Bakenov Z, Gosselink D and Chen P 2012 *J. Power Sources* **216** 222
[12] Zhang H, Wu X, Yang T, Liang S and Yang X 2013 *Chem. Commun.* **49** 9977
[13] Zhang L, Chen L, Zhou X and Liu Z 2015 *Adv. Energy Mater.* **5** 1400930
[14] Trócoli R and La Mantia F 2015 *View issue TOC* **8** 481
[15] Wang R Y, Wessells C D, Huggins R A and Cui Y 2013 *Nano Lett.* **13** 5748
[16] Xu C, Li B, Du H and Kang F 2012 *Angew. Chem. Int. Ed.* **51** 933
[17] Lee B, Yoon C S, Lee H R, Chung K Y, Cho B W and Oh S H 2014 *Sci. Rep.* **4** 6066
[18] Karyakin A A 2001 *Electroanalysis* **13** 813
[19] Pasta M, Wang R Y, Ruffo R, Qiao R, Lee H W, Shyam B, Guo M, Wang Y, Wray L A, Yang W, Toneye M F and Cui Y 2016 *J. Mater. Chem. A* **4** 4211
[20] Sun X, Duffort V and Nazar L F 2016 *Adv. Sci.* **3** 1600044
[21] Wang R Y, Shyam B, Stone K H, Weker J N, Pasta M, Lee H W, Toney M F and Cui Y 2015 *Adv. Energy Mater.* **5** 1401869
[22] Lee C and Jeong S 2016 *Chem. Lett.* **45** 1447
[23] Pavlov M, Siegbahn E M and Sandstrolm M. 1998 *J. Phys. Chem. A* **102** 219
[24] Hartmann M, Clark T and Eldik R V 1997 *J. Am. Chem. Soc.* **119** 7843
[25] Kameda Y, Saitoh H and Uemura O 1993 *Bull. Chem. Soc. Jpn.* **66** 1919
[26] Hou Z, Zhang X, Li X, Zhu Y, Liang J and Qian Y 2017 *J. Mater. Chem.* **5** 730

Acknowledgments
This work (Grants No.C0502212) was supported by Business for Cooperative R&D between Industry, Academy, and Research Institute funded Korea Small and Medium Business Administration in 2017. This research was supported by the Basic Science Research Program through the National Research Foundation of Korea (NRF) funded by the Ministry of Education (No. NRF-2017R1A2B4010544).

Enhanced visible-light-driven photocatalytic activities for assembled hierarchical Bi_2WO_6 microspheres by fluorine substitution

Yu Hongguang[1, 2] **and Chu Chenglin**[1, 2]

1 School of Materials Science and Engineering, Southeast University, Nanjing, China
2Jiangsu Key Laboratory for Advanced Metallic Materials, Southeast University, Nanjing, China
E-mail: yuhongguang@vip.qq.com; clchu@seu.edu.cn (corresponding author)

Abstract. The different compositions of Bi_2WO_6 by fluorine substitution have been successfully synthesized via a two-step hydrothermal process in the absence of a surfactant. The morphology of all F-doped Bi_2WO_6 samples was similar with pure Bi_2WO_6 samples. On the basis of maintaining good morphology, the introducing fluorine element only replaced the O atoms of WO_6 octahedron in Bi_2WO_6. The effects of F- substitution on the crystal structure and photocatalytic activity of Bi_2WO_6 were investigated. For the Bi_2WO_6 samples with the different Fluorine content, their band gaps have changed obviously. All F-doped Bi_2WO_6 samples exhibited good photocatalytic activity in degradation of RhB under 500W Xe lamp light irradiation. It shows that photocatalytic activities of F-doped Bi_2WO_6 samples are relevant to band gap and F content.

1. Introduction

On account of the rising level of environmental pollution, the researchers paid more and more attentions to environmentally friendly materials of photodegradation catalysts [1, 2]. Since Bi_2WO_6 has many excellent intrinsic physical and chemical properties such catalytic behavior, oxide anion conduction and luminescent properties [3-8], it can be used as an excellent photocatalyst splitting water and mineralization of organic pollutants under visible-light irradiation [1, 2, 9].
Although the semiconducting photocatalysts have been carried out extensive research by the researchers, they were restricted by their structure, morphology and immanence propertys over the past decade. The semiconductors photocatalytic activity depends on photogenerated carriers generation and separation [10]. Therefore, it is important for maximizing the photocatalytic efficiency to maximize the separation efficiency and increase the oxidation power of photogenerated carriers as far as possible. A few researchers have reported that the photocatalytic activity of semiconductor photocatalysts can be improved by surface fluorine modification or fluorine doping [11-16]. The fluorine ions could increase the coordination sphere around the W atom in a WO_6 octahedron and cause the distortion of the WO_6 octahedron in a Bi_2WO_6 crystal, resulting in an increased transfer rate of photogenerated electrons to the photocatalyst surface and the enhanced photoactivity for rhodamine B degradation. Meanwhile, F-doped Bi_2WO_6 presented enhanced photoactivity for RhB degradation, which could be a synergetic effect of the surface fluorination and the doping of the crystal lattice [17]. On the basis of these results, it can be inferred that the introduction of fluorine, irrespective of the approaches of fluorine doping, really enhanced the photocatalytic activity of semiconductor

photocatalysts. On account of the special morphologies, higher surface areas and so on, the microspheres containing nanosheets usually had better photocatalytic activities. Experimental observations and theoretical calculations also indicate that anisotropic metal nanostructures exhibit shape-dependent optical properties [9, 18]. So, it is very important for photocatalysts to control the morphology of Bi_2WO_6 microspheres [19–21]. Meanwhile, Bi_2WO_6 microspheres should be maintained when fluorine was doped in the different samples.

Herein, in these cases above, we exhibit the F-doped Bi_2WO_6 samples of the different RF value (RF =0, 0.125, 0.25, 0.5, 1) synthesized by the two-step hydrothermal process. The visible light (λ>400nm) photocatalytic activities have also been investigated, and we found that the photocatalytic performances of the F-doped Bi_2WO_6 samples are greatly different and are strongly dependent on the concentration of fluorine substitution. Our principal aim is not only to investigate the role of F-substitution on the structure and photocatalytic activity of Bi_2WO_6 crystal but also how to get high crystallinity and good morphology of Bi_2WO_6 samples. Moreover, our work may be further extended to the design F-doped samples and understand on the photocatalytic mechanisms of F substitution.

2. Experimental

2.1. Synthesis of Bi_2WO_6 hierarchical microspheres

All chemicals were analytical grade and used as received without further purification. The one-step hydrothermal process was similar to that in the Ref. [2, 9]. On the basis of the literature, the different concentrations of NaF were added into the precursor. Then it was heated at 200 ºC for 24 h.

A two-step hydrothermal process: First, 3 mmol of $Bi(NO_3)_3 \cdot 5H_2O$ and 1.5 mmol of citric acid were added to 60 mL deionized water. After 20 min of magnetic stirring, $Na_2WO_4 \cdot 2H_2O$ and $NaHCO_3$ were added into the mixed solution apiece 3 mmol. After being further stirred for10 min, the above mixed solution was poured into a stainless steel autoclave with the 100 ml Teflon liner. It was heated at 200 ºC for 24 h. The as- synthesized samples were separated centrifugally and washed three times separately with deionized water and absolute ethanol when the autoclave cooled to room temperature. Subsequently, the samples were dried under vacuum at 80 ºC for 10 h.

2.2. Synthesis of Bi_2WO_6 hierarchical microspheres by fluorine substitution

Following the above procedures, high purity of Bi_2WO_6 hierarchical microspheres were synthesized on a large scale. Then, 0.5mmol as-synthesized samples were dispersed into a mixture of 40mL ethanol and 20mL deionized water. After 20 min of magnetic stirring, different concentrations of NaF were added to the above solution. In the process, the value of R_F was used to describe the molar ratio of NaF to Bi_2WO_6, these were 0, 0.125, 0.25, 0.5 and 1 nominal ratios, respectively. After being further stirred for 20 min, the suspension was transferred into a stainless steel autoclave with a 100 ml Teflon liner, and heated at 180ºC for 20h. After the autoclave had cooled to room temperature, the products were separated centrifugally and washed three times separately with deionized water and absolute ethanol, respectively. Subsequently, the products were dried under vacuum at 60 ºC for 10 h.

2.3. Characterization

The purity and crystallinity of the as-synthesized samples were characterized by X-ray diffractometer (XRD, Bruker D8) using Cu Kα radiation (λ = 1.5418 Å). The morphology and characterization of the products were performed by field emission scanning electron microscopy (FE-SEM, JSM-6700F) and transmission electron microscopy (TEM, HITACH H-600). The optical diffuse reflectance spectra were recorded on a Japan Shimadzu UV-2550 (Japan) using $BaSO_4$ as reference.

2.4. Photocatalytic experiments

The visible light photocatalytic activities activity of Bi_2WO_6 samples containing the different R_F value (R_F = 0, 0.125, 0.25, 0.5, 1) were evaluated by the decomposition of rhodamine B (RhB) under a 500 W Xe arc lamp light (λ>400 nm). The photocatalytic reaction procedures were similar to that in the Ref. [1,2] In every experiment, RhB solutions (30 mL, 10^{-5} mol/L) containing the different F-doped Bi_2WO_6 samples (100mg) were added into test tube. Before illumination, the above solutions were magnetically stirred in the dark for 2 h to ensure an adsorption-desorption equilibrium between the as-synthesized samples and RhB. After that, the test tubes were taken at the given time intervals, and then centrifuged to remove the photocatalyst particles. Then, the filtrates were analyzed by recording variations of the absorption band maximum (553 nm) by a Shimadzu UV-2550 spectrophotometer.

3. Results and discussion

3.1. Effect of F^- substitution on the morphology of Bi_2WO_6 samples

As shown in Figure 1, the FE-SEM images of Bi_2WO_6 was synthesized for different R_F values via one-step hydrothermal process. With the increasing of fluorine content, the construction of Bi_2WO_6 hierarchical microspheres caused more and more destruction. It indicates that one-step hydrothermal process is disadvantaged to photocatalysts because the bad morphology was unsuited to the scattering and reflecting of the light in the 3D construction of Bi_2WO_6 samples. So, two-step hydrothermal process was employing for F-doped Bi_2WO_6 microspheres.

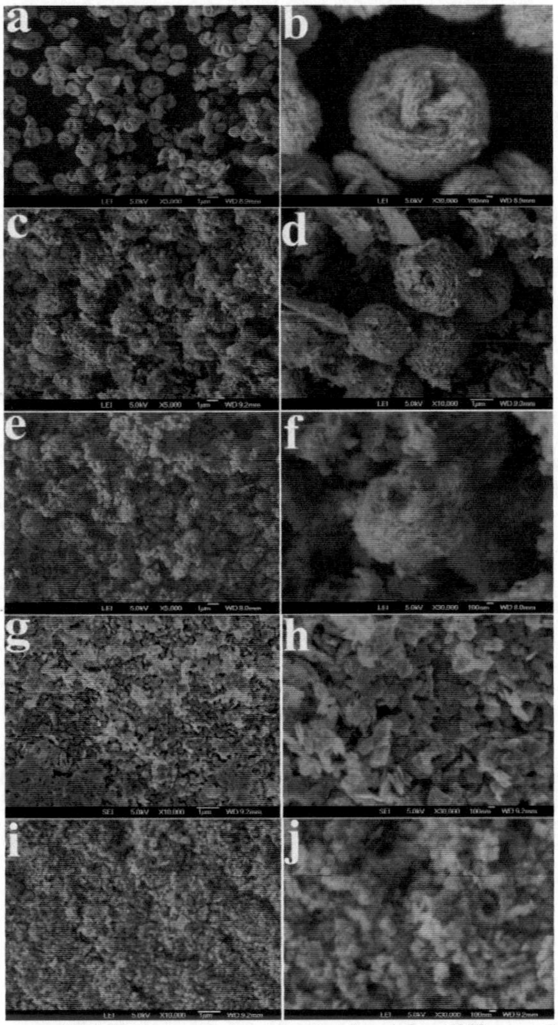

Figure 1. FE-SEM images of Bi_2WO_6 samples synthesized for different R_F values via one-step hydrothermal process. a-b) R_F=0; c-d) R_F =0.125; e-f) R_F =0.25; g-h) R_F =0.5; i-j) R_F =1.

3.2. Effect of F^- substitution on the structure of Bi_2WO_6 hierarchical microspheres

Figure 2a shows XRD patterns of the samples with different R_F values prepared by the two-step hydrothermal process at 200 °C for 20 h. The diffraction peaks of all of the as-prepared samples were in accordance with phase-pure orthorhombic Bi_2WO_6 (JCPDS No. 39-0256). Further observation showed that the peak intensities of the F-doped Bi_2WO_6 samples were almost not change with increasing R_F value. It indicates that the substitution of different R_F values doesn't result in the development of new crystal orientations or changes in preferential orientations. In other words, nothing to do with the presence or absence of F^-, the diffraction peak of the as-synthesized samples with different R_F values were in keeping with the pure Bi_2WO_6. No other peaks can be detected in Figure 2a, indicating the high purity of the samples. As shown in Figure 2b, with an increase of F^- content, a comparison of the (131) diffraction peaks in the range of $2\theta = 26\sim30°$ shows that the peak position of different F^- doped Bi_2WO_6 samples with increasing R_F values shifts slightly toward a higher 2θ value. According to Bragg's law, the increase in 2θ value should result in the decrease in lattice parameters (d(131) value). Therefore, the shift of the diffraction peak with increasing R_F values toward higher angle could be due to the substitution of one O^{2-} by two F^- in order to maintain an electroneutrality condition.

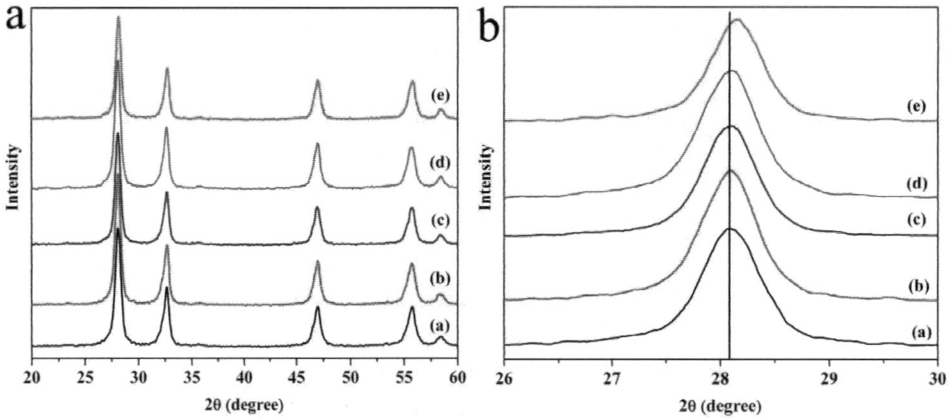

Figure 2. XRD patterns of the Bi_2WO_6 microspheres with different R_F values. (a) $R_F = 0$, (b) $R_F = 0.125$, (c) $R_F = 0.25$, (d) $R_F = 0.5$, and (e) $R_F = 1$. (b) Comparison of 2θ peaks from 26° to 30° for the as-synthesized samples of different RF values.

3.3. Effect of F- substitution on the morphology of Bi_2WO_6 hierarchical microspheres

Figure 3 gives the FE-SEM images of Bi_2WO_6 hierarchical microspheres synthesized for different R_F values. As shown in Figure 3, the samples are composed of many microspheres which are regular in shape and uniform in size with the average diameter length of 2~4μm. With the increase of the F content, it can be found that the shape and superstructure of Bi_2WO_6 microspheres with different R_F values almost has no changes by contrasting from Figure 3a to Figure 3j. Comparing with one-step hydrothermal process (as shown in Figure 1), the morphology of Bi_2WO_6 microspheres may maintain better by two-step hydrothermal process (as shown in Figure 3). It indicates that two-step template-free approach employing a hydrothermal process is beneficial to photocatalysts because high crystallinity and good morphology generally means fewer traps, bigger specific surface area and stronger photocatalytic activity.

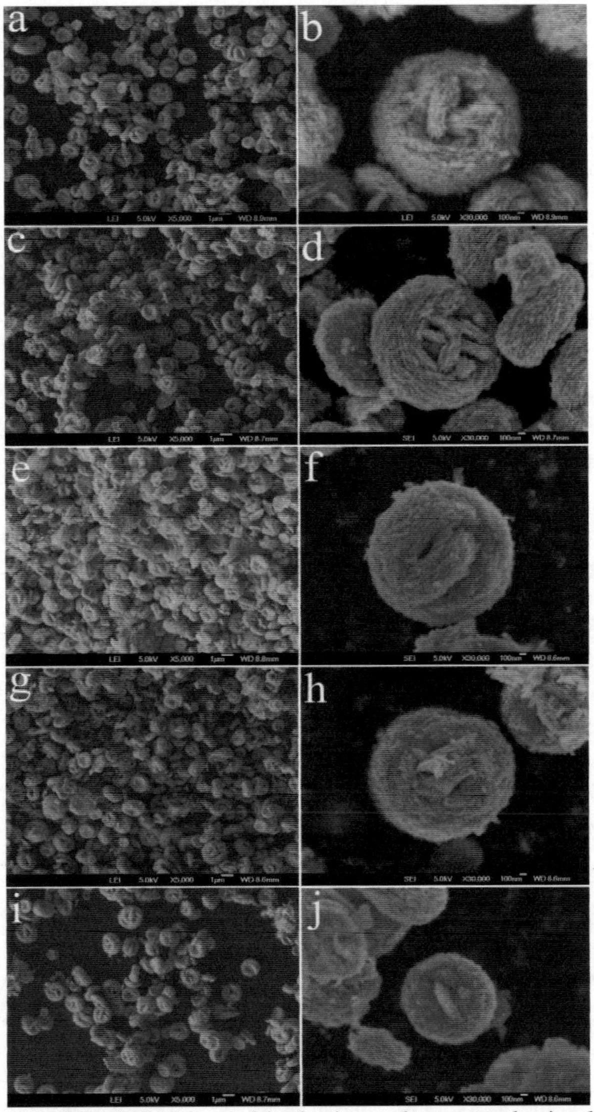

Figure 3. FE-SEM images of Bi_2WO_6 hierarchical microspheres synthesized for different R_F values. a-b) R_F=0; c-d) R_F =0.125; e-f) R_F =0.25; g-h) R_F =0.5; i-j) R_F =1.

3.4. Chemical compositions of Bi_2WO_6 microspheres

The oxidation states and surface chemical composition of the F-doped Bi_2WO_6 were characterized by X-ray photoelectron spectroscopy (XPS). Typical survey XPS spectrum and Bi 4f, W 4f, O 1s XPS spectra of F-doped Bi_2WO_6 could all be detected as shown in Figure 4. The high resolution spectra of W 4f, Bi 4f, O 1s are provided in Figure 4b-d. Compared to that of pure Bi_2WO_6, the binding energies of the W peaks of F-doped Bi_2WO_6 samples have significant changes (as shown in Figure 4c) that should be ascribed to the alteration of the chemical coordination environment of W surroundings, meanwhiles it indicated that F atom succeeded in the substitution of O atom in WO6 octahedron.[22] The binding energies of Bi peaks of F-doped Bi_2WO_6 samples changed almost nothing (as shown in Figure 4b) which indicated that the coordination environment of Bi surroundings was not influenced by F substitution. [23] Therefore, the F atoms only replaced the O atoms of WO_6 octahedron in Bi_2WO_6 and it suggested that the method successfully introduced the fluorine element on the basis of maintaining good morphology.

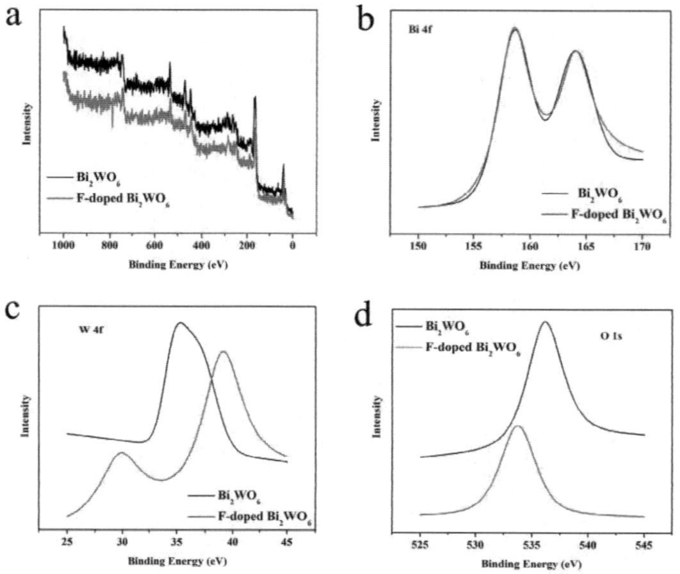

Figure 4. XPS of Bi_2WO_6 and F-doped Bi_2WO_6 samples. (a) survey, (b) Bi 4f, (c) W 4f and (d) O 1s.

3.5. Optical properties of the F-doped Bi_2WO_6 microspheres

Figure 5 displays the diffuse reflectance spectra of the F-doped Bi_2WO_6 samples. When the wavelengths are shorter lower than about 470nm, the Bi_2WO_6 samples with different R_F values presented the photoabsorption abilities, which is in good agreement with the previous report (3). It is outstanding that the edges of the catalysts are dependent strongly on the R_F value. Although all Bi_2WO_6 samples show the similar absorption edges, the absorption edges of the Bi_2WO_6 samples are red shift apparently with the increase of R_F values from 0 to 1. The band gap of the Bi_2WO_6 samples with different R_F values is estimated to be 2.81, 2.79, 2.75, 2.72, and 2.69 eV from the onset of the absorption edge, corresponding to $R_F = 0$, 0.125, 0.25, 0.5, and 1, respectively. We concluded that the difference of the absorbance edges could arise from the fluorination of Bi_2WO_6 catalyst. The band gap of the F-doped Bi_2WO_6 samples can be related closely with two factors [24, 25]: (i) the degree of F 4s and W 5d orbitals being involved in the conduction band of the F-doped Bi_2WO_6 samples, (ii) the degree of delocalization of excitation energy due to lattice deformation arising from F substitution. With increasing of the R_F value, the change of the band gap energy was not necessarily proportional with the R_F value, but the band gap had the outstanding red shifts. Therefore, the results indicated that Bi_2WO_6 samples containing the different R_F value had a suitable band gap for photocatalytic decomposition of organic contaminants under the visible-light irradiation.

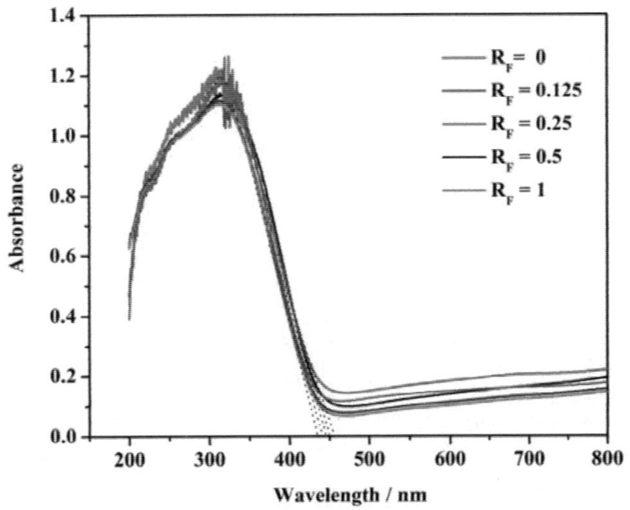

Figure 5. Uv-vis diffusive reflectance spectra of the Bi_2WO_6 samples with different R_F values.

3.6. Photocatalytic performance of the F-doped Bi_2WO_6 samples

The photocatalytic activities of RhB mediated by the Bi_2WO_6 samples with different concentration of fluorine substitution as well as without photocatalyst (the blank test in the inset of Figure 6) under visible light irradiation (λ >400 nm) are displayed in Figure 6. The blank test demonstrates that the degradation of RhB is extremely few without a photocatalyst under visible light irradiation. As shown in Figure 6, the visible-light irradiation of the aqueous RhB/photocatalyst dispersions lead to vastly different photodegradation rates of RhB, and it is obviously that the photodegradation efficiency of RhB is increasing with the concentration of fluorine substitution in the Bi_2WO_6 samples. After 105 minutes irradiation, it is found that the photocatalytic performances of all samples are greatly different and are strongly dependent on the concentration of fluorine substitution. The fluorine is integrate into the Bi_2WO_6 crystal, it may improve the photocatalytic performances of Bi_2WO_6 samples. The enhanced photocatalytic activity came from the following factor. Comparing with O^{2-} ions, F^- ions have the stronger attraction for electrons. Therefore, the electronic cloud can deviate from the original position in the Bi-O polyhedron, meanwhile, the bonds could be weakened between Bi and O in F-doped Bi_2WO_6 samples. So, plenty of oxygen vacancies can be easily produced under visible light irradiation, and they could induce the OH• radicals to take part in the photocatalytic process. Undoubtedly, it could improve the photocatalytic efficiency. In addition, the fluorinated process could probably increase defects which are beneficial to the photocatalytic activity.

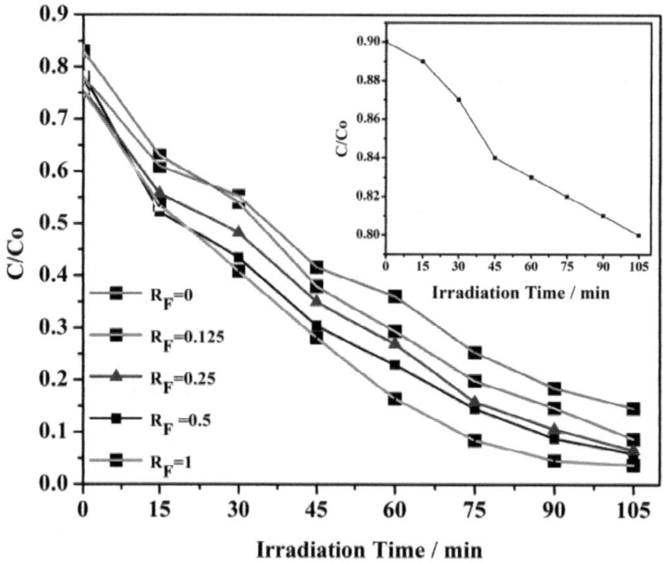

Figure 6. Photocatalytic degradation of RhB solution by the F-doped Bi_2WO_6 samples. inset: the degradation of RhB solution in the absence of Bi_2WO_6 samples.

4. Discussions

The Bi_2WO_6 samples containing the different R_F value (R_F = 0, 0.125, 0.25, 0.5, 1) have been successfully synthesized through a two-step hydrothermal process in the absence of a surfactant. With increasing of the R_F value from n_w = 0 to n_w =1, the morphology of all F-doped Bi_2WO_6 samples are similar. The results of XRD and FE-SEM indicate that the morphology of F-doped Bi_2WO_6 microspheres may maintain better via a two-step hydrothermal process comparing with one-step process. On the basis of maintaining good morphology, the two-step hydrothermal process successfully introduced the fluorine element replaced the O atoms of WO_6 octahedron in Bi_2WO_6. For the Bi_2WO_6 samples with the different R_F value, their band gaps have changed obviously. Under 500W Xe lamp light irradiation, the F-doped Bi_2WO_6 samples exhibit good photocatalytic activity in degradation of RhB. Comparing with Bi_2WO_6, the photocatalytic activity of the samples with different R_F value (R_F = 0.125, 0.25, 0.5, 1) increased outstandingly. Therefore, our work provides new possibilities to keep better morphology with different doped elements for future materials.

5. References

[1] Hongguang Y, Zhenfeng Z, Jianhong Zhou, Jing W, Junqi L, Yanli Z 2013 Appl. Surf. Sci. **265** 424
[2] Zhenfeng Z, Hongguang Y, Junqi L 2012 J. Nano. R. **17** 1
[3] Chuan Z, Yongfa Z 2005 Chem. Mater. **17** 3537
[4] Neil A, Kevin S, Philip L 2006 Chem.Eur. J. **12** 1493
[5] Mei L, Heimo S and Stephen M 1999 Nature. **402** 393
[6] Park S, Lim JH, Chung SW, Mirkin CA 2004 Science. **303** 348
[7] Liu B, Zeng HC 2004 J. Am. Chem. Soc. **126** 8124
[8] Zhang T, Dong W, Mary KB, Roland N, Njabonand Z, Ryan T 2006 J. Am. Chem. Soc. **128** 10960
[9] Dekun M, Shaoming H, Wenxiu C, Shengwei H, Fangfang S and Kelei F 2009 J. Phys. Chem. C. **113** 4369
[10] Rui S, Guangli H, Jie L and Yongfa Z 2004 J. Phys. Chem. C. **113** 19633
[11] Minero C, Mariella G, Maurino V and Pelizzetti E 2000 Langmuir. **16** 2632
[12] Minero C, Mariella G, Maurino V and Pelizzetti E 2000 Langmuir. **16** 8964
[13] Vohra M, Kim S, Choi W 2003 J. Photochem. Photobiol. A. **160** 55

[14] Jimmy Y, Jia G, Wing K, Zi T and Li Z 2002 Chem. Mater. **14** 3808
[15] Di L, Hajime H, Nitin K, Shunichi H, Naoki O 2005 Chem. Phys. Lett. **401** 579
[16] Huang G, Zhu Y 2007 J. Phys. Chem. C. **111** 11952
[17] Hongbo F, Shicheng Z, Tongguang X, Yongfa Z and Jianmin C 2008 J.EnViron. Sci. Technol. **42** 2085
[18] Shushan Yao, Jiyong W, Baibiao H, Shengyu F, Xiaoyang Z, XIaoyan Q, Peng W,Zeyan W, Qi Z, Xiangyang J, Jie Z 2009 J. Solid State Chem. **182** 236
[19] Jun W, Choi JS, Cheon J 2006 Angew. Chem. Int. Ed. **45** 3414
[20] Sanghun C, Eunggyu K, Jongnam P, Kwangjin A, Nohyun L, Sungchu K and Taeghwan H 2005 J. Phys. Chem. B. **109** 14792
[21] Youngwook J, Jaehyun L, Jinsil C and Jinwoo C 2005 J. Phys. Chem. B. **109** 14795
[22] Huang HW, Liu K, Chen K 2014 J. Phys. Chem. C. **118** 14379
[23] Liu H, Cao WR, Su Y, Wang Y, Wang XH 2012 Appl. Catal. B **111** 271
[24] Zhang LW, Man Y, Zhu YF 2011 ACS Catalysis. **1** 841
[25] Yoshino M, Kakihana M, Cho WS, Kato H and Kudo A 2012 Chem. Mater. **14** 3369-3376

Acknowledgments

This work was jointly supported by the National Natural Science Foundation of China (Grant No. 31570961), the State Key Program of National Natural Science Foundation of China (Grant No. 51631003), the Fundamental Research Funds for the Central Universities and Research Innovation Program for College Graduates of Jiangsu Province (Grant No. KYLX15_0120).

IOP Conf. Series: Materials Science and Engineering **284** (2017) 012003 doi:10.1088/1757-899X/284/1/012003

Based on the rainfall system platform raindrops research and analysis of pressure loss

Gang Cao[1] and Jian Sun[2]

1 Senior engineer , the 28th research institute of China electronics technology group , Nanjing, China
2 Engineer, the 28th research institute of China electronics technology group , Nanjing, China
E-mail: sj517794780@163.com

Abstract. With the rapid development of China's military career, land, sea and air force all services and equipment of modern equipment need to be in the rain test, and verify its might suffer during transportation, storage or use a different environment temperature lower water or use underwater, the water is derived from the heavy rain, the wind and rain, sprinkler system, splash water, water wheel, a violent shock waves or use underwater, etcTest the product performance and quality, under the condition of rainfall system platform in the process of development, how to control the raindrops pressure loss becomes the key to whether the system can simulate the real rainfall [1], this paper is according to the rainfall intensity, nozzle flow resistance, meet water flow of rain pressure loss calculation and analysis, and system arrangement of the optimal solution of rainfall is obtained [2].

1. Introduction
This article in view of the different types of rainfall shower nozzle, through comparing the experimental data, choose this rainfall system is the most suitable nozzle [3]. Through the different types of nozzle experiment, with the aid of computer optimization technique, the optimization design of nozzle layout, brings forward the laboratory artificial rainfall shower space layout of the optimal solution.

2. Rainfalsystem composition and the function
By the n * m nozzle, n, n root branch line of speed regulating valve, n, n flow meter pressure gauge, n * m electromagnetic switch valve and n * m a manual throttle valve. Its main function is to achieve uniform rainfall rainfall l is continuously adjustable [4].

3. Nozzle area computation
3 the nozzle diameter, respectively d1=0.95mm， d2=1.5mm， d3=2.1mm

$$S_{喷}=S1+S2+S3=\pi((\frac{d_1}{2})^2+(\frac{d_2}{2})^2+(\frac{d_3}{2})^2)=5.93\times10^{-4}\ m^2 \quad (1)$$

So, a list of nozzle (10) of the pipe diameter d1 finance content [5]:

$$d_1 \geq 2\sqrt{\frac{10S_1}{\pi}}=8.7\ m \quad (2)$$

Diameter d2 points should satisfy large area in charge of the road [6]:

$$d_2 \geq 2\sqrt{\frac{10\times 19S_1}{\pi}} = 38\text{ m} \tag{3}$$

Small diameter d points should satisfy area in charge of the road [7]:

$$d_3 \geq 2\sqrt{\frac{10\times 7S_1}{\pi}} = 23\text{mm} \tag{4}$$

Nozzle spray process and the rain to Competent road calculation, As shown in Figure 1. Take outlet pipe diameter d = 125 mm (line allows velocity can be chosen as: import v = 1.5 m/s,export v = 2.5 m/s),so
Inlet pipe (large pump) d = 150 mm in diameter [8]:

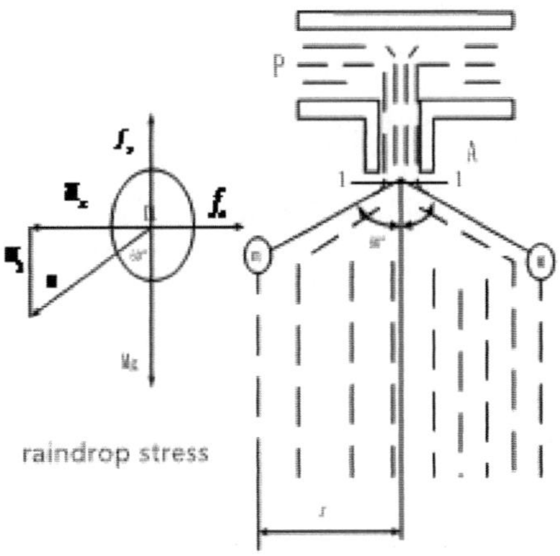

Figure 1. Nozzle spray process and the rain to Competent road calculation.

$$V_{出} = \frac{Q_B}{\pi(\frac{d}{2})^2} = 1.6\text{m/s} \tag{5}$$

All the main routine Reynolds number, $\mathrm{Re} = v.d/\upsilon = 2.26\times 0.125/1.007\times 10^{-6} = 2.8\times 10^5$
The tube running Cheng Liu resistance calculation formula:

$$\Delta p = \lambda \frac{l}{d} \frac{v^2}{2}\rho \tag{6}$$

The turbulent flow along the Cheng Liu resistance coefficient:
$\lambda = 0.032 + 0.221\mathrm{Re}^{-0.237} = 0.043$

$$\Delta p = \lambda \frac{l}{d} * \frac{v^2}{2}\rho = 0.043\frac{l}{0.125}\frac{2.26^2}{2}\times 1000 = 888\times l\text{ Pa} \tag{7}$$

If line calculated at 100 m, the path loss is 0.09 Mpa.

4. In charge of the way to calculate

By the assigned to each branch of the main road, shunt d = 125 mm. Water enter the shunt into three road, all the way to the corresponding branch, the other two road on both sides of the flow. To one

side of the flow rate is 100/19 (100 -) / 2 = 47.4 m3 / h, each column, through traffic 100/19 = 5.3 m3 / h.
After the first column, Q=47.4 m3/h

$$V_1 = \frac{Q}{\pi(\frac{d}{2})^2} = 1.07\ \mathrm{m} \tag{8}$$

Along the Cheng Liu shunt resistance $Re = v.d/\upsilon = 1.07 \times 0.125 / 1.007 \times 10^{-6} = 1.33 \times 10^5$
The tube running Cheng Liu resistance calculation formula:

$$\Delta p = \lambda \frac{l}{d} \square \frac{v^2}{2} \rho \tag{9}$$

The turbulent flow along the Cheng Liu resistance coefficient

$$\lambda = 0.032 + 0.221 \square Re^{-0.237} = 0.045 \tag{10}$$

$$\Delta p_1 = \lambda \frac{l}{d} * \frac{v^2}{2} \rho = 0.045 \frac{1.2}{0.125} \frac{1.07^2}{2} \times 1000 = 252\ \mathrm{Pa} \tag{11}$$

After the second column Q=（47.4 -5.3） m3/h=42.1 m3/h

$$V_2 = \frac{Q}{\pi(\frac{d}{2})^2} = 0.95\ \mathrm{m} \tag{12}$$

Along the Cheng Liu shunt resistance $Re = v.d/\upsilon = 0.95 \times 0.125 / 1.007 \times 10^{-6} = 1.18 \times 10^5$
The tube running Cheng Liu resistance calculation formula

$$\Delta p = \lambda \frac{l}{d} \square \frac{v^2}{2} \rho \tag{13}$$

The turbulent flow along the Cheng Liu resistance coefficient

$$\lambda = 0.032 + 0.221 \square Re^{-0.237} = 0.045$$

$$\Delta p_2 = \lambda \frac{l}{d} * \frac{v^2}{2} \rho = 0.045 \frac{1.2}{0.125} \frac{0.95^2}{2} \times 1000 = 200\ \mathrm{Pa} \tag{14}$$

After the third column Q=（47.4 -5.3*2） m3/h=36.8m3/h

$$V_3 = \frac{Q}{\pi(\frac{d}{2})^2} = 0.83\ \mathrm{m} \tag{15}$$

Along the Cheng Liu shunt resistance, $Re = v.d/\upsilon = 0.83 \times 0.125 / 1.007 \times 10^{-6} = 1.03 \times 10^5$,
The tube running Cheng Liu resistance calculation formula

$$\Delta p = \lambda \frac{l}{d} \square \frac{v^2}{2} \rho \tag{16}$$

The turbulent flow along the Cheng Liu resistance coefficient

$$\lambda = 0.032 + 0.221 \cdot Re^{-0.237} = \tag{17}$$

$$\Delta p_3 = \lambda \frac{l}{d} * \frac{v^2}{2} \rho = 0.045 \frac{1.2}{0.125} \frac{0.83^2}{2} \times 1000 = 154\ \mathrm{Pa} \tag{18}$$

Is for the end of the frictional pressure loss

$$\Delta p = \Delta p_1 + \Delta p_2 + \Delta p_3 = 606Pa \tag{19}$$

5. Nozzle control algorithm

Two-dimensional rainfall distribution expectations by analog computer is given, and the actual measurement of rainfall, rainfall intensity data for two-dimensional rainfall error values, through the error diffusion algorithm, nozzle control of quantity, and then control the two-dimensional rainfall distribution. Distributed rainfall sensor measurement of rainfall data continuous rainfall distribution is obtained by interpolation calculation, for the use of closed loop control of rainfall [9].

Error diffusion algorithm: control the amount of each nozzle, Its control algorithm is as follows:

$$p_4 = W_1 \times pid(e_1) + W_2 \times pid(e_2) + W_3 \times pid(e_3) + W_4 \times pid(e_4)$$

Among them, W_1, W_2, W_3, W_4, For the weight, $pid(\bullet)$ is PID operation.

6. Rainfall intensity uniformity simulation

In order to simulate different aperture of nozzle in different pressure, we make a simulation software which simulate the artificial rainfall, it can set the value, such as nozzle pressure, nozzle diameter, nozzle center distance, nozzle in distance. And then we can look at the combination figure clearly, a three-dimensional figure of a single nozzle rain intensity distribution, a three-dimensional figure of all areas rainfall intensity distribution, a section figure of all areas rainfall intensity distribution, a 85% section figure of rainfall intensity distribution, and rainfall intensity distribution of the contour [10]. The operation panel of simulation software interface is shown in Figure 2.

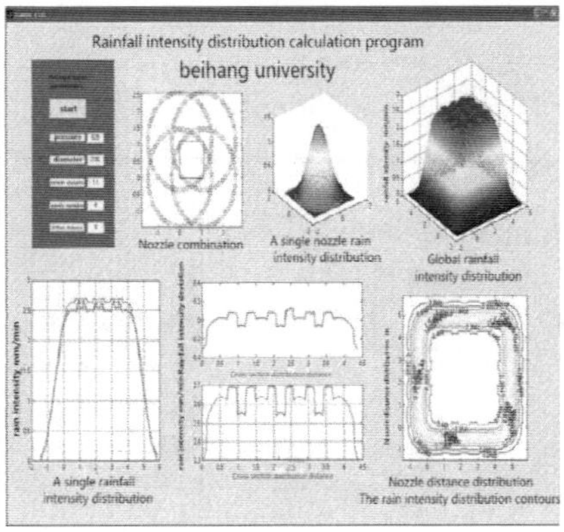

Figure 2. Software analysis.

7. The rain intensity distribution calculation and analysis

Nozzle diameter is small, the Reynolds number is larger, so the shower nozzle model for the round hole turbulent jet model. The rainfall intensity distribution accord with normal distribution, satisfy:

$$u = u_{\max} \exp(-\frac{r^2}{be^2})$$

Among them: Be as strong rain when r = be in equal, is proportional to the jet distance, of the relationship between the radius and jet as follows:

$$b_e \propto actg 60^0 x_{\max}$$

Based on the constant flow are:

$$Q=\int_0^{X\max}2\pi urdr=2\pi u_{\max}\int_0^{X\max}re^{(-\frac{r^2}{be^2})}dr \quad (20)$$

$$Q=\pi b_e^{\ 2}u_{\max}[1-e^{(-\frac{x\max^2}{b_e^2})}]$$

$$u_{\max}=\frac{Q}{\pi b_e^2[1-e^{(-\frac{x\max^2}{b_e^2})}]}$$

8. Conclusion

In the process of rainfall realize the change of rainfall intensity, the pressure with the additional requirements for water supply, need water supply pipeline pressure adjustment is done in a very short period of time, to make changes in rainfall intensity steady rainfall process can in the shortest time. In view of this, in view of the water supply pipeline pressure control stability and uniformity of rainfall influence research. Through experimental study and theoretical calculation, put forward in the artificial rainfall rainfall intensity changes to ensure that the nozzle discharge pressure constant solution.

9. References

[1] The command information system and technology, 2011, 02 (2): 28-31
[2] The command information system and technology, 2010, 01 (5): 28 and 33
[3] The command information system and technology, 2012, 3 (1): 25 to 28
[4] Cui Peng. The mechanism of debris flow initiation [Ph.D. Thesis] [D]. Beijing; Beijing Forerstry University, 1990.
[5] Cui P. Studies on condition and mechanism of debris flow initiationby means of experiment [J]. Chinese Science Bulletin, 2014, 37(9), 759–763.
[6] Anderson S A, Siter N. Analysis of rainfall-induced debris flows [J]. J. Geotech. Eng., ASCE, 2009, 121(7): 544–552.
[7] Study on the landslides change into debris flows [A]. In: Proceedings of Landslide [C]. Chengdu: Sichuan Science and Technology Press, 2016. 69–76.
[8] Cui Peng. The mechanism of debris flow initiation [Ph.D. Thesis] [D]. Beijing: Beijing Forerstry University, 2011
[9] Tian FeiFei Ma Tao. Explore innovative communication, the sixth China aviation institute youth science and technology BBS corpus (part ii) [C]. 2014
[10] Zhang Shuan, Zhang Tong, more sun. Explore innovation - the sixth China aviation institute youth science and technology BBS corpus (part ii) [C]. 2014

IOP Conf. Series: Materials Science and Engineering **284** (2017) 012004 doi:10.1088/1757-899X/284/1/012004

Synthesis, structure and temperature dependent luminescence of Eu^{3+} doped hydroxyapatite

Xiaobing Luo*, Xiaoxia Luo, Hongwei Wang, Yue Deng, Peixin Yang, Yili Tian

College of Science, Chongqing University of Posts and Telecommunications, Chongqing, 400065, P. R. China
E-mail: luoxb@cqupt.edu.cn

Abstract. A series of Eu^{3+} substituted hydroxyapatite (HA) were prepared by co-precipitation reactions. The phase, fluorescence and temperature dependent luminescence of the phosphors were investigated by X-ray diffraction (XRD) and photoluminescence (PL). It is found that the doped Eu^{3+} ions have entered the hexagonal lattice with no obvious secondary phase were detected by XRD. The $^5D_0 \rightarrow ^7F_0$ transition was clearly split into two even at room temperature. The predominate 573 nm peak illustrates Eu^{3+} ions occupy more Ca(II) sites. The temperature dependent luminescent results show HA:xEu might be applied as one potential optical thermometry material.

1. Introduction

Hydroxyapatite (HA) widely exists in the bones and teeth of animals. Owing to its chemical similarity to natural HA, artificial HA ($Ca_{10}(PO_4)_6(OH)_2$, Ca/P = 1.67) has good biocompatibility and biodegradation. Therefore, artificial is extensively employed in biomedical sector, particularly in dentistry and orthopedics [1], gene vector [2], drug carrier [3], fluorescent labeling [4] and so on. Fluorescent labeling can realize the continuous and nondestructive observations, which is helpful for monitoring the implanted subject, the delivering progression of drug carries. Thus, luminescent HA, which can be obtained by substituting Ca^{2+} ions in the crystal lattice by rare earth ions, is challenging. In fact, because the components of the natural hydroxyapatite are complicated, people have studied the substitution of various cations and anions in the hydroxyapatites [1]. Y. Han et al. have found that Eu-HA nanocrystallines could be used to label Bel-7402 human liver cancer cells [5].

The crystal structure of HA is unique. Two different crystal phases of HA have been observed in previous reports. One is hexagonal with the lattice parameters $\boldsymbol{a} = \boldsymbol{b} = 9.432$ Å, $\boldsymbol{c} = 6.881$ Å and $\gamma = 120^o$, the other one is monoclinic with the lattice parameters $\boldsymbol{a} = 9.421$ Å, $\boldsymbol{b} = 2\boldsymbol{a}$, $\boldsymbol{c} = 6.881$ Å and$\gamma = 120^o$ [6]. In the case of hexagonal structure with the space group $P6_3/m$, it is possible to identify two sites for calcium ions, i.e., Ca(I) and Ca(II): Ca(I) is called as the columnar Ca, and Ca(II) the screw axis Ca [7]. Ca(I) is in columns parallel to the $\boldsymbol{c}$-axis and are surrounded by 9 oxygen atoms, among which 6 oxygen atoms from PO_4^{3-} groups are at the first nearest neighboring sites, while further 3 oxygen atoms are located at the second nearest neighboring sites. Ca(I) is in high C_3 symmetry. However, Ca(II) is seven-coordinated with 6 PO_4^{3-} groups and 1 OH^-, forming two equilateral triangles along $\boldsymbol{c}$ axis. Specifically, for Ca(II) site, 6 oxygen atoms are located at the first nearest neighboring sites, while a further one oxygen atom is placed at the second nearest neighboring site Ca(II) is in low C_s symmetry.

2. Experimental

2.1. Synthesis of Eu^{3+} doped hydroxyapatite.
The nanocrystalline powders of Eu^{3+} doped hexagonal hydroxyapatites ($Eu:Ca_{10}(PO_4)_6(OH)_2$) were synthesized by co-precipitation method. The concentration of Eu^{3+} ions has been set to 1, 3, 5, and 7 mol% in respect to the appropriate Ca^{2+} molar content. The starting materials used for the preparation were Eu_2O_3, $NH_4H_2PO_4$ and Eu_2O_3. According to the molar ratios of (Eu+Ca)/P=1.67, Eu_2O_3 was dissolved in dilute HNO_3 to obtain $Eu(NO_3)_3$. The mixture of $Ca(NO_3)_2$ aqueous solution and $Eu(NO_3)_3$ aqueous solution was poured into the $NH_4H_2PO_4$ aqueous solution under vigorous stirring. After the pH value was adjusted to about 10 by $NH_3{\cdot}H_2O$, the precipitation was washed by distilled water at least three times and then dried by oven-dried, Finally, the dried precipitation was calcined at 600°for 6h to obtain the samples.

2.2. Characterization
The powder X-ray diffraction (XRD) measurement was recorded by an X-Ray Diffractometer (Persee, XD-2) with Cu-Kα (λ = 1.5406 Å) at 36 KV and 20 mA. The scan rate was 2°/min and the scan range was between 10° and 60°. The UV-visible reflectance spectrum was obtained on a Φ 130 mm integrating sphere attachment in a double-beam spectrometer (Cary 530). $BaSO_4$ was employed as the reference for 100% reflectance. The photoluminescence excitation (PLE), photoluminescence (PL) spectra and the fluorescence decay were recorded by a fluorescence spectrophotometer (Edinburgh, FLS920) equipped with a 450 W xenon lamp or a pulsed xenon lamp as the light source, and a Shimidazu R9287 photomultiplier as the detector. The samples were mounted in one home-made temperature controller to obtain the temperature-dependent excitation and emission spectra.

3. Results and discussion

3.1. Phase formation
The XRD patterns of different Eu^{3+}- doped HA phosphors were shown in Figure 1. As shown, all the diffraction peaks are weak and the baselines are noisy, which might illustrate that the crystallization of the prepared Eu^{3+} doped HA is not perfect. But it is similar with the reported results where the samples were synthesized by hydrothermal method [8], thermal decomposition method, or ultrasound assisted precipitation method [9]. On the other hand, the position and the relative intensity of the diffraction peaks of each sample are in agreement with the hexagonal HA in $P6_3/m$ space group with the JCPDS No. 74-0566. No significant signal related to other phases was detected, which indicated that Eu^{3+} ions had successfully doped into HA crystals. As stated above, the doped Eu^{3+} ions could have two different sites since Eu^{3+} ions substituted Ca^{2+} ions in the lattice.

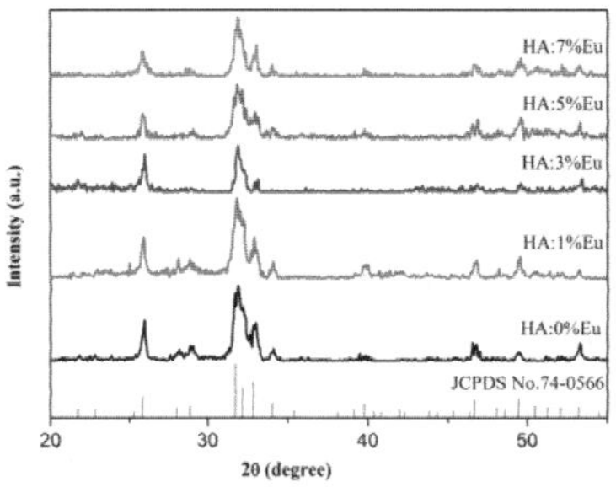

Figure 1. XRD powder diffraction patterns of Eu^{3+} doped HA phosphors.

3.2. Diffuse reflectance spectroscopy
The diffuse reflectance spectra of powders are often used to investigate the absorption properties of the samples. Figure 2 shows the diffuse reflectance spectra of HA: xEu (x = 1%, 3%, 5% and 7%). From the figure, it is see that all the profiles present high and constant reflectance (above 90% s) at wavelength longer than 600 nm, which illustrate there is no absorption in 600-800 nm spectra range. The weak sharp

absorption peaks due to Eu^{3+} ions at 394, 466, and 531 nm were observed. The assignments of the f-f transitions are noted in the figure. One intense wide band centered at about 305 nm could be attributed to the charge transfer band of Eu-O, which is consistent with the documented data [10].

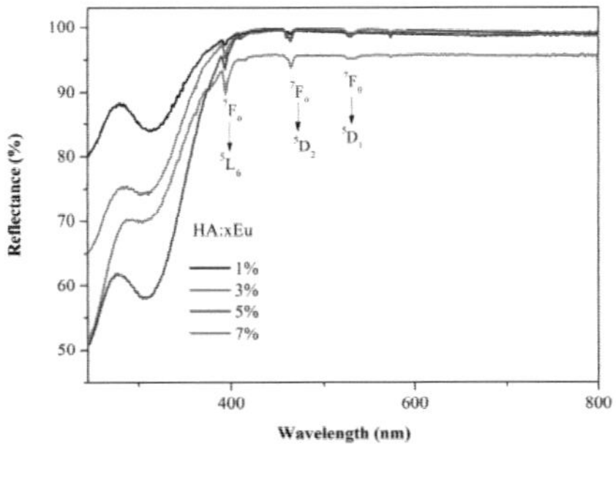

Figure 2. UV-vis diffuse reflectance spectra of HA:xEu.

3.3. Luminescence of HA:Eu at room temperature

Figure 3 shows the PL and PLE spectra of one representative sample HA: 5% Eu. Both the sharp structured intra-configurational f-f transition of Eu^{3+} ions and the broad charge transfer band of Eu-O are observed in the PLE spectra. The assignments of the peaks are marked in the figure. From the excitation spectrum, it is seen that the Eu doped HA has strong excitation property. In the emission spectra, the transition from the excited state of $^5D_0$ to the ladder energy levels of ^7F_J (J = 0, 1, 2, 3, 4) are clearly observed. The emission intensity under the excitation of 391 nm (the characteristic transition of Eu^{3+}:$^7F_6 \rightarrow ^5L_6$) is much higher than that of 284 nm (the charge transfer band of Eu-O) as expected.

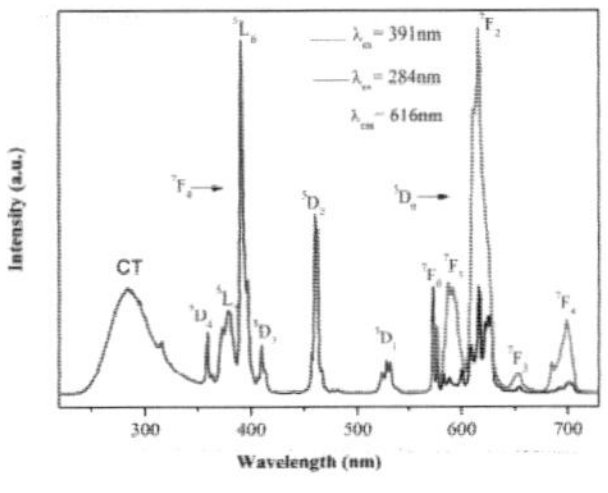

Figure 3. PL and PLE spectra of HA:5%Eu sample at room temperature.

It is interesting to noted that the transition of $^5D_0 \rightarrow ^7F_0$ is clearly to be slipped into two sharp peaks even at room temperature under the excitation of 391 and 284 nm. It is known that $^5D_0 \rightarrow ^7F_0$ transition of Eu^{3+} is generally parity forbidden, but when Eu^{3+} ions are located in C_s, C_n and C_{nv} sites, the parity forbidden will be partially permitted and thus a $^5D_0 \rightarrow ^7F_0$ transition can be observed. Furthermore, the number of $^5D_0 \rightarrow ^7F_0$ transition peaks will correspond to the number of lattice site in the luminescent materials. The observation of the split of $^5D_0 \rightarrow ^7F_0$ transition illustrates two lattice sites of Eu^{3+} ions in the lattice, which is in good agreement with the crystal structure. Furthermore, the two peaks in $^5D_0 \rightarrow ^7F_0$ transition located at 573 and 576 nm recorded in this work corresponds to Ca(II) and Ca(I) sites respectively. In addition, the intensity of 573 nm is much higher than that of 576 nm (see the black curve excited by 284 nm in Figure 3), which hints that there are more Eu^{3+} ions occupy Ca(II) sites than Ca(I) sites in the lattice. The result is consistent with the previous study of site preference of rare earth doped in hydroxyapatite [11]. At the same time, it is found in the emission spectrum excited by 284 nm that the overall intensity of $^5D_0 \rightarrow ^7F_0$ is as high as that of $^5D_0 \rightarrow ^7F_2$, which is quite interesting.

The occupying of Eu^{3+} ions in Ca(II) or Ca(I) sites depends on the preparation method, the morphology, the size of the luminescent materials and the concentration of doped Eu^{3+} ions. Figure 4 (a) and (b)

IOP Conf. Series: Materials Science and Engineering **284** (2017) 012004 doi:10.1088/1757-899X/284/1/012004

displays the excitation and emission spectra of HA:xEu (x = 1%, 3%, 5% and 7%) samples at room temperature. It is found out that the emission intensity of these series of samples increases at first until the doping concentration of Eu^{3+} is 5%, the intensity reaches a maximum and it decreases with the concentration. But the shape of the spectrum profiles does not show obvious change with the doping concentration in this work. The relative emission intensity of $^5D_0 \rightarrow {}^7F_0$ Ca(II) at 573 nm and that of $^5D_0 \rightarrow {}^7F_0$ Ca(I) at 576 nm does not change either, which illustrates that the energy migration as reported in ref. [5] does not occur.

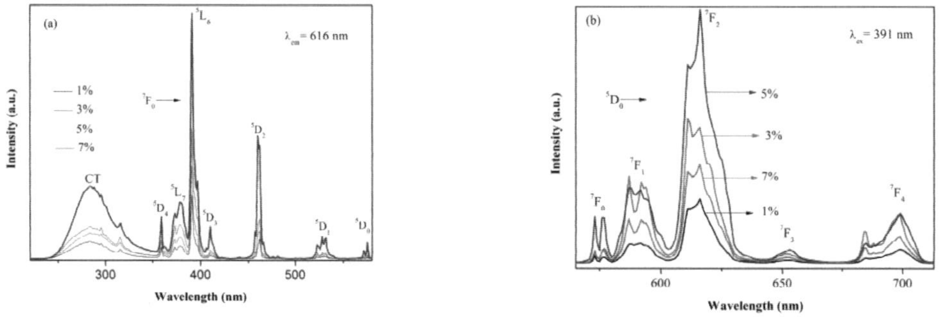

Figure 4. PLE (a) and PL(b) spectra of HA:xEu at room temperature .

3.4. Temperature dependent excitation and emission spectra of HA:5%Eu

From the above results it is known that the HA:5%Eu phosphor presents the maximum emission intensity among the HA:xEu (x = 1-7%) samples. Figure 5 and Figure 6 present the temperature dependent excitation and emission spectra along with the fitting results of this optimal sample. It could be found that the intensity declines rapidly when temperature increases from 25 K to 275 K both in excitation and emission spectra as expected, because the non-radiactive transition increases with temperature. Figure 5 (a) shows the 3-dimentional excitation spectra of this representative sample HA:5%Eu at different temperatures, and Figure 5(b) are the data fitting results based on temperature dependent excitation spectra, from which it is seen that the excitation intensity ratio between Eu^{3+} f-f transition at 391 nm and that of the charge transfer band at 284 nm, the relative sensitivity and the excitation at 391 nm exhibit linear relationship with the temperature.

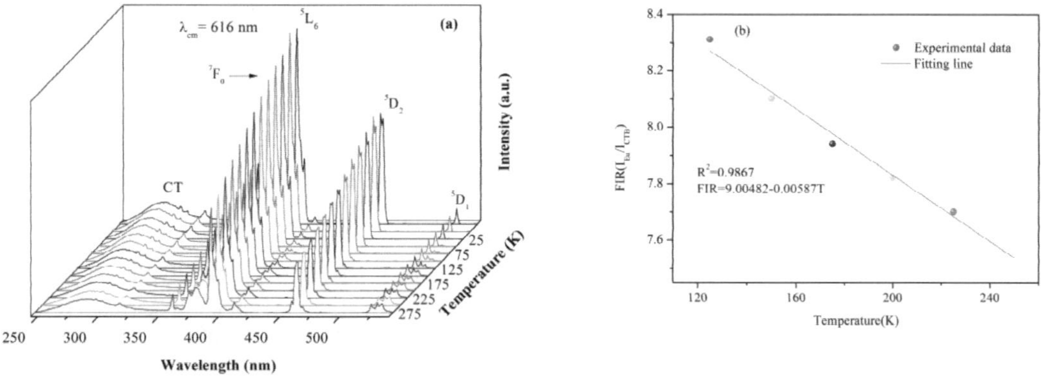

Figure 5. (a) The 3-dimentional PLE spectra of HA:5%Eu at different temperatures. (b) The intensity ratio between Eu^{3+} f-f transition at 391 nm and that of CTB at 284 nm as a function of temperature.

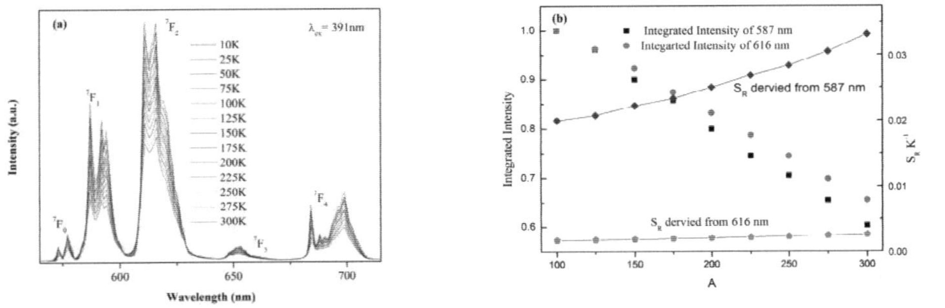

Figure 6. (a) The PL spectra of HA:5%Eu at various temperatures. (b) Temperature dependence of emission @616 nm 587 nm.

Figure 6 (a) is the emission spectra of HA:5%Eu phosphor at various temperatures. Figure 6 (b) shows the temperature dependence of the emission intensity of 616 nm /587 nm. Similar with the results obtained from excitation spectra, the linear relationship is observed.

4. Summary

The Eu^{3+} doped hydroxyapatite HA:xEu (x =0, 1%, 3%, 5% and 7%) were synthesized by co-precipitation reactions. XRD and PL were employed to investigate the phase, fluorescence and temperature dependent luminescence of the samples. It is found that the doped Eu^{3+} ions have successfully entered the hexagonal lattice. The powders could be pumped both by the charge transfer band (CTB) of Eu-O and the characteristic transition of Eu^{3+}. It presents red emission under the excitation of 391 nm (Eu^{3+} characteristic transition) and orange emission under the excitation of 284 nm (CTB of Eu-O). The splitting of $^5D_0 \rightarrow ^7F_0$ transition illustrates the substituted Eu^{3+} ions occupy two different sites, and Ca(II) sites were dominate in this study. The optimal doping concentration of Eu^{3+} is determined to be 5%. The temperature dependent luminescence shows the samples might be served as one potential optical thermometry material.

5. References

[1] I. Cacciotti, *Cationic and Anionic substitutions in hydroxyapatite, Handbook of Bioceramics and Biocomposites,*(DOI)10.1007/978-3-319-09230-0-7-1.

[2] H. Nie, C.H. Wang, *Fabrication and characterization of PLG/Hap composite scaffolds for delivery of BMP-2 plasmid DNA*(J. Control. Release) 120 (2007) 111-121.

[3] P. Yang, Z. Quan, C. Li, X. Kang, H. Lian J. Lin, Bioactive, *luminescent and mesoporous europium-doped hydroxyapatite as a drug carrier* (Biomater). 29 (2008) 4341-4347.

[4] R. Singh, H. S. Nalwa, *Medical applications of nanoparticles in biological imagine, cell labeling, antimicrobial agents, and anticancer nanodrugs*(J. Biomed. Nanotech)7 (2011) 489-503.

[5] Y. Han, X. Wang, H. Dai, S. Li, *Synthesis and luminescence of Eu^{3+} doped hydroxyapatite nanocrystallines: Effects of calcinations and Eu^{3+} content*(Lumin) 135 (2013) 281-287.

[6] A. S. Posner, A. Perloff, A. F. Diorio, *Refinement of the hydroxyapatite structure*(Acta Crystallogr)11(1958)308-309.

[7] M. H. Uddin, T. Matsumoto, M. Okazaki, A. Nakahira, T. Sohmura, *Biomimetic fabrication of apatite related biomaterials. In: Mukherjee A (ed) Biomimetics learning from nature*(InTech, Rijeka) 2010, pp 289-303

[8] S. Huang, J. Zhu, K. Zhou, *Effects of Eu^{3+} ions on the morphology and luminescence properties of hydroxyapatite nanoparticles synthesized by one-step hydrothermal method* (*Mater. Res. Bull)* 47 (2012) 24-28.

[9] Y. Han, X. Wang, H. Dai, S. Li, *Synthesis and luminescence of Eu^{3+} doped hydroxyapatite nanocrystallines: effects of calcinations and Eu^{3+} content*(J. Lumin.)135 (2013) 281-287.

[10] A. K. Parchur, R. S. Ningthoujam, *Behavior of electric and magnetic dipole transitions of* Eu^{3+}, ${}^5D_0 \rightarrow {}^7F_0$ *and Eu-O charge transfer band in* Li^+ *co-doped YPO4:*Eu^{3+}(RSC Advances) 2 (2012) 10859.
[11] M. E. Fleet, X. Liu, Y. Pan, *Site preference of rare earth elements in hydroxyapatite* $[Ca_{10}(PO_4)_6(OH)_2]$(J. Solid State Chem.) 149 (2000) 391-398.

Acknowledgement
The project was funded by National Natural Science Foundation of China (11674044, 11604037), Natural Science Foundation Project of Chongqing (CSTC2015jcyjA50005), Chongqing Key Laboratory Improvement Plan (cstc2014pt-sy40001) and University Innovation Team Construction Plan Funding Project of Chongqing (CXTDG201602009).

IOP Conf. Series: Materials Science and Engineering **284** (2017) 012005 doi:10.1088/1757-899X/284/1/012005

Electric field numerical simulation of disc type electrostatic spinning spinneret

L Wei[1], ZL Deng[2], XH Qin[1] and ZY Liang[2]

1 Key Laboratory of Textile Science & Technology, Ministry of Education, College of Textiles, Donghua University, Shanghai, China
2 College of Science, Donghua University, Shanghai, China
E-mail: xhqin@dhu.edu.cn

Abstract. Electrospinning is a new type of free-end spinning built on electric field. Different from traditional single needle spinneret, in this study, a new disc type free surface spinneret is used to produce multiple jets, this will greatly improve production efficiency of nanofiber. The electric-field distribution of spinneret is the crux of the formation and trajectory of jets. In order to probe the electric field intensity of the disc type spinneret, computational software of Ansoft Maxwell 12 is adopted for a precise and intuitive analysis. The results showed that the whole round cambered surface of the spinning solution at edge of each layer of the spinneret with the maximum curvature has the highest electric field intensity, and through the simulation of the electric field distribution of different spinneret parameters such as layer, the height and radius of the spinneret. Influences of various parameters on the electrostatic spinning are obtained.

1. Introduction

Electrostatic spinning is a new method for the preparation of polymer nanofibers. Because of its simple equipment, less cost and convenient operation, it is widely used in mass production of polymer nanofibers. An electric field was adopted to produce thin charge jet and fabricate fibers [1]. Fong et al. [2] investigated the effect of the viscoelasticity of solution, jet charge density and the surface tension of the solution on the formation of the beaded fibers. Spivak et al. [3] started with the theory of fluid mechanics analysis of the electrostatic spinning. Deitzel et al. [4] studied the voltage and the solution concentration's influences on the fibers. Theron et al. [5] simulated the morphological characteristics of the electrostatic spinning spinneret. Due to the limited production, single needle electrostatic spinning method is not appropriate for practical production which has prompted a number of multi-needle electrostatic spinning methods to spring up. Electric field simulation method could be found to investigate spinneret design and nanofiber preparation [6-9]. Jiang et al. [10] calculated and analyzed the electric field distribution of typical electrospinning device based on finite element calculation theory of electric field. Electric field structure of needle-plate type electrospinning machine was simulated using ANSYS software. Zheng et al. [11] designed the appropriate spinneret and collector geometry to achieve uniform electric field distribution in multi-jet electrospinning based on the three-dimensional electric field simulation. The result showed that a flat spinneret creates a more uniform electric field distribution than the multi-needle spinneret with an auxiliary plate. Kim et al. [12] used the five metal ring electrostatic spinning to simulate the relationship between electric field and the stability of the jet. But these spinning methods to prepare nanofiber are still very limited. In order to further improve the efficiency of spinning and the productivity of nanofiber, free surface electrostatic spinning method is a very good solution to this problem. The first innovation point of the research is the adoption of three layers of round free surface spinneret which is designed by the authors. Another innovation point of the research is that computational software Ansoft Maxwell 12 is used to simulate electric field intensity of spinneret.

2. The model establishment of the disc type electrostatic spinning spinneret

In this model, the spinning solution uses the shape which is in actual freedom of the liquid surface. The influence of the high-voltage electrostatic field is ignored here. Because the spinning solution on the copper plate flows, the contact angle becomes smaller. The contact angle is 75 °, and the thickness is 3.6 mm. The first layer to the third layer of copper radius is 46mm, 38 mm and 30 mm, the thickness of each layer of the copper plate is 8 mm, and the model is shown in Figure 1 (a) and 1(b). Because the model is a center symmetric figure, the three-dimensional model is simplified as the two-dimensional model for the research.

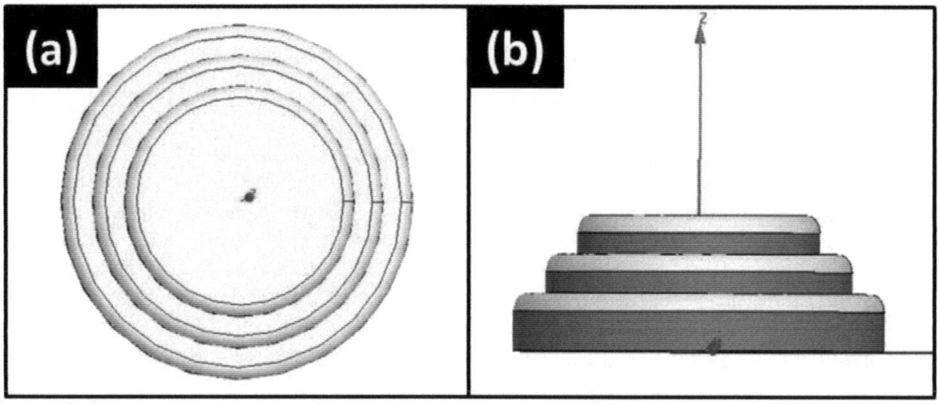

Figure 1. Disc type electrostatic spinning spinneret model of (a) top view; (b) mian view

3. The electrostatic spinning electric field intensity simulation analysis

3.1. Different number of layers of disc type electrostatic spinning electric field intensity analysis

Figure 2 (a)-(c) show that different numbers of layers of the disc type electrostatic spinning electric field intensity contours are broadly similar. The red part of the figure means the maximum electric field strength, and the blue part means the minimum electric field strength. It can be seen from Figure 2 (a)-(c) that the red parts are mainly concentrated in each layers of spinning solution of curve edges and adjacent areas. The area outspreads farther, the color turns from red to yellow and green. This suggests that the electric field intensity is the strongest near the edge of the spinning solution surface. This is because since the edges have the largest curvature, they experience maximum accumulation of the surface charge which accordingly, gives an equally big surface charge density. Therefore this area's electric field intensity is the largest.

The nearer to the receiver, the deeper the blue color, which shows that the farther from spinning solution, the smaller the electric field intensity. To further observe the spinning solution and its nearby electric field intensity, the internal spinning solution is blue, indicating that the internal electric field intensity in the spinning solution is almost zero. The reason is that an extra electric field exists within the internal spinning solution which has been induced by the inductive charge. The superposition of these two electric fields makes the internal electric field of the spinning solution far less than the surface electric field. The top layer of the spinning solution is a free surface, and its distance to the receiver is the nearest thus under the same voltage, its electric field intensity is the biggest. Table 1. shows the different numbers of layers of disc type electrostatic spinning maximum electric field intensity value.

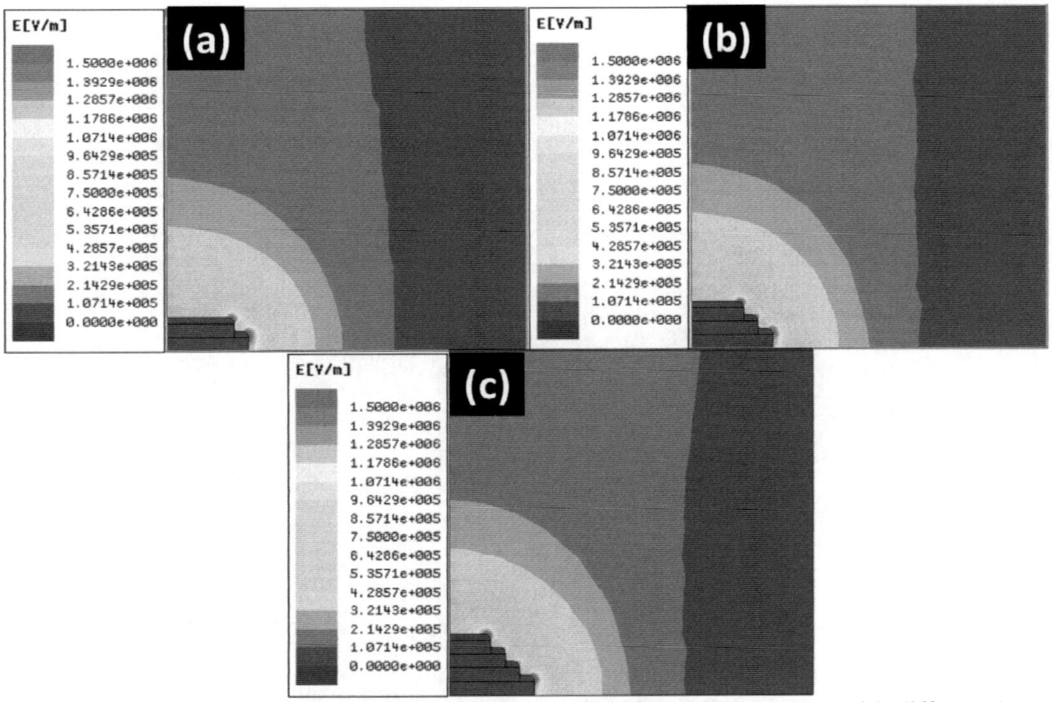

Figure 2. Disc type electrostatic spinning electric field intensity contours with different layers of (a) two; (b) three; (c) four

Table 1. Different numbers of layers of disc type electrostatic spinning maximum electric field intensity value

Number of layer of the spinneret	2	3	4
The maximum electric field intensity value ($\times 10^6$V/m)	2.1829	2.1091	2.0375

Further comparisons of the three different layers of disc type electrostatic spinning of electric field intensity contours show that the electric field intensity decreases with the increasing number of layers. Specifically, by comparing the maximum electric field intensity value; the increase of in the number of layers reduced the maximum electric field intensity accordingly. It can be seen from Table 1 that with the increase of the number of layers, the corresponding yellow and green areas are also gradually reduced. This suggested that the increase in layers decreases the overall regional electric field intensity value. It can also be seen in Fig.2 (a)-(c) that in the middle tier of the spinning solution of curve edges, the red areas are less, and the electric field intensity is smaller. This is because the middle tier of the spinning solution will be affected by the electric fields in the top and bottom layers of spinning solution. Since the effect of electric field's interference is small, at the top curve edge of the spinning solution's top layer and the bottom curve edge of the bottom layer, the red areas are still large. This shows that the electric field intensity in these two areas is the largest and most likely to pump out the fibers.

3.2. Different height of layers of disc type electrostatic spinning electric field intensity analysis

Figure 3 (a)-(c) show that different heights of layers of disc type electrostatic spinning electric field intensity contours are broadly similar. With the increase of the height, the red, yellow and green areas of the figure have some increase. This suggests that the increase of the height increased the whole electric field intensity. Because under the same voltage, the increase of the height reduces the distance from the spinning solution to the receiver, correspondingly the electric field intensity increases. Through comparison we found that with the increase of the height, the electric field interference effect between the two layers of the spinning solution is weakened. Comparing the data in Table 2, with the increase of the height, correspondingly the maximum electric field intensity value also increases. To sum this up, increasing the height will do good to weaken electric field interference effect between two adjacent layers of the spinning solution, it also can improve the maximum electric field intensity value, which will be conducive to the formation of the jets and the fibers.

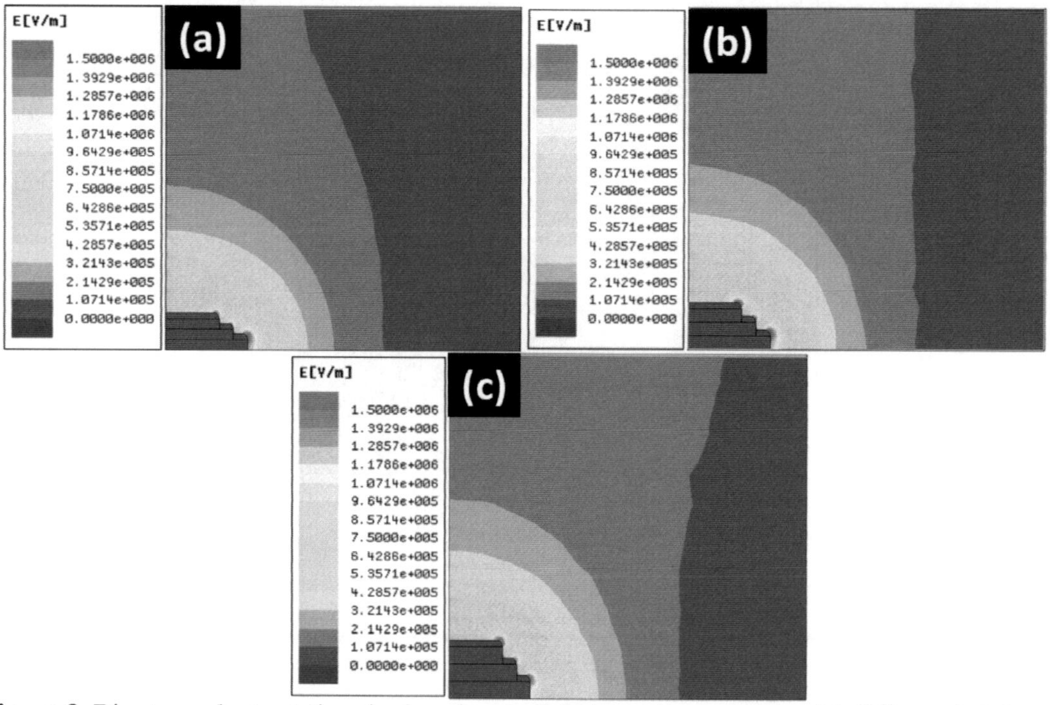

Figure 3. Disc type electrostatic spinning electric field intensity contours with different heights of layers of (a) 6mm; (b) 8mm; (c) 10mm

Table 2. Disc type electrostatic spinning maximum electric field intensity value with different heights of layers

Height of layers of the spinneret(mm)	6	8	10
The maximum electric field intensity value ($\times10^{6}$V/m)	1.9549	2.1091	2.2361

4. Conclusions

In the article, numerical simulation was used to analyze the disc type electrostatic spinning spinneret model. By changing the number and height of layers in the disc type electrostatic spinning spinneret, electric field intensity was investigated. Numerical simulation of the electric field intensity of the disc type electrostatic spinning spinneret shows that the maximum area of the electric field intensity is distributed at the edge of the surface of each layer of the spinning solution and the nearby area. The phenomenon of jet and nanofiber formation is highly consistent with the experimental results. The electric field intensity distributions of the two-layered, three-layered and four-layered nozzles were analyzed by changing the number of the disc type electrostatic spinning spinneret. With the increase in the number of layers, the electric field intensity of the whole calculation domain decreased, and the maximum electric field intensity value got smaller and smaller. A numerical simulation analysis done by varying the disc spinneret heights and analyzing the electric field intensity distributions show that along with the increase of height, the electric field intensity of the whole calculation domain increased the maximum electric field intensity value grows accordingly.

5. References

[1] Doshi J and Reneker D H 1995 *J. Electrostat.* **35** 151
[2] Fong H, Chun I and Reneker, D 1999 *Polymer* **40** 4585
[3] Spivak A, Dzenis Y and Reneker D 2000 *Mech. Res. Commun.* **27** 37
[4] Deitzel J M, Kleinmeyer J, Harris D and Tan N B 2001 *Polymer* **42** 261
[5] Theron S, Yarin A, Zussman E and Kroll E 2005 *Polymer* **46** 2889
[6] Xu L, Zhao J H and Liu H 2015 *Therm. Sci.* **19** 1255
[7] Park C H, Pant H R and Kim C S 2014 *Text. Res. J.* **84** 1044
[8] Kim I G, Lee J H, Unnithan A R, Park C H and Kim C S 2015 *J. Ind. Eng. Chem.* **31** 251
[9] Wang X, Liu Y, Zhang C, An Y, He X and Yang W 2013 *J. Nanosci. Nanotechno.* **13** 4680
[10] Jiang J G, Duan H W, He T H and Li B 2015 *J. Comput. Theor. Nanosci.* **12** 2016
[11] Zheng Y, Zhuang C, Gong R H and Zeng Y 2014 *Ind. Eng. Chem. Res.* **53** 14876
[12] Kim G, Cho Y S and Kim W D 2006 *Eur. Polym. J.* **42** 2031

Acknowledgements

This work was partly supported by the Chang Jiang Youth Scholars Program of China and grants (51373033 and 11172064) from the National Natural Science Foundation of China to Prof. Xiaohong Qin. As well as "The Fundamental Research Funds for the Central Universities" and "DHU Distinguished Young Professor Program" to her. It also has the support of the Key grant Project of Chinese Ministry of Education (No 113027A). This work has also been supported by "Sailing Project" from Science and Technology Commission of Shanghai Municipality (14YF1405100) to Dr. Hongnan Zhang. This work was also supported by the Fundamental Research Funds for the Central Universities (CUSF-DH-D-2016008) to Mr. Liang Wei.

Effect of incident kinetic energy on a-Si:H structure: A molecular dynamics simulation study

Yaorong Luo[1], Naigen Zhou[1 *], Hongyong Gong[2], Haibin Huang[3] and Lang Zhou[3]

[1] School of Materials Science and Engineering, Nanchang University, Nanchang 330031, China
[2] School of New Energy Science and Engineering, Xinyu College, Xinyu 338004, China
[3] Institute of Photovoltaics, Nanchang University, Nanchang 330031, China
* E-mail: ngzhou@ncu.edu.cn

Abstract. This paper utilized molecular dynamics simulation to investigate the influences of the incident kinetic energy on the structure of hydrogenated amorphous silicon (a-Si:H) thin films. The SiH_3 radical impinged on the Si (001) surface with substrate temperature of 500 K in the incident kinetic energy range from 0.04 to 5.81 eV. The results showed that high incident kinetic energy could smooth the surface, promote the densification of internal structure, and increase the adsorption rate of SiH_3 radical and the crystalline volume fraction. The SiH_x and dangling bonds, which are the key factors that affecting the film quality, have been further analyzed, the results reflected that the increase of the incident kinetic energy will increase the content of SiH combination pattern and decrease the dangling bonds in the films, thus improving the quality of the thin films. There is a critical value, 0.64 eV, 1.45 eV or 4.04 eV, of this variation rate. Steep variation trends is corresponding to the incident kinetic energy less than this critical value, and variation trends is gentle for that above it.

1. Introduction

Hydrogenated amorphous silicon (a-Si:H) thin films have played a crucial role as intrinsic absorber layers with doped layers in heterojunction (HIT) solar cells [1, 2]. The quality of a-Si:H thin films will determinate the conversion efficiency of HIT solar cells. The structures of thin films, including dangling bond and the proportion of SiH_x are one of the key factors which influencing the quality of a-Si:H thin films. Full understanding microstructural information is helpful to building the process-structure-property relationship, even rising high-efficieney HIT solar cells. However, the deposition conditions are interlaced each other. It is difficult to investigate the relationship between atomic structures of a-Si:H and single condition experimentally. The molecular dynamics (MD) simulation is a useful tool for investigating atomic arrangement [3] and the structural evolution [4]. Therefore, we utilizing molecular dynamics (MD) simulation to explore the microstructural information behind different incident kinetic energy. It is expected to find the correlation between the process condition and microstructure and provide theoretical basis for the experiment.

2. Computational model and methods

2.1. Computational details

The interatomic interactions are described by an extended form of the Tersoff potential, which was first proposed by Tersoff [5] and developed by Murty and Atwater [6], includes Si-Si, Si–H and H–H interactions.

A single crystal silicon of 22 atomic layers is used as the substrate. The simulation cell consisting of three regions is represented in figure 1: (i) the five bottom layers are fixed to their perfect lattice positions, (ii) the middle nine layers above the fixed are utilized Langevin method to control the temperature of 500 K, (iii) the eight top layers can move freely. Periodic boundary conditions (PBCs) are applied in both the x- and y- directions. The initial size of the model is Lx × Ly × Lz = 81.8 Å × 81.8 Å × 28.9 Å.

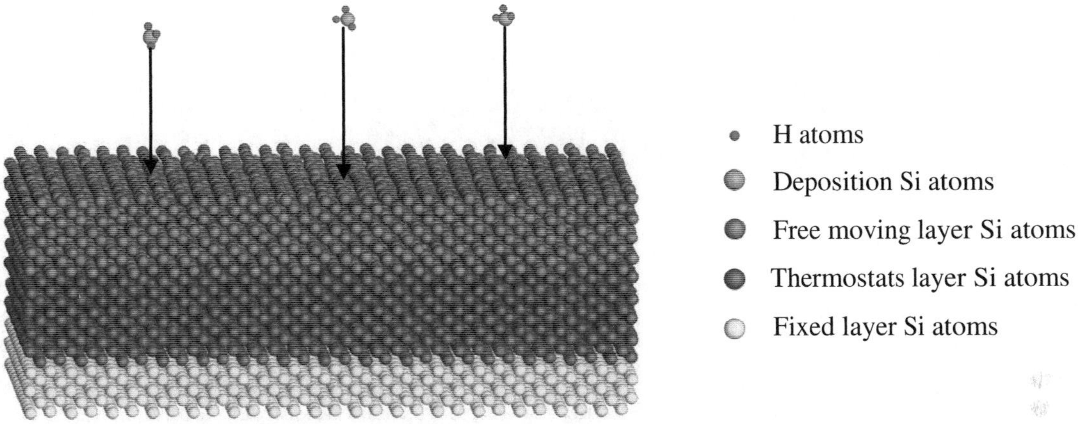

Figure 1. The schematic diagram of the deposition of a-Si:H/c-Si thin films system.

In the process of thin films deposition, the radicals deposit on the crystal Si (001) surface, and the detail parameters are as follows: Substrate temperature is 500 K, incident frequency is one radical per picosecond, the total number of the incident radicals are 8000. The incident radical is SiH_3, which is believed to be the main films precursors for high-quality films. The time step used in this simulation is 1.0 fs. The incident kinetic energies are selected as 0.04, 0.64, 1.45, 2.58, 4.04 and 5.81 eV, respectively. The final configuration obtained at different incident kinetic energy is quenched to 0.01 K or lower (eliminating thermal vibrations) before analyzing its characteristics.

2.2. Characterization methods

Surface-roughness [7] is a scalar that represents the surface-roughness of a-Si:H thin films. Surface-roughness (Rq) is defined as

$$R_q = \left(\frac{\sum_{i=1}^{N} (Z_i - \overline{Z})^2}{N} \right)^{1/2} \tag{1}$$

where: Z_i-the height of the i-th exposed atom on the film surface; $\overline{Z}$-the mean height of all the surface atoms; N-the number of the surface atoms.

Uttormark etc [8, 9] proposed a characterization index (A, equation (2)) to determine whether an atom is at crystalline state or not. The crystalline volume fraction is the ratio of the number of crystalline atoms to the total atomic number in the thin film.

$$A = \sum_{i=1}^{6} \left(\cos\theta_i + 1/3 \right)^2 \tag{2}$$

3. Results and discussion

Figure 2 (a) and (b) have shown the structures of the a-Si:H/c-Si thin films system with the incident kinetic energy of 0.04 and 5.81 eV, respectively. As can be seen, the crystal structure is at the bottom and the amorphous is at the top. For the incident kinetic energy of 0.04 eV, the atoms on the surface of the thin films emerge island structure, loosely connection and very rough appearance. For the incident kinetic energy of 5.81 eV, the surface of the film is smoother than the former.

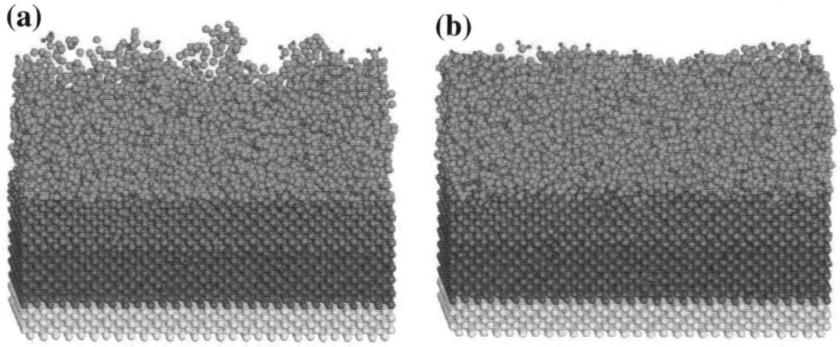

Figure 2. The structure of the a-Si:H/c-Si thin films depositing at the incident kinetic energy of 0.04 eV (a) and 5.81 eV (b), respectively.

In order to obtain the quantitative differences, the surface-roughness and relative atomic mass density of the films with different kinetic energy are analyzed, the results are shown in figure. 3. With the incident kinetic energy increasing, the surface-roughness decreased. In the range of lower incident kinetic energy (0.04-1.45 eV), the downward trend is steep, and then, as the incident kinetic energy continues to increase, the downward trend become a little of gentle. The surface structure is smoother obtained at higher incident kinetic energy. As we can see from figure 3b, the relative atomic mass density increases with increasing the incident kinetic energy and gradually approach to 1. Its variation trend is opposite to that of surface-roughness. The above results show that: Increasing incident kinetic energy, the surface of the films is more smooth and flat, and its internal structure is more compact. The better performance of a solar cell is predictable.

We further analyzed the adsorption rate of SiH_3 radical with different incident kinetic energy. As shown in figure 4a, the adsorption rate increases as the incident kinetic energy increasing. When the incident energy is below 0.64 eV, the radical cannot be effectively adsorbed onto the surface. It is easy to combine with the surface H atoms of the film and then return to the gas phase as the SiH_4. When the incident energy is more than 0.64 eV, the adsorption rate tend to stable at about 0.97.

Figure 3. The surface-roughness (a) and relative atomic mass density (b) of the films depositing at different incident kinetic energy.

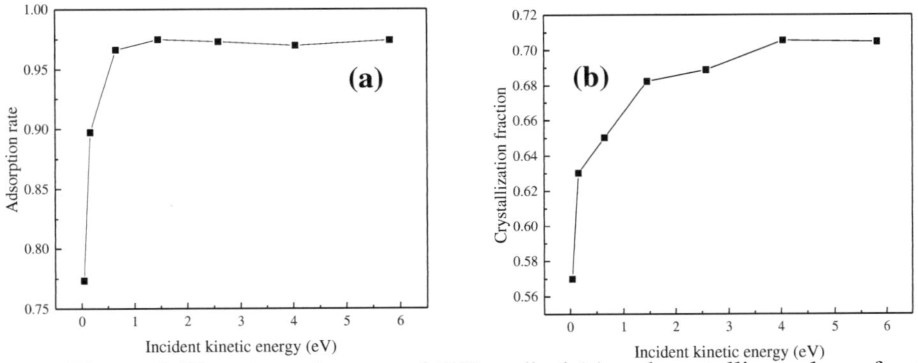

Figure 4. The adsorption rate of SiH_3 radical (a) and crystalline volume fraction (b) of the films depositing at different incident kinetic energy.

The crystalline volume fraction of the thin film layer at different incident kinetic energy is shown in figure 4b. The crystalline volume fraction rose gradually when the incident kinetic energy increased before 4.04 eV. After that value, the crystalline volume fraction tended to stable. It means that the incident kinetic energy can induce crystallization of atoms and improve the order degree of the atomic structure in the range of 0.04 to 4.04 eV. However, the local paracrystalline structure arrangement is not obvious varied after the energy over 4.04 eV.

The incident kinetic energy affects not only the structure of the silicon thin films, but also the combination pattern of hydrogen and silicon atoms. Besides the Si-Si bond and Silicon dangling bond, there are many SiHx radical, include SiH, SiH_2 and SiH_3. All of them will influence the conversion efficiency of a-Si:H/c-Si thin films heterojunction solar cells [10]. However, it has been reported [11] that only the SiH can reduce the density of defect states in the band gap and improve properties of the thin films.

The evolutions of dangling bond, SiH, SiH_2 and SiH_3 in the thin films with different incident kinetic energy are shown in figure 5. No matter what the incident kinetic energies are, the SiH combination pattern is the dominated, the contents of SiH_2 and SiH_3 are very less. Moreover, the content of SiH combination pattern increases with the increase of the incident kinetic energy, the contents of SiH_2 and SiH_3 decrease. This phenomenon can be understood from two aspect. One is that the initial SiH_3 radical changed easily to SiH due to more H atoms escaping with high kinetic energy. The other is that the higher incident kinetic energy will induce the more crystalline volume. The content of SiH_2 and SiH_3 will decrease because the more Si atoms have replaced the sites of H atoms during the process of crystalline. The content of dangling bond decreases with increasing the incident kinetic energy (figure 5a). The above two results show that the increase of the incident kinetic energy will increase the content of SiH combination pattern and decrease the dangling bonds in the films, thus improving the quality of the thin films.

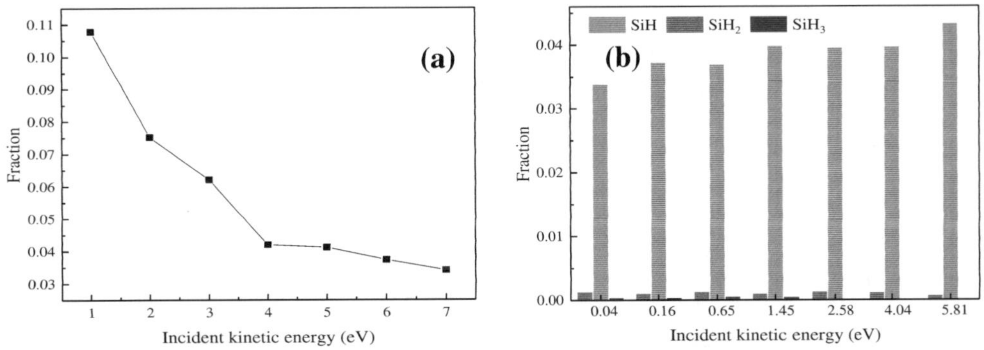

Figure 5. The evolutions of dangling bond (a), SiH, SiH_2 and SiH_3 (b) in the thin films with different incident kinetic energy.

4. Summary

In the present work, the influence of the incident kinetic energy on the structure of hydrogenated amorphous silicon thin film is investigated on atomic scale using molecular dynamics simulation. The results indicated that the surface-roughness decreases, internal structure is more compact and the probability of appearing the cavity structure decreases as the incident kinetic energy is from 0.04 to 5.81 eV. The content of SiH combination pattern increases and the dangling bonds in the films decreases with increasing the incident kinetic energy, and consequently improving the quality of the thin films. There is a critical value of these variation rate. Steep variation trends is corresponding to the incident kinetic energy less than this threshold value, and gentle variation trends for that above it. For the adsorption rate, this critical value is 0.64 eV; For surface-roughness and relative atomic mass density, the value is 1.45 eV; For crystalline volume fraction is 4.04 eV.

5. References

[1] Stuckelberger M, Biron R, Wyrsch N, Haug F J and Ballif C 2016 *Renewable Sustainable Energy Rev.* **76** 1497–1523
[2] Polman A, Knight M, Garnett E C, Ehrler B and Sinke W C 2016 *Science.* **352** aad4424
[3] Sheng H W, Luo W K, Alamgir F M, Bai J M, and Ma E 2006 *Nature.* **439** 419-425
[4] Wu Z W, Li M Z, Wang W H and Liu K X 2015 *Nat. Commun.* **6** 6035
[5] Tersoff J 1988 *Phys. Rev. B* **37** 6991-7000
[6] Murty M V R and Atwater H A 1995 *Phys. Rev. B* **51** 4889-4893
[7] D Huang, Pu J, Lu Z and Xue Q 2012 *Surf. Interface Anal.* **44** 837-843
[8] Uttormark M J, Thompson M O and Clancy P 1993 *Phys. Rev. B* **47** 15717-15726
[9] Albenze E J, Thompson M O and Clancy P 2004 *Phys. Rev. B* **70** 094110
[10] Olibet S, Vallat-Sauvain E and Ballif C 2007 *Phys. Rev. B* **76** 035326
[11] Robertson J 2000 *J. Non-Cryst. Solids.* **266–269** 79-83

Acknowledgements

This work is supported by the National Natural Science Foundation of China (No. 51561022, 51361022 and 61306084) and the Natural Science Foundation of Jiangxi Province (No. 20151BAB206001).

Physical models for MoS_2 and their application to simulations of MoS_2 MSM device

Shan Zhang[1], Hongbin Pu[1] and Yong Yang[1]
[1] Department of electronic engineering, Xi'an university of technology, Xian, 710082, PR China
E-mail: zhsirving@hotmail.com

Abstract. MoS_2 is a kind of two-dimensional semiconductor material with excellent electrical and optical properties, which can be used to develop new types of electronic devices and optoelectronic devices. Here, we established the mobility model of the MoS_2. It is found that the simulation results agree with the experimental data. And our model are better than the theoretical results at low temperature. At the same time, we used the model to simulat the MoS_2 MSM device. The I – V curve are basically identical with the measurement results, which proves the practicability of the model.

1. Introduction

In the last few years, bidimensional materials have driven the attention of the scientific community as one of the best alternatives to the traditional Si-based technology, among which MoS_2 is a typical materials. The bandgap of MoS_2 can be modified by changing its thickness or applying stress and interesting optoelectronic properties of MoS_2 have also been demonstrated because of the direct bandgap behavior of monolayer films[1]. Comparing to one-dimensional materials, MoS_2 materials are easy to fabricate complex structures and attractive for using in next-generation electronic devices.

In order to analyze the performance of MoS_2-based devices, the mobility of MoS_2 need to be studied. Recently, Sunkook Kim[2] has obtained the mobility by measuring the Ids - Vgs curve of MoS_2 FET at room temperature. Others like J.M. Gonzalez-Medina[3] has theoretically calculated the mobility of MoS_2, which are in good agreement with Sunkook Kim's data. In this paper, the analysis model of MoS2 carrier mobility will be established and investigated by comparing with other groups experimental and theoretical result. Further, the MoS_2 MSM device has been fabricated and test the analysis model.

2. Physical models

The mobility is affected by several scattering mechanisms, including lattice scattering, ionized impurity scattering, carrier-carrier scattering and so on. The structure of MoS_2 is very similar to the graphene, S and Mo atoms are covalently bonded in the monolayer. And there is no dangling bond on the surface. So we focus on three different scattering mechanisms such as acoustic phonons, optical intervalley phonons and ionization impurities, and establish the mobility models for each scattering mechanism. Further, the combined mobility, which includes the three mechanisms above, is[4,5]

$$u_T^{-1} = u_{ac}^{-1} + u_b^{-1} + u_{im}^{-1} \quad (1)$$

Where u_{ac} is the mobility from acoustic phonons interaction, u_b is the mobility from optical intervalley phonons interaction and u_{im} is the mobility from ionization impurities interaction.

2.1. Acoustic phonons

In the two-dimensional deformation potential theory of surface phonon scattering and for a

IOP Conf. Series: Materials Science and Engineering **284** (2017) 012007 doi:10.1088/1757-899X/284/1/012007

nondegenerate surface, the electron mobility is given by [5]:

$$u_{ac} = q\hbar\rho\, u_l^2 / m^* m_u Z_A^2 \tag{2}$$

Where q is the elementary charge, ħ is the Dirac constant, u_l is the sound velocity, m* and mu are the effective mass and the mobility mass, respectively, Z_A is the deformation potential, k is the Boltzmann constant, and T is the absolute temperature. The areal mass density of MoS_2, ρ(g/cm^2), is given by

$$\rho = \rho_{bulk} z_{av} \,. \tag{3}$$

Where ρ_{bulk} is the mass density (g/cm^3) of MoS_2 and Z_{av} is the effective thickness of the inversion layer. Z_{av} was assumed to be independent of temperature and normal electric field. Here, Z_{av} adopts the approximation suggested by Schwarz and Russek[6]

$$z_{av} = z_{CL} + z_{QM} \,. \tag{4}$$

$$z_{CL} = 3kT/2qE_{eff} \,. \tag{5}$$

$$z_{QM} = \left(\frac{9\hbar^2}{4m_\perp qE_{eff}} \right)^{1/3} . \tag{6}$$

$$E_\perp = E\sin\phi = |E \times J| / |J| \leftrightarrow E_{eff} \,. \tag{7}$$

So the acoustic phonons (LA and TA) is expressed as:

$$u_{ac} = \frac{a}{(E_\perp/E_1)^b} + \left(\frac{c(N/N_1)^d}{(E_\perp/E_1)^e} \right) \cdot \frac{1}{T_n} \,. \tag{8}$$

$$T_n = \left(\frac{T_L}{300} \right)^f . \tag{9}$$

Where T_L is the lattice temperature, $E_\perp$ is the perpendicular electric field. a, b, c and d are the empirical constants, respectively. f is the temperature dependence of the probability of surface phonon scattering and N is the total doping concentration. The E_1 and N_1 are the constant with a value of 1 V/cm and 1cm^{-3}, respectively.

2.2. *Optical intervalley phonons*

At low temperature, the acoustic phonon vibration is the main influence factor to the mobility. however, when the temperature increased, optical intervalley scattering play a major role.

The mobility (u_b) from optical intervalley phonons interaction adopts the empirical model in a slightly modified form which preserves their model parameters fitted at room temperature[5,6].

$$u_b = u_0 + \frac{u_{max}(T) - u_0}{1 + (N_A/C_r)^\alpha} - \frac{u_1}{1 + (C_S/N_A)^\beta} \,. \tag{10}$$

$$u_{max}(T) = u_{max} \left(\frac{T}{300} \right)^{-\gamma} . \tag{11}$$

Where T is the temperature, u_l is the sound velocity, N_A is the acceptor concentration. Others parameters are the empirical constants.

So this component is given by:

$$u_b = f \times e^{-\frac{h}{N}} + \frac{\left[g \times (T_L/300)^{-a} - f\right]}{1+(N/b)^c} - \frac{f}{1+(d/N)^e} \tag{12}$$

Here, N is the total density of impurities and T_L is the lattice temperature, a, b, c, d, e and f are the empirical constants, respectively.

2.3. *Ionization impurities*

Ionized impurities are mainly affected by temperature and impurity concentration. The mobility from ionization impurities interaction, u_{im} can be described by:

$$u_{im} = Z^{1-\alpha_1} u_{i.n} \left(\frac{T}{300}\right)^{3\alpha_1 - 1.5} \left(\frac{N_1}{Z^2 N_1}\right) + u_{i.c} \left(\frac{300}{T}\right)^{0.5} \left(\frac{c}{Z^2 N_1}\right) . \tag{13}$$

Where N_1 is the constant with a value of $1cm^{-3}$, T is the temperature, Z and c are the empirical constants, respectively. In the calculations, $Z^{1-\alpha_1} \cong 1$, the parament α_1 stems from the description of the majority carrier mobility as a function of impurity concentration[7].

It can be seen that the second part plays a dominant role in the case of high doping, so the impurity scattering components is obtained using the following equation:

$$u_{im} = \frac{a^2}{a-b} \times \left(\frac{T_L}{300}\right)^{(3\times c - 1)} . \tag{14}$$

Where T_L is the lattice temperature, a, b and c are the empirical constants, respectively.

3. Results and discussion

3.1 *Mobility model*

Figure 1 shows the simulation results of MoS_2 carrier mobility model. We can see that the u_{ac}, u_b and u_{im} fitting well with the Kim's theoretical results in Fig. 1(a), (b) and (c), which can be explained that the model is reasonable. And in Fig. 1(d), we can find that the combined mobility is very consistent with the J.M.'s experimental data. Meanwhile, the model is better than the Kim's theoretical calculation at low temperature.

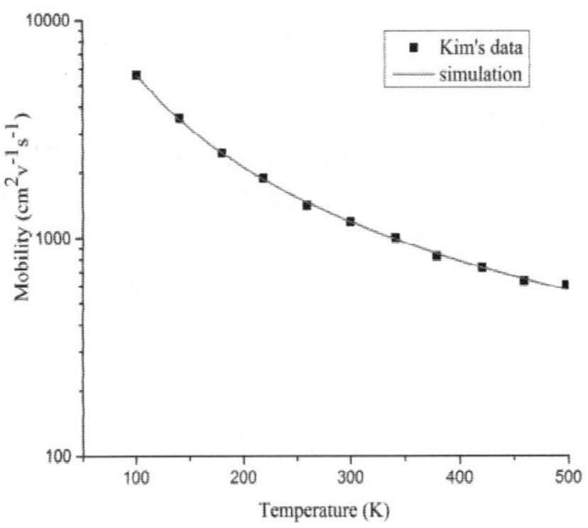

(a)

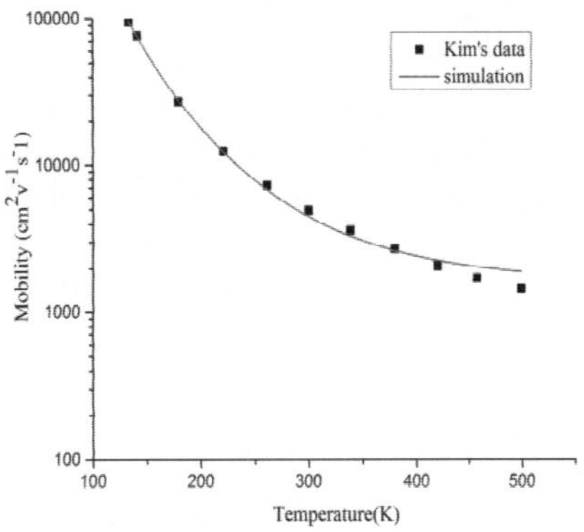

(b)

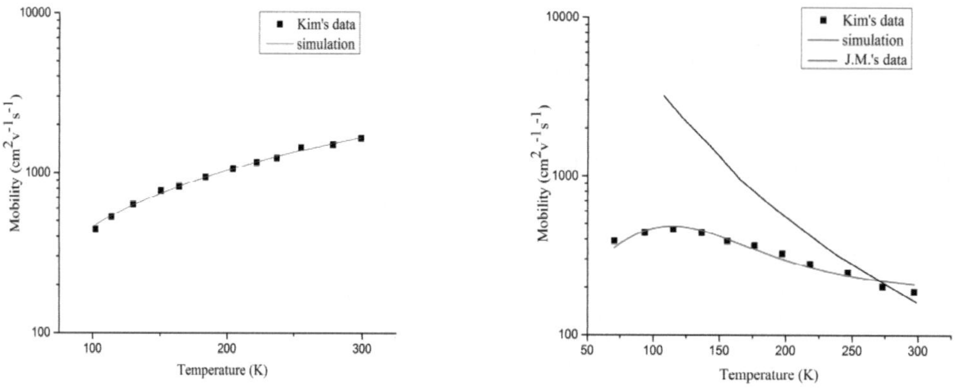

(c) (d)

Figure 1 the simulation results of MoS_2 mobility model, (a) acoustic phonon mobility (b) optical intervalley phonons (c) impurity scattering and (d) comparison of the combined mobility.

3.2. Mobility model test

The MSM device adopted Ni electrodes with Hexagon MoS_2/Al_2O_3 substrates and measured the I – V curve. The mobility model is applied to simulating the same structure device and compared with the experimental results in Figure 2. The results are in good agreement with the experimental data.

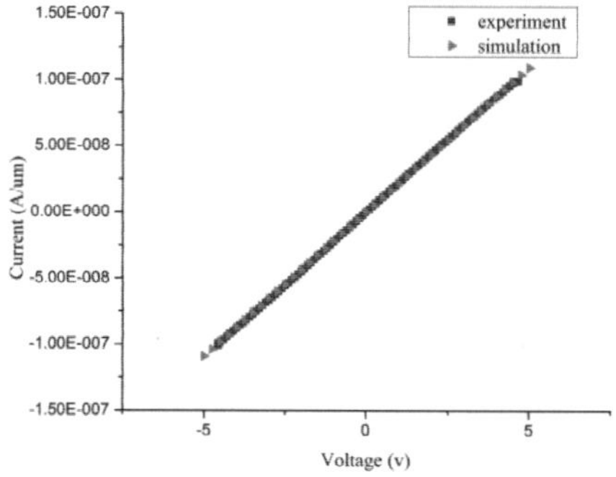

Figure 2. MSM simulation used the mobility model.

4. Summary

Physical models for MoS_2 have been established and investigated. And we have stimulated the characteristic of the model and compared with the experimental data and the theoretical calculation. According to the results, it is can been seen that u_{ac}, u_b and u_{im} is the important part of the MoS_2 mobility. Finally, the MoS_2 MSM device has been fabricated and simulated to verify this model. the results indicate that the new model is reasonable.

5. References

[1] Radisavljevic B, Radenovic A, Brivio J, et al. Single-layer MoS_2 transistors[J]. *Nature Nanotechnology*, 2011, **6(3)**:147.
[2] Kim S, Konar A, Hwang W S, et al. High-mobility and low-power thin-film transistors based on multilayer MoS_2 crystals[J]. *Nature communications*, 2012, **3**: 1011.
[3] González-Medina J M, Ruiz F G, Marin E G, et al. Simulation study of the electron mobility in few-layer MoS_2 metal–insulator-semiconductor field-effect transistors[J]. *Solid-State Electronics*, 2015, **114**: 30-34.

[4] Lombardi C, Manzini S, Saporito A, et al. A physically based mobility model for numerical simulation of nonplanar devices[J]. *IEEE Transactions on Computer-Aided Design of Integrated Circuits and Systems*, 1988, **7(11)**: 1164-1171.
[5] Reggiani S, Valdinoci M, Colalongo L, et al. A Unified Analytical Model for Bulk and Surface Mobility in Si n-and p-Channel MOSFET's[C]//Solid-State Device Research Conference, 1999. Proceeding of the 29th European. IEEE, 1999, 1: 240-243.
[6] Schwarz S /, Russek S E. Semi-empirical equations for electron velocity in silicon: Part II—MOS inversion layer[J]. *IEEE Transactions on Electron Devices*, 1983, **30(12)**:1629-1633.
[7] Klaassen D B M. A unified mobility model for device simulation[C]// Electron Devices Meeting, 1990. IEDM '90. Technical Digest. International. IEEE, 1991:357-360.

Acknowledgments
This work was supported by the National Nature Science Foundation of China [grant numbers 51677149].

IOP Conf. Series: Materials Science and Engineering **284** (2017) 012008 doi:10.1088/1757-899X/284/1/012008

Influence of surface rectangular defect winding layer on burst pressure of CNG-II composite cylinder

H X You[1], L Peng[2], C Zhao[3], K Ma[4] and S Zhang[5]

[1,2,3,4,5] School of Chemical Machinery and Safety Engineering, Dalian University of Technology, Dalian 116023, Liaoning, China
E-mail: youhx@sina.com pl2012znx@163.com.

Abstract. To study the influence of composite materials' surface defect on the burst pressure of CNG-II composite cylinder, the surface defect was simplified as a rectangular slot of certain size on the basis of actually investigating the shape of cylinder's surface defect. A CNG-II composite cylinder with a rectangular slot defect (2mm in depth) was used for burst test, and the numerical simulation software ANSYS was used to calculate its burst pressure. Through comparison between the burst pressure in the test and the numerical analysis result, the correctness of the numerical analysis method was verified. On this basis, the numerical analysis method was conducted for composite cylinders with surface defect in other depth. The result showed that surface defect in the form of rectangular slot had no significant effect on the liner stress of composite cylinder. Instead, it had a great influence on the stress of fiber-wrapped layer. The burst pressure of the composite cylinder decreased as the defect depth increasing. The hoop stress at the bottom of the defect in the shape of rectangular slot exceeded the maximum of the composite materials' tensile strength, which could result in the burst pressure of composite cylinders decreasing.

1. Introduction

High-strength fiber is used to wrap pressure vessel, and pressure vessel made of composite materials can be obtained after the fiber being solidified. The composite materials, characterized as high strength, large elasticity modulus, good design property, excellent fatigue resistance, outstanding failsafe performance and superior manufacturability [1], significantly improve the overall performance of the vessel. Therefore, pressure vessels made of composite materials have been widely applied in a variety of fields [2,3,4]. The fiber-wrapped composite cylinder (CNG-II composite cylinder) is a typical type of composite pressure vessel. However, the composite materials also have some drawbacks, such as having a low surface hardness, being vulnerable to scratch, and having poor wearability [5] which cause the fiber-wrapped layer to be unavoidably to generate defect during the production and the usage of CNG-II composite cylinders. It has also been observed that the damage is generally comprised of small-scale scratches, pits or wearing. According to the provisions [6] of Chinese National Standards *GB24162-2009 Periodic Inspection and Evaluation of Metal-lined and Fiber-Wrapped Composite Cylinders for Compressed Natural Gas Used by Automotive Vehicles*, the allowable and repairable defect depth on composite materials' surface is 1.25mm. However, inspectors usually perform the procedures based on experience instead of with a mind to strict compliance to this standard in the periodic inspection and scrapping determination for cylinders. At the same time, the absence of strict computational and experimental verification during the standard execution makes it highly possible that the composite cylinders still have great residual strength [7, 8] even when the defect is deeper than 1.25mm. The investigation of the damage tolerance and residual strength of composite cylinders with surface defect of certain size is necessary, and it benefits the safety management of composite cylinders. To obtain the damage tolerance of composite cylinders, it is essential to study the influence of surface defect on composite cylinders' stress.

The research needs to be carried out in the form of experiment and numerical simulation for researching the influence of surface rectangular defect on composite cylinders. Hydraulic burst test was conducted to obtain the ultimate burst pressure of the defective cylinders in the experiment, and numerical simulation was used to study the reason why the defect influenced the burst pressure. Numerical simulation analysis can be applied for defect in different depth, but the accuracy of numerical simulation needs to be verified by experiment. Through comparing the test result and the consequence of numerical analysis, the correctness of the modeling and analysis in numerical simulation was verified. On this basis, calculation was made for composite cylinders with other defect depth by means of numerical simulation. The ANSYS software has won recognition for related researcher with its powerful functions and can be applied to the analysis of composite materials with defect [9, 10].

On the basis of hydraulic burst test for the composite cylinders with surface rectangular defects, the paper used ANSYS software to analyze the stress field of defective cylinders. After verifying the correctness of finite element model by means of experiment, underlying causes were analyzed for the stress changes in the liner and the fiber-wrapped layer as well as changes in burst pressure.

2. Test methods

Hydraulic burst method was adopted to obtain the burst pressure of the composite cylinders with surface defects, and the test apparatus was capable to record pressure automatically. A G300LVE gas-driven pump was used for boosting pressure; a pressure sensor was used for measuring pressure during the pressure-boosting process. The measuring range of the pressure sensor was 0~150MPa, and the accuracy of the pressure sensor was Grade 0.5. The measuring range of the pressure gauge was 0~160MPa and the accuracy of the pressure gauge was Grade 1.6. Corresponding software was used to record the pressure changes during the process of hydraulic burst test.

The CNG-II composite cylinder used in the test was 406 mm in diameter and the size of the liner is shown in Figure 1. The fiber-wrapped layer was 8.4mm in thickness and the cylinder was 100L in volume. The liner was made of 30CrMo, and the composite materials comprised ECR560-2400 glass fiber and epoxy resin. The mechanical properties of the liner and the composite materials is shown in Table 1. The performance parameters of the composite materials are shown in Table 2. And the burst pressure of the defect-free composite cylinders with the same size in the test is 50-55MPa.

The defect on the surface of the composite cylinder were in the shape of a rectangular slot, and the defect was in the center of the composite cylinder. The scratch, which is along the axial direction of the cylinder, was 2mm in depth, 50mm in length, and 3.5mm in width, is shown in Figure 2.

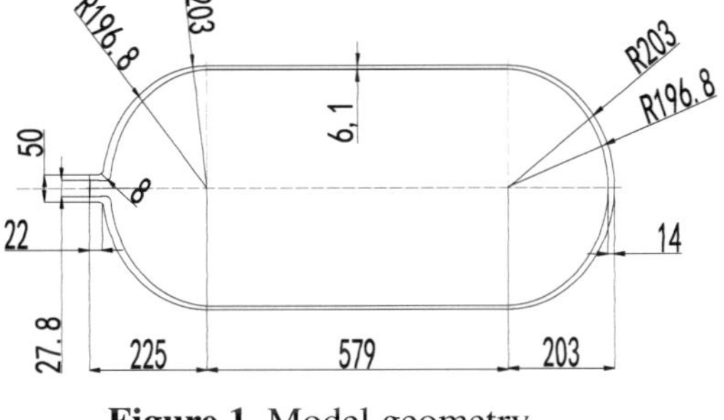

Figure 1. Model geometry

Figure 2. The shape of the defect

Table 1. The mechanical properties of the liner and the composite layer

Material	Modulus of elasticity (GPa)	Poisson's ratio	Yield strength (MPa)	Tensile strength(MPa)	Elongation (%)
30CrMo	200	0.29	≥720	≥820	≥14
ECR560-2400 glass fiber/epoxy resin composite materials	≥81.49			≥2625	

Table 2. The performance parameters of the composite materials

E_X (GPa)	E_Y (GPa]	E_Z (GPa)	V_{XY}	V_{YZ}	V_{XZ}	G_{XY} (GPa)	G_{YZ} (GPa)	G_{XZ} (GPa)
33	8.97	8.97	0.26	0.05	0.26	6.2	2.0	2.0

The specific test procedures were as follows: The cylinder with scratch defect was fully filled with water first. Then it was fitted with a connector. A blasting machine was used for boosting water pressure. The pressure was boosted after air was exhausted. As the pressure reached approximately 43MPa, it was observed that the fiber-wrapped layer of the composite cylinder was initially broken at the zone of rectangular slot defect with the pressure dropping. Then the pressure was fluctuated and the fiber-wrapped layer continued to rupture with crack. Finally, the liner ruptured at approximately 26MPa. The shape of the burst composite cylinder is shown in Figure 3. The data of the hydraulic burst test was recorded and the time-pressure curve is shown in Figure 4.

Figure 3. The shape of the burst composite cylinder

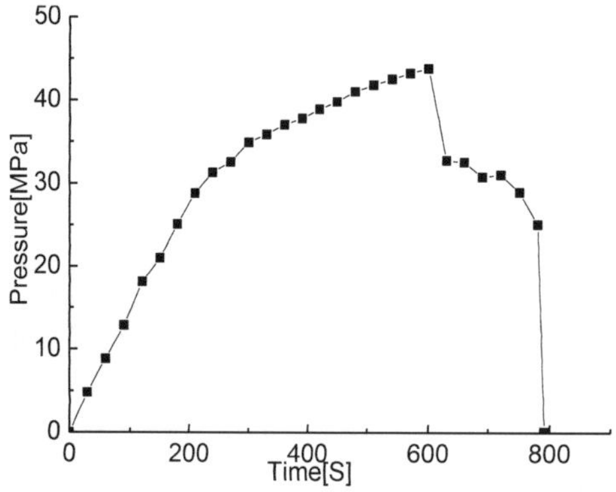

Figure 4. The time-pressure curve

From the test it is evident that the burst pressure of the composite cylinder with the rectangular slot (2mm in depth, 50mm in length and 3.5mm in width) was 43.87MPa. Compared with the burst pressure of the defect-free composite cylinders of the same size, there was a significant decrease in burst pressure for composite cylinders with rectangular slot defect on their surfaces.

3. Finite element analysis

3.1 Model establishment

Numerical analysis method was conducted for composite cylinder with the rectangular slot defect (2mm in depth, 50mm in length and 3.5mm in width). Half of the CNG-II composite cylinder model was established because the defect, which was in the center of the cylinder and along the axial, only affected the local part of the cylinder. The geometric structure and material properties of the CNG-II composite cylinder used in ANSYS was as same as the cylinder used in the foregoing burst test. The geometric model of the defective composite cylinder is shown in Figure 5.

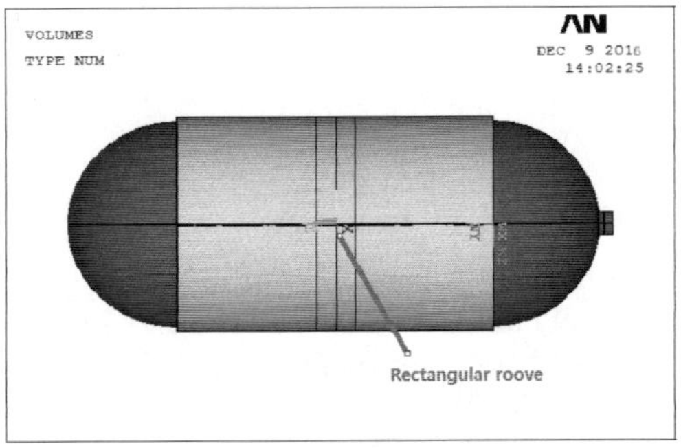

Figure 5. The geometric model of the defective composite cylinder

Finite element software ANSYS was used to make simulated calculation, and appropriate unit was selected according to the different model materials. The liner was made of isotropic material while the fiber-wrapped layer was made of transversely anisotropic material, which therefore meant that an 8-node isotropic unit solid45 was used for simulating the liner, and an 8-node anisotropic unit solid64 was used for simulating the fiber-wrapped layer. Since the cylinder is wrapped circumferentially, the composite fiber-wrapped layer was considered as a whole piece of anisotropic layer laminate, and the X axis of the fiber-wrapped layer unit coordinate system was adjusted along the fiber direction when establishing the model. The finite element model is shown in Figure 6.

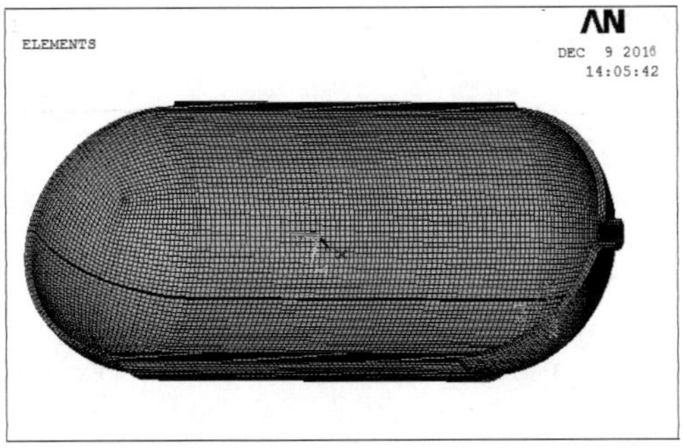

Figure 6. The finite element model

3.2 Constraint and load

The boundary conditions of the cylinder model were set as follows: Symmetry constraint was applied on the profile of the cylinder model. The equivalent tensile stress was applied on the top of the cylinder model and zero of axial displacement was applied at the bottom of the cylinder model.

Both normal load conditions and auto-frettage process should be taken into consideration during the ANSYS numerical simulation. The specific load conditions were as follows: the auto-frettage pressure was set as 34MPa, the operating pressure was set as 20MPa, and the maximal pressure was set as 65MPa to calculate the burst pressure easier. The load was set as: 0MPa→34MPa

→0MPa→20MPa→65MPa during the ANSYS numerical calculation. In addition, certain sub-steps were set for each load-step to ensure that the load was gradually increased.

3.3 Finite element calculation of burst pressure

It was first necessary to determine the burst criteria to calculate the burst pressure of the composite cylinder using numerical calculation method. The maximum stress criterion and the maximum strain criterion were used most popular. In the paper, it was concluded [11] that the maximum stress criterion had higher accuracy than the maximum strain criterion. At the same time, in light of that fact that the defect could give rise to discontinuity of the cylinder structure, stress singularity might occur to the local part of defect and consequently, calculation based on stress could potentially result in inaccuracy. Maximum strain criterion was therefore adopted for judgment. The basis of judgment for maximum strain criterion [12, 13] was the following.

$$\varepsilon_t = \varepsilon_f^h \tag{1}$$

Where, ε_t represents the maximum strain of the circumferential fiber layer; ε_f^h represents strain limit.

Multiple burst tests for the composite cylinders indicated that the hoop facture strain of the fiber-wrapped layer at the time of cylinder burst was 85% of that of pure fiber. The fracture strain of the pure glass fiber provided by the glass fiber manufacture r was 3%, and that therefore the fracture strain of the circumferential fiber was 0.0255.

Finite element calculation was carried out for the composite cylinder with the rectangular slot defect (D×L×X=2×50×3.5mm), and the curve of relation between the stress changes of the winding layer and the load changes is shown in Figure 7.

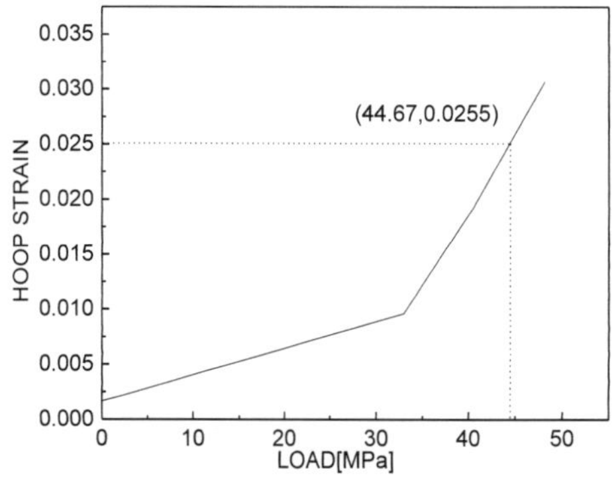

Figure 7. The curve of relation between the stress changes of the winding layer and the load changes

It can be seen from Figure 7 that while the facture strain was 0.0255, the corresponding load was 44.67MPa. Therefore, the burst pressure of the composite cylinder with rectangular slot defect (2mm in depth, 50mm in length and 3.5mm in width) was 44.67MPa. The starting point of the Y-axis in the Figure was not 0 owing to the fact that the previous auto-frettage treatment caused partial deformation to the fiber-wrapped layer.

Through comparison, the deviation between test result and the consequence of numerical calculation was 1.82%. This indicates that the finite element modeling was correct and that numerical calculating result was relatively accurate. So the same method can be adopted to calculate the burst pressure of the composite cylinders with the rectangular slot defect at different depth.

3.4 Burst pressure of the cylinder with various defect depths

The cylinder model above-mentioned was used for numerical calculation to evaluate the influence of rectangular slot defect with different depth on the burst pressure of CNG-II composite cylinder. The size of rectangular slot defect on the surface is shown in Table 3.

Table 3. The size of rectangular slot defect on the surface

The number of the defect	Length (mm)	Width (mm)	Depth (mm)	Analysis type
1	50	3.5	1.25	Numerical analysis
2	50	3.5	1.50	Experiment and Numerical analysis
3	50	3.5	1.75	Numerical analysis
4	50	3.5	2.00	Numerical analysis
5	50	3.5	2.25	Numerical analysis

Through ANSYS numerical calculation for the composite cylinder with rectangular slot defect (50mm in length, 3.5mm in width, and 1.25mm, 1.5mm, 1.75mm and 2.25mm in depth respectively), the hoop strain curve of fiber-wrapped layer with the increasing of load is shown in Figure 8.

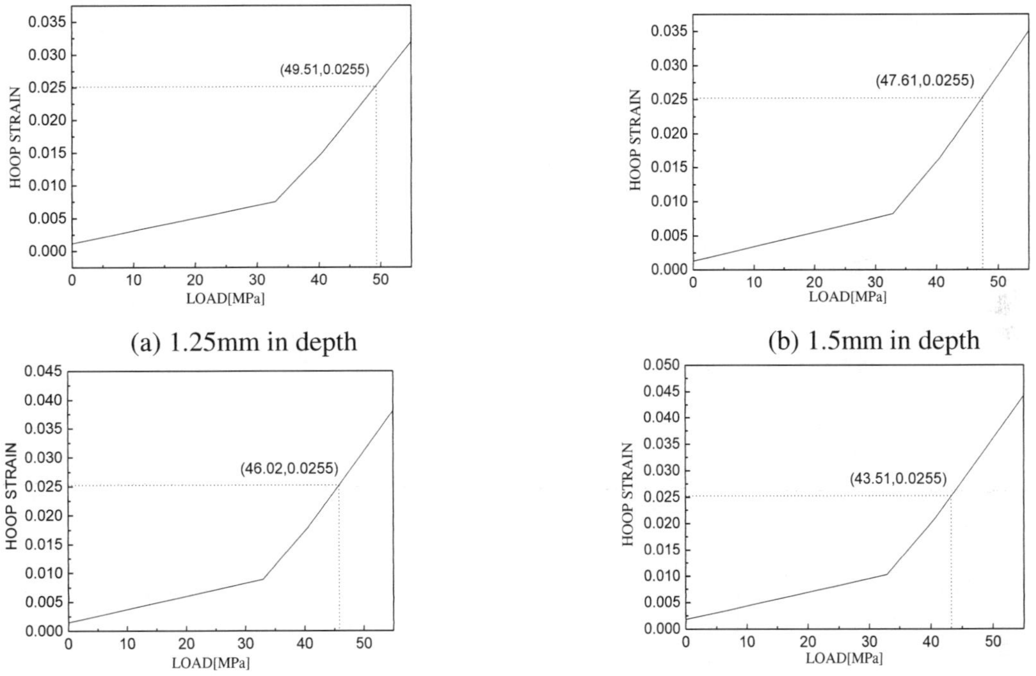

(a) 1.25mm in depth (b) 1.5mm in depth

(c) 1.75mm in depth (d) 2.25mm in depth

Figure 8. The curve of relation between the stress changes of the winding layer and the load changes

Figure 8. shows that the burst pressure of composite cylinders with rectangular slot defect (50mm in length, 3.5mm in width, and 1.25mm, 1.5mm, 1.75mm and 2.25mm in depth respectively) was 49.51MPa, 47.61MPa, 46.02MPa and 43.51MPa, respectively. And the burst pressure curve of defective cylinder with the increasing of defect depth is shown in Figure 9.

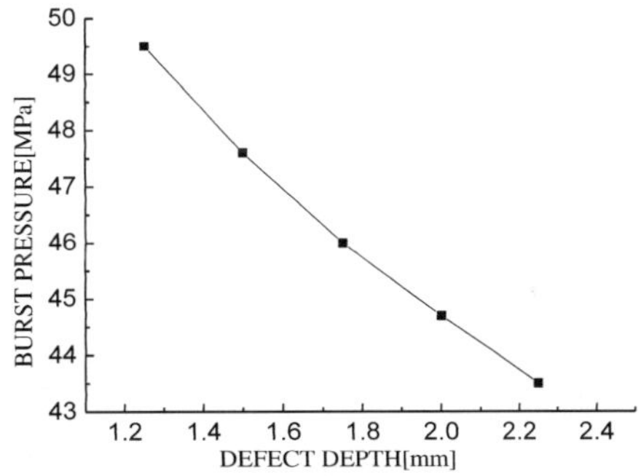

Figure 9. The burst pressure curve of defected cylinder with the increasing of defect depth

As seen from Figure 9, the burst pressure of defective composite cylinder decreased as the defect depth increasing. The stress analysis of liner and fiber-wrapped layer was taken into consideration to explain the fact that the burst pressure of defective composite cylinders decreased with the increasing of defect depth.

3.5 Stress analysis

Both the test result and the consequence of ANSYS numerical simulation indicated that the burst pressure of cylinder decreased because of the defect. Through analyzing the stress of the composite cylinder with rectangular slot defect (50mm in length, 3.5mm in width and 2mm in depth), it was calculated that the burst pressure of the defective cylinder above-mentioned was 44.67MPa.

When the cylinder was under burst pressure 44.67MPa, the Mises stress nephogram of the cylinder liner is shown in Figure 10, and the hoop stress nephogram of the fiber-wrapped layer is shown in Figure 11.

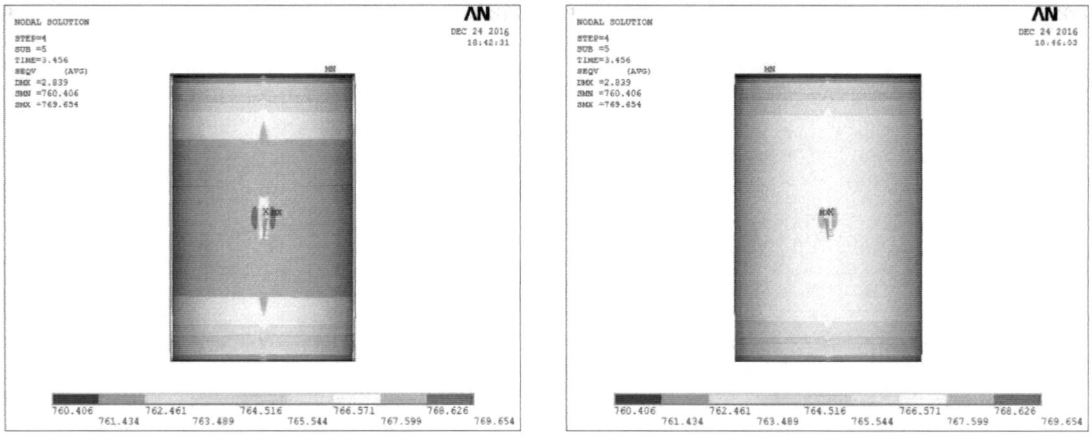

(a) The inner surface of the liner (b) The outer surface of the liner

Figure 10. The Mises stress nephogram of the cylinder liner under burst pressure 44.67Mpa

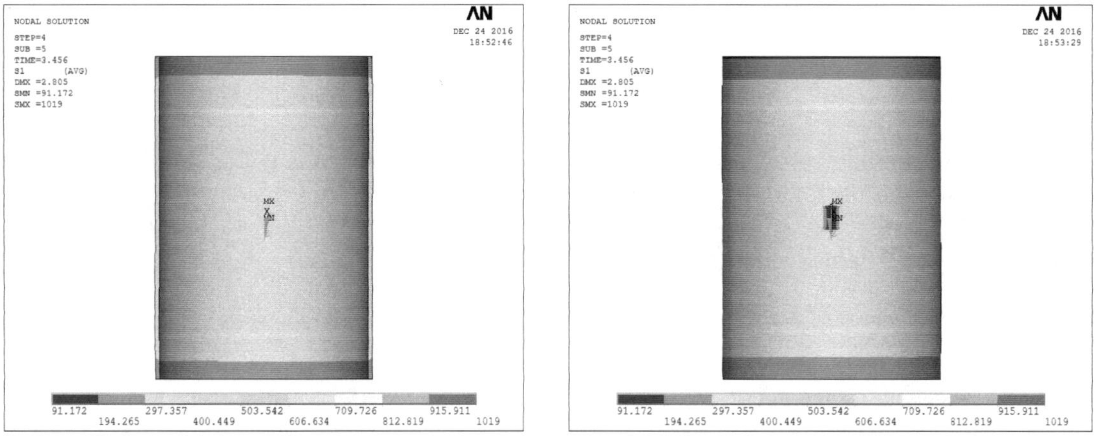

(a) the inner surface of the fiber-wrapped layer (b) the outer surface of the fiber-wrapped layer

Figure 11. The hoop stress nephogram of the fiber-wrapped layer

Figure 10 shows that the maximal Mises stress of the liner was in the center of the liner's inner surface when the loading pressure reached the burst pressure, namely 44.67MPa. Furthermore, the maximal Mises stress was 769.65MPa, which was less than the tensile strength of the liner materials, namely 820MPa. Therefore, the liner had no danger of bursting under 44.67MPa. Besides, it can be found that there was a little difference, approximate 1-2MPa, between the maximal Mises stress and the other Mises stress. That is, the surface defect of fiber-wrapped layer had little effect on the liner stress. Figure 11 indicates that the maximal hoop stress of fiber-wrapped layer, which was 1019.00MPa, is located at the defective area. And the maximal hoop stress exceeded the guaranteed value 995MPa of the tensile strength of fiber-wrapped layer, which had resulted in the decreasing of cylinder's burst pressure. The stress at the maximal stress point was, moreover, significantly greater than the stress at locations far away from the defective area. Therefore, the fiber-wrapped layer started to break from the maximal stress point, which conformed to the test observation. It can be concluded that the rectangular slot defect had significant influence on the stress distribution of fiber-wrapped layer.

The correctness of numerical analysis method was verified by the conformance between the test result and the consequence of numerical analysis. To study the influence of different defect depth on cylinder, the analysis method using for defect in 2mm depth was adopted for other defect depth. Figure 12 plots maximal Mises stress of the liner against five different defect depth. And Figure 13 plots maximal hoop stress of the fiber-wrapped layer against five different defect depth.

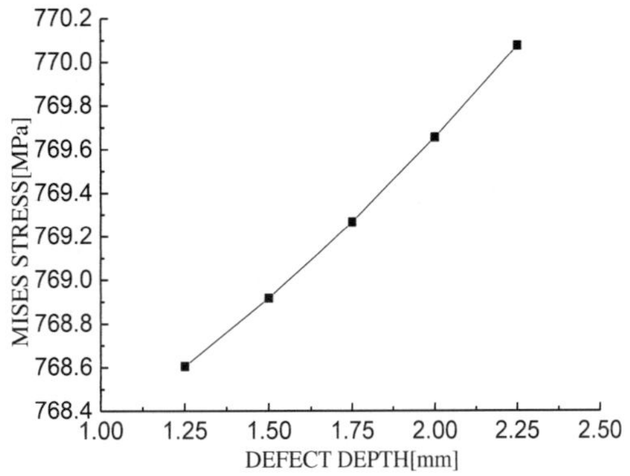

Figure 12. The relation maximal Mises stress of the liner against five different defect depth

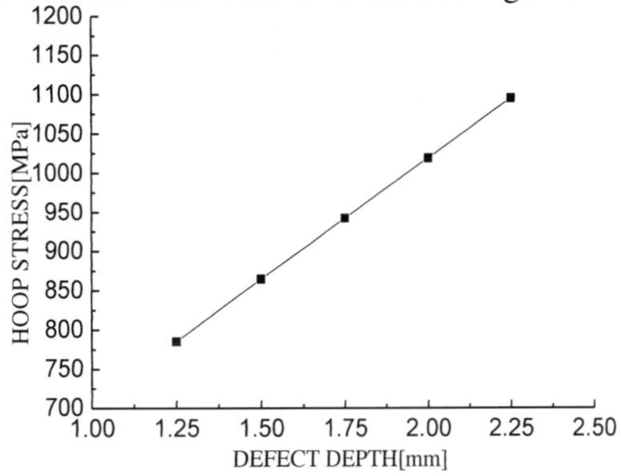

Figure 13. The relation maximal hoop stress of the liner against five different defect depth

Figure 12 indicates that the maximal Mises stress of liner rose, but slowly, as the defect depth increasing. The Mises stress of liner increased by 1.47MPa as the defect depth increasing from 1.25mm to 2.25mm, which further proved the conclusion that the defect had little influence on the liner. Figure 13 illustrates that the maximal hoop stress of fiber-wrapped layer rose, approximately 80MPa, as the defect depth increasing from 1.25mm to 2.25mm. This implies that the defect had larger effect on the fiber-wrapped layer. The maximal hoop stress of fiber-wrapped layer rose with the increasing of defect depth, and it could result in lower burst pressure of composite cylinder.

4. Conclusions

The method of finite element numerical simulation was used to study the influence of the surface defect, which was on the fiber-wrapped layer of composite cylinder, on the burst pressure. Hydraulic burst test was used to verify the correctness of the finite element model. Through analysis of stress changing law for each part of composite cylinder, it was concluded that the burst pressure of composite cylinder decreased because of the existence of surface defect. Surface defect had little influence on the liner stress of composite cylinder. The burst pressure of composite cylinder with surface defect decreased as the defect depth increasing, because the maximal hoop stress of fiber-wrapped layer rose with the increasing of defect depth.

5. Reference

[1] Zhang Z M 1993 Beijing. *Composite structures mechanics*. Beihang University Press
[2] Wang R G and Jiao W C Research progress of lightweight composite pressure vessel *J. Aeronautical Manufacturing Technology*. 2009(15):86-89
[3] Luo X L and Liu L L Performance analysis for composite pressure vessel *J. Machinery Design and Manufacture*. 2011(05):220-222
[4] Liu P F and Zheng J Y Progressive failure analysis of carbon fiber/epoxy composite laminates using continuum damage mechanics *J* 2008. *Materials Science And Engineering a-Structural Materials Properties Microstructure And Processing*. 485(1-2):711-717
[5] Huang J K and Yue H J 1999Beijing. Composites molding techniques. *Chemical Industry Press*
[6] *GB 24162-2009*. Periodic inspection and evaluation of hoop wrapped fibre reinforced composite gas cylinders with metal liners of compressed natural gas for automotive vehicles
[7] You H X, Ma K and Li B The method of security analysis and evaluation to the surface damage on the CNG-Ⅱ composite cylinder's winding layer J. *Pressure Vessel* 2013(10):54-60
[8] Yu W W and Gao B J 2006 Beijing The application of ANSYS in the mechanical and chemical equipment. *China Water Power Press*
[9] Zhang H S, Hu RX 2011 Beijing. ANSYS13.0 finite element analysis from entry to the master. *China Machine Press*
[10] Zhang G Z, Wang H H and Guan K S Autofrettage analysis and prediction of burst pressure of carbon fiber-winding hydrogen storage vessels by FEM *J. Pressure Vessel*. 2011(08):27-34
[11] Duan C H, Wu X and Luo X P Finite element analysis of burst for carbon fiber wrapped composite cylinder *J. FRP/Composite Materials*. 2012(02):17-19
[12] JC Shu, ST Chiu, JB Chug An enhanced analysis vessels *R*. AIAA-95-1024-CP
[13] Sneddon K, Sllunders S and Dever R The Eurostar 2000 Helium COPV *R*. AIAA-97-3033

Hydrophilicity, pore structure and mechanical performance of CNT/PVDF materials affected by carboxyl contents in multi-walled carbon nanotubes

Yanxia Zhang[1,a], Ce Jiang[1,b], Run Tian[1,c] and Guangfen Li[1,d *]

1 State Key Laboratory of Separation Membranes and Membrane Processes, School of Materials Science and Engineering, Tianjin Polytechnic University, Tianjin 300387, P. R. China
E-mail: [a]272259193@qq.com, [c]519301263@qq.com, [d]liguangfen@tjpu.edu.cn

Abstract. Poly (vinylidene fluoride) (PVDF) membranes have been prepared by loading different type of MWCNTs-COOH as the dispersed phase via phase inversion method. The chemically functionalized MWCNTs with increasing carboxyl content were chosen for achieving a better dispersion in PVDF and altering the membrane hydrophilicity. The effect of the carboxyl content in MWCNTs on crystal structure, thermal behavior, membrane morphology, hydrophilicity, and water flux of blended membranes were investigated. Due to the addition of carbon nanotubes, various performances of the hybrid membrane had obvious changes. The most prominent was that thermal stability could be enhanced and the pore morphology was more preferable, also that the hydrophilicity were improved, further that water flux could be increased to some extent.

1. Introduction

PVDF is a semi-crystalline polymer [1], which exhibits more hydrophobic and can easily dissolve in common solvent [2]. The application of PVDF membranes have been extensively reported since the 1980[3], the remarkable progress has been made in achieving higher performance of the PVDF membrane for diverse applications [4]. Despite of the many advantages, the PVDF materials are easily fouled due to its hydrophobicity. Many efforts have been made in improving the hydrophilicity and anti-fouling performance by different methods such as surface coating [5, 6], the addition of hydrophilic additives [7] and inorganic material [6]. Among these materials, carbon nanotubes were regarded as ideal nanofillers for polymeric ultrafiltration membranes as they could provide extraordinary mechanical and thermal properties, improved water flux as well as anti-fouling behavior. However, the poor dispersion of CNTs in solvent and polymer has restricted its application. Therefore, some tactics were taken by surface modified such as carboxylation and hydroxylation to improve their hydrophilicity and the dispersion of CNTs in the polymer membranes. Moreover, the CNTs with different content of functional groups may result complicated effects on the resultant hybrid membranes, thus it is necessary to further study the unknown effect. In this study, three carboxylated MWCNTs with carboxyl content of 0.49, 2, 3.86 wt% and pristine MWCNTs were introduced to prepare PVDF/MWCNTs blended membranes by the phase inversion method. The characterization means included scanning electron microscopy (SEM), static contact angle, DSC. The synergism effects of carboxyl content of MWCNTs in membranes on the crystallization, surface morphology, hydrophilicity, and thermal properties were elaborated in details.

2. Experimental

2.1. Materials

Powdered polyvinylidene fluoride (904, Mw=450KDa) was purchased from Arkema Inc. N,N-dimethylformamide (DMF, purity ≥99.5%) used as solvent was supplied by Tianjin Guangfu technology development co., LTD. Pristine multi-walled carbon nanotubes (p-MWCNTs, length 10-30 μm) and MWCNT-COOH (carboxyl content 0.49 wt.%, length 0.5-2μm, 2 wt.% and 3.86 wt.%, length 10-30 μm) were obtained from Chengdu institute of organic chemistry.

2.2. Preparation of membranes and Membranes characterization

A certain amount of MWCNTs was dispersed in 30 ml DMF for 40 min by ultrasound assisted (KQ3200DB) until achieving a uniform dispersion. PVDF powder was dried at 60℃ for 2h prior to use and then dissolved in DMF at 70℃ for approximately 4 h to form a homogeneous casting solution, which was then cooled down and stored at room temperature overnight to remove the trapped air bubbles. The degassed solution was cast onto a glass plate by a rod with thickness 300 μm and dried for 7 min, which was immersed into a DI water coagulation bath at room temperature by a water/DMF exchange. After coagulation, the membranes were transferred to fresh DI water bath and stored for 2 days to remove the residual solvent. Finally, the membranes were dried in air at room temperature and stored. The PVDF/MWCNT blended membranes were prepared via phase inversion and labeled that contained carboxyl 0.49wt.%, 2wt.%, 3.86wt.% as F1-PVDF, F2-PVDF and F3-PVDF, respectively.
The functional groups and crystalline structure of membranes were analyzed by wide angle X-ray diffraction (WXRD). The thermal behavior of the membranes and the effects of carbon nanotubes contained different carboxyl content on the glass transition temperature (Tg) were analyzed by differential scanning calorimetry (DSC, Perkin-Elmer, U.S.). The surface and cross-section morphology of the prepared membranes were investigated by scanning electron microscopy (SEM, Hitachis-4800, Hitachi Limited, Japan). The hydrophilicity of membrane was analyzed via water contact angle measurements (DSA-100), using the sessile drop method of water on dry membranes. And the measurement method of the pure water flux of the prepared membranes can be found in ref [8].

3. Results and discussion

3.1 Membrane properties

In order to determine crystal structure of prepared membranes and measure the phase qualitatively, we studied the five types of membranes by wide angle X-ray diffraction. Figure.1 (left) presents the XRD spectra for pure PVDF and all prepared MWCNTs/PVDF blended membranes. Two major diffraction peaks at 2θ values of 18.4 ° and 26.3 ° are observed in all membranes, which correspond to PVDF characteristic α-phase (020) and (021), respectively. A diffraction peak at 20.4° presents to the (110) and (200) of β-phase. For MWCNTs/PVDF membranes, the diffraction peak at 20.3° enhances, which hints that the incorporation of MWCNTs in PVDF matrix slightly alters the crystal transformation of PVDF. The formation of β-polymorph was induced by the interaction between carboxyl groups presented in MWCNTs and the -CF_2 segments in PVDF molecular chains. This further proves a better dispersibility of MWCNTs-COOH in PVDF membranes.
Differential scanning calorimetry was performed to distinguish the effect of adding various types of MWCNT to PVDF matrix on the Tg. A series of TG/DTG curves of pure PVDF membrane and four different MWCNTs/PVDF membranes tested were obtained, as shown in Figure 1.(right). The pure PVDF membrane starts to decompose at 450 ℃, whereas, the decomposition temperatures of the other membranes shift to around 455 ℃. The enhanced thermal stability of PVDF membranes contained MWCNTs is attributed to the presence of MWCNTs regardless of the type of MWCNTs and the carboxyl content in MWCNTs, in this case, as the variation of carboxyl content in MWCNTs is too small to dramatically change the overall thermal stability of the membranes. In comparison to p-MWCNTs/PVDF membranes, the slightly decreased thermal stability of all MWCNTs-COOH/PVDF blended membranes is directly correlated to the attached carboxyl, which

actually lowers the decomposition temperature. A significant decrease is observed in the composite with 3.86 % MWCNTs.

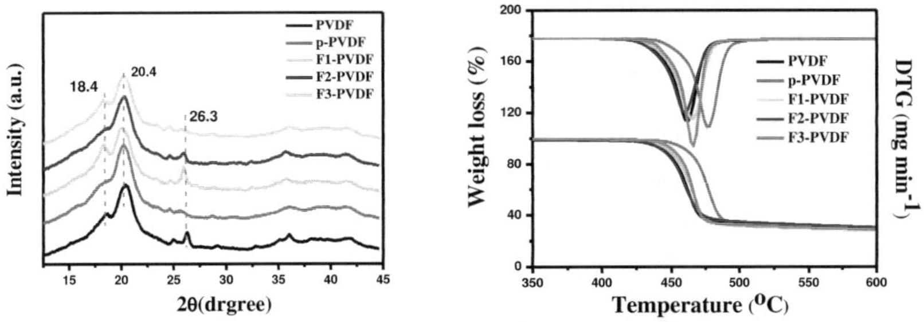

Figure 1. XRD spectra (left) and TGA/DTG curves (right) of blended membrane PVDF, p-PVDF, F1-PVDF, F2-PVDF and F3-PVDF.

3.2. Membrane morphology

The surface and cross-section morphologies of the various membranes were investigated by SEM, as shown in Figure 2. Three carboxylated MWCNTs with carboxyl content as 0.49, 2, and 3.86 wt% were selected for a comparison. It can be viewed from surface morphologies of all membranes that the type of MWCNTs has predominant effect on the porous structure. The membranes appear more macrovoids and aggregates on the surface, probably due to the fast phase separation process caused by higher evaporation speed of lower viscose solution. Compared with surface morphology of pure PVDF, more aggregates present on the surfaces of p-PVDF, which caused by worse dispersibility of p-MWCNTs, whereas, the surface morphologies of F1-PVDF exhibits rather smooth surface only a few macrovoids are found. The same tendencies are also observed in the blended membranes with higher carboxyl content in MWCNTs. These results verify that the hydrophilic carboxyl content in MWCNTs can actually promote its dispersibility in PVDF, enhance water diffusion, and thus inhibits the aggregation.

To the cross-section images, the membranes display a typical asymmetric structure with a thinner top-layer and a thicker sub-layer. The formation of such asymmetric structure originates from the solvent diffusivity in the membranes and water in the coagulation bath. For the membrane with different carboxyl content of MWCNTs, obvious discrepancies in cross-section morphology as follows: the pure PVDF membranes possess finger-like structure due to the low solution viscosity, but the membranes of p-PVDF, F2-PVDF and F3-PVDF have wider finger-like structures than pure PVDF. Besides, the finger-like pores of F1-PVDF membrane were obviously more homogeneous and shape more regularly. In order to explain the pore distribution, the membranes were characterized by BET method. From the analysis data of the bar graph, the major concentration of pore size distributed in 0-20nm, and pore structure of PVDF, F1-PVDF, F3-PVDF membranes was more concentrated and smaller than the other two, and the number of pores in F1-PVDF, F2-PVDF, F3-PVDF membranes had been on the rise, therefore, the MWCNT-COOH can enhance the formation of pore structures.

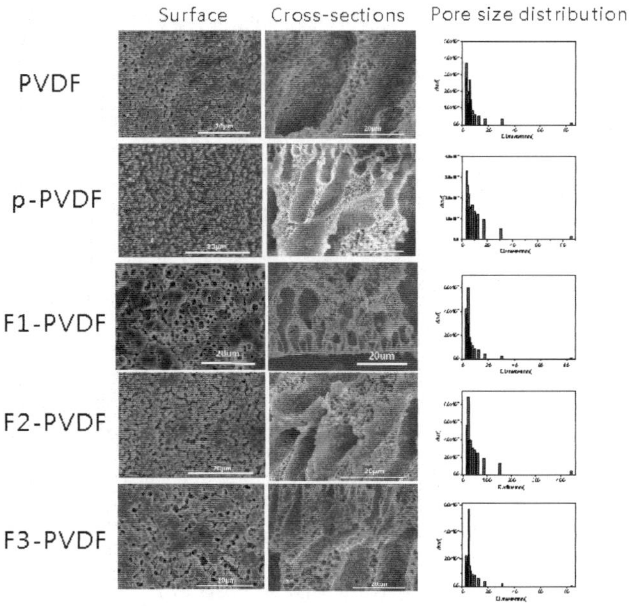

Figure 2. Surfaces and cross-sections morphologies of PVDF, p-PVDF, F1-PVDF, F2-PVDF, and F3-PVDF membranes (left). The pore size distribution of five membranes analyzed by BET is shown in the right column.

3.3. Hydrophilicity of membrane

To characterize the hydrophilicity of the different types of membranes, water contact angles on the membrane surfaces were measured and plotted in Figure 3.(A). All of the membranes exhibit a similar contact angle around 80°. With the increases of carboxyl content in MWCNTs, the hydrophilicity of membranes shows a trend of rising, which proves that the carboxyl content in MWCNTs affect the hydrophilicity of the MWCNTs/PVDF membranes to some extent. However, an exception is observed for the membrane with 2 wt.% carboxyl in MWCNTs. It has a water contact angle of above 90°, which displays slightly increase compared to the other membranes. Therefore, the membranes hydrophilicity is determined by not only the carboxyl contents of MWCNTs, but also the surface roughness of membranes. Permeability of the prepared membranes was characterized by measuring pure water flux, as presented in Figure 3.(B). Many factors may determine water flux, for example, morphologies of the membranes, pore properties and the dispersion of MWCNT. Compared with the water flux of PVDF membrane, the data of p-PVDF, F1-PVDF and F3-PVDF membranes were significantly enhanced, especially, the water flux of F1-PVDF membrane reached the maximum about 342 L/m^2*h. However, the exception was also F2-PVDF membrane, the water flux abnormally reduced, which might be caused by hydrophilicity.

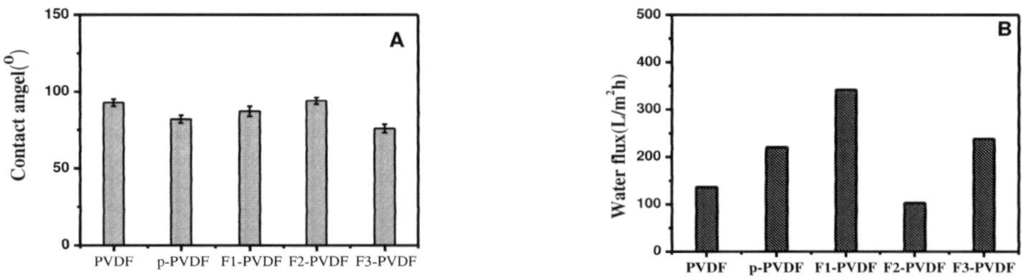

Figure 3. Water contact angle analysis and Water flux of PVDF, p-PVDF, F1-PVDF, F2-PVDF and F3-PVDF membranes

4. Conclusion

Carboxyl MWCNTs/PVDF flat hybrid membranes were prepared via the phase inversion method. The various performances of blended membranes had been markedly improved due to the addition of MWCNT-COOH. The crystal structure of F1-PVDF membrane enhances significantly, such as the diffraction peak at 18.4^{o} and 26.3^{o} more strengthen. It is observed the highest degradation temperature was 477℃ for the p-PVDF on account of the presence of MWCNTs. For the SEM, the p-PVDF and F2-PVDF membranes showed more large pores and aggregates, which made the pore size reduced and the pore distribution expanded. As for cross-section morphologies, all membranes presented both finger-like pores and dense sponge-like pores around the finger pores except for pure PVDF, which are attributed to the addition of MWCNT. Besides, adding MWCNTs-COOH to the membranes can indeed promote hydrophilicity of membranes and the present of carboxyl can enhance the binding forces between MWCNTs and PVDF matrix. The water flux of p-PVDF, F1-PVDF and F3-PVDF membranes were also significantly improved, especially, the water flux of F1-PVDF membrane reached the maximum.

5. References

[1] Rekik H, Ghallabi Z, Royaud I, et al. Dielectric relaxation behaviour in semi-crystalline polyvinylidene fluoride (PVDF)/TiO_2, nanocomposites[J]. Composites Part B Engineering, 2013, 45(1):1199-1206.

[2] Cha S N, Kim S M, Kim H J, et al. Porous PVDF As Effective Sonic Wave Driven Nanogenerators[J]. Nano Letters, 2011, 11(12):5142.

[3] Kang G D, Cao Y M. Application and modification of poly(vinylidene fluoride) (PVDF) membranes – A review[J]. Journal of Membrane Science, 2014, 463(1):145-165.

[4] Kim I, Choi D C, Lee J, et al. Preparation and application of patterned hollow-fiber membranes to membrane bioreactor for wastewater treatment[J]. Journal of Membrane Science, 2015, 490:190-196.

[5] Shi H, Xue L, Gao A, et al. Fouling-resistant and adhesion-resistant surface modification of dual layer PVDF hollow fiber membrane by dopamine and quaternary polyethyleneimine[J]. Journal of Membrane Science, 2016, 498:39-47.

[6] Barsan M M, Ghica M E, Brett C M. Electrochemical sensors and biosensors based on redox polymer/carbon nanotube modified electrodes: a review. [J]. Analytica Chimica Acta, 2015, 881:1.

[7] Silva T L S, Morales-Torres S, Figueiredo J L, et al. Multi-walled carbon nanotube/PVDF blended membranes with sponge- and finger-like pores for direct contact membrane distillation [J]. Desalination, 2015, 357:233-245.

[8] Chu K H, Yoo S S, Ahn J Y, et al. Determining flux behavior via a modified flux-step method for surface water treatment: pilot-scale ultrafiltration membrane operation[J]. Desalination, 2014, 341(1): 19-26.

Acknowledgements

The research was supported by the Science and Technology Plans of Tianjin (No.15YFXQGX00070 and 15PTSYJC00240)

Effect of heat-setting on UV protection and antibacterial properties of cotton/spandex fabric

M N Pervez[1, 2, 3], M E Talukder[1, 3], F Shafiq[1], K M F Hasan[1, 3], M A Taher[1, 3], M M Meraz[1, 3], Y Cai[1, 4] and Lina Lin [4, 5] *

1 School of Chemistry and Chemical Engineering, Wuhan Textile University, Wuhan 430073, China
2 Research Institute of Flexible Materials, School of Textiles & Design, Heriot-Watt University, Galashiels TD1 3HF, UK
3 Department of Textile Engineering, Southeast University, Dhaka-1213, Bangladesh
4 Hubei Provincial Engineering Laboratory for Clean Production and High Value Utilization of Bio-based Textile Materials, Wuhan Textile University, Wuhan 430073, China
5 College of Textile Science and Engineering, Wuhan Textile University, Wuhan, 430200, China
* Corresponding author, E-mail: linalin@wtu.edu.cn

Abstract. An unexampled approach for simultaneous heat setting process with optimized condition at C3 (140°C, 45 s) and functional finishing, i.e. UV protection and antibacterial properties of cotton/spandex fabric were studied in this research. Experimental results disclosed that, ameliorative antibacterial efficacy and perdurable UV protection of heat-treated cotton/spandex fabrics with best sample A3 among all samples was achieved and mechanical properties also improved as the temperature rose from 120 to 140ºC. In addition, Ultraviolet (UV) radiation protection and antibacterial properties are becoming increasingly necessary for human health, and textiles play an important role and this report will be appurtenant to meet regular demand.

1. Introduction

Due to the semi-crystalline structures orientation of synthetic polymeric fibers, the macromolecules are not widely distributed in their equilibrium state. Further instabilities are imparted if the fibers are adapted to yarns and the yarns to fabrics. Heat setting is an important industrial method, since it rids them of their instabilities [1]. The accepted method for the heat setting of synthetic material is the continuously acclimated through stenter frame. The stenter frame heat sets by way of the usage of warm air and important parameters like temperature, dwell time, width and overfeed are considered. For natural fibers relaxation dryers, crimping and steaming units are used [2]. The heat-setting has two important functions-it imparts dimensional balance and it relaxes the inter-yam forces at the crossover points in the fabric. The relaxation of inter-yarn forces profoundly modifies the mechanical behavior of the fabric in shear and in bending and, in allowing the conclusion of the whole high-recuperation capacity of the synthetic fibers from bending traces, makes feasible the brilliant smooth-care overall performance of cloth constituted of these materials [3-5]. Heat-setting procedure, seen as an essential issue, is undertaken during the initial phase of the textile industry, considering the fact that it proffers the appropriate product's output to make it appropriate for additional steps and for the customer to utilize it. Under-setting after-effects in eventual loss of fabric dimensions; over heat-setting lowers

power and can discolor the spandex and companion fibers. Relaxation remedy is used to lessen capacity distortion or deformation of the material from residual uneven tension. It develops the power and recovery of the fabric. The fabric should be secure prior to heat-setting to avoid rope marks and puckering during dyeing to ensure good dimensional balance of the final garment [6]. Special interest has been focused these days at the ultraviolet (UV) transmission of textile due to the developing demand in the marketplace for lightweight apparel that offers to human's protection from UV radiation (UVR). In addition, some antimicrobials may additionally motive allergic reactions amongst purchasers, and the practical corporations liable for these reactions aren't well set up in many cases. Degradation of antibiotics as a result of heat treatments may also result in a loss of the antibacterial potential of those antibiotics and also in a likely lack of their allergenic ability [7].
Therefore, we were interested to elucidate the effects of heat-treatment on the UV protection and antibacterial properties of cotton/spandex fabrics and no study has been found in literature about assessing functional properties of heat setting processed fabric.

2. Experimental

2.1. Materials and methods
For experimental work knitted single jersey fabrics with (1-5%) spandex and 95% cotton have been used. The fabric samples were knitted on Mayer and Cie (Germany) single knit machine by using 30/1 Ne cotton yarn and 24 denier spandex filament yarn and produced fabric had 74 course/inch, 38 wales/inch and fabric density of 276 g sq. m^{-1}.
The stenter machine (BRUCKNER, Germany) was used to carry out the heat setting process. At first, samples were placed between parallel bars at 10 cm length and stretched out to different degrees on an aluminum frame with pre-determined tension. There were six levels of temperature and time such as (120, 130, 140, 150, 160, 170°C) and (15, 30, 45, 60, 75, 90 s) respectively, while all other machine variables were held constant. Then, the process conditions were denoted as C1, C2, C3, C4, C5 and C6. After heat setting, the samples were cooled before being removed from the aluminum frame. Then the samples were conditioned under standard conditions of 65±2 % relative humidity and 21±1°C for 24 h.

2.2. Measurement
The tensile strength of samples was calculated according to our previous report [8]. Transmittance (%) of the fabric samples was measured at the wavelength changing from 200 to 1000 nm using a UV–Vis spectrophotometer. The antibacterial activity of samples was evaluated qualitatively according to the Australian/New Zealand AS/NZS 4399:1996 in presence of Staphylococcus aureus (gram-positive) and Escherichia coli (gram-negative) bacteria.

3. Results and discussion

3.1. Tensile strength
The study of physical properties of fabrics is important due to apparel usually undergoes small extension and relaxation during wear and tensile strength is one of them [9]. Figure 1 shows the tensile strength values of the samples before and after under different heat-setting conditions. It can be seen that the tensile strength is increased after heat setting except for the sample number A1, as while it is decreased after heat setting. The highest tensile strength is shown in case of sample A3 and process condition C3 about 5 % improvement compared to untreated sample and also there is little improvement about 3% observed from sample A4 and A5. This result was attributed due to the fiber adhesion and fabric becomes stiffer during the heat setting and also may be affiliated to the rearrangement of fibers after heat treatment which improved the tensile strength of the blended fabric [10]. Finally from these results it can be conclude that: (1) heat treatment at 140°C results to increase tensile strength and (2) degradation in fabric starts at about 160°C which results to decrease in tensile strength. Sample A3 was chosen for further study.

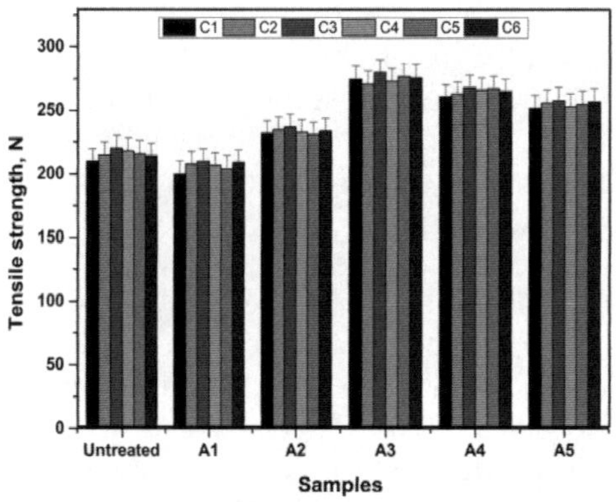

Figure 1. Tensile strength values.

3.2. UV protection

Clothing is in actuality one of the best means of attention humans adjoin solar UV radiation [11], but not all apparel is abundantly careful adjoin UV radiation. From the angle of textile chemicals, dyes and UV absorbers, UV-protective clothing has not yet become a breakthrough. As it is known, The UVR band consists of three regions: the UV-A band (320–400 nm), UV-B band (290–320 nm), and UV-C band (200–290 nm). The UV-C band is captivated by oxygen and ozone in the high atmosphere. Of the solar UVR extensive the Earth's surface, 6% is in the UV-B region and 94% in the UV-A region. Light radiation of wavelengths 280-400 nm permits tanning of the epidermis. Rays of wavelength 290-320 nm (UVB) causes erythemas and skin burns, which can inhibit skin tanning. Radiation of wavelength 320-400 nm (UVA) is known to induce pores and skin tanning, but can also purpose skin damage, especially to touchy skin which is uncovered to sunlight for a long time frame. Examples of such harm include lack of pores and skin elasticity, the advent of wrinkles, advertising of the onset of erythemal response, and the inducement of phototoxic or image allergic reactions [12, 13]. The amount of UV protection supplied through the samples turned into evaluated in phrases of ultraviolet protection factor (UPF value) that is contrary feature of transmission and there's simply the other correlation between the transmission and coloration values of a smaller depth [14]. UPF is the ratio of the minimum erythemal dose of sun UV radiation for skin covered by fabric to that unprotected [15]. Therefore, the transmittance of UVR, including UV-A and UV-B, through the fabrics turned into evaluated in this experiment.

Figure 2 shows the UV transmission curve of untreated and different heat-setting conditions treated sample (A3). It was noticed that in case of untreated sample indicating that almost 45% of the UV-A and UV-B can penetrate into the fiber which is insufficient for providing UV protection. In comparison, different conditions like C1, C2, C3, C4, C5 and C6 noticeable UPF were increases with the transmittance values decreases but C4 condition provided less protection than C5 and C6. Figure 2 indicates that the best UPF results condition C3 was obtained with the heat setting temperature at 120°C for 45 s. This UPF behavior can be classified as excellent and very good UV protection, respectively, according to according to the Australian/New Zealand AS/NZS 4399:1996 for ultraviolet radiation (UVR) protection categories of fabrics [16]. In UPF rating, fabric shade performs a crucial function. Dark shaded fabric gift a higher protection towards ultraviolet radiation than light colored fabrics. Black colored fabrics recommend not simplest suitable, but even amazing UV protection. Light colors transmit greater UVR than darkish ones and shade and color depth of the cloth may be related to UV transmittance [17]. In addition, figure 2 indicates that, the presence of different heat setting conditions inside the treated fabric sample increases the UPF values compared to the untreated fabric. Also, the degree of upgrading in the UV-protection properties is estimated and follows the order: C3 > C5 > C6 > C4 > C2 > C1 > untreated. These results show clearly that heat-setting treated fabric can greatly enhance the UV-protection properties with C3 heat setting condition.

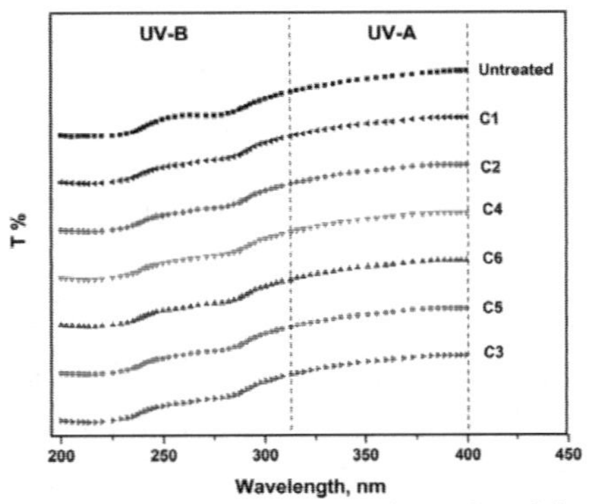

Figure 2. Transmittance (%) spectra in the UV of sample (A3) with different heat setting condition.

3.3. Antibacterial property

Antimicrobial finishes prevent the growth of microorganisms on fabrics utilized in wide kind of apparel, home furnishing, business and business merchandise. Fabrics for tents, tarpaulins, and car fabric could have an extended life when dealt with a few type of antimicrobial finish that reduces or prevents damage from rot and mildew [18]. Worldwide efforts are continuing to develop antimicrobial substances and heat treatment is the normally practiced era to feature value to products and complements the resistance of materials to weathering and rot fungi [19]. In general, the most suitable heat-setting condition for antimicrobial activity is C3 (140°C for 45 s) among all treated conditions. When temperature is lower than 140°C, the antibacterial activity against *E.coli* increases with the increase of temperature plotted in figure 3. However, as heat-setting temperature is higher than 140°C, the antimicrobial activity decreases. It is because the melting point of cotton is around 149°C, it is easier for cotton getting detached from the fabric surface at high temperatures, which may result in poor fixation of cotton, meaning it may dissolve again in water during the subsequent finishing process. Finally, the content of cotton and spandex decreases with the increase of the heat-setting temperature and the antimicrobial activity decreases [20]. Therefore, the optimal heat setting temperature for the antibacterial activity was 140°C at 45 s in this research.

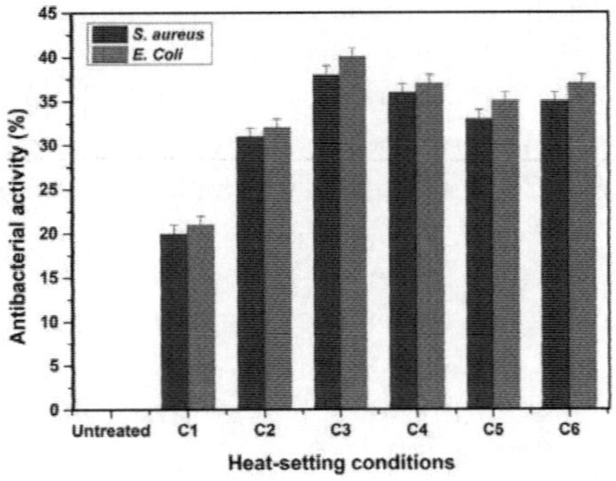

Figure 3. Antibacterial activity of sample (A3) with different heat setting condition.

4. Conclusion

Based on a series of experimental studies in this work, an analytical probe was performed to research the heat-setting effect on UV protection, antibacterial and mechanical properties of cotton/spandex knitted fabric. Then samples were annealed at 120 to 170°C range and subjected to assorted tests at room temperature. The aforementioned properties of cotton/spandex were significantly improved by annealing treatment with best condition C3 (140°C, 45 s) on sample A3. We have developed the procedure where simultaneous heat-setting and functional finishing of cotton/spandex fabrics with prominent antibacterial and perdurable UV protective properties can be evaluated.

5. References

[1] Gupta V B 2001 Heat setting *J. Appl. Polym. Sci.* **83** 586
[2] Besler N, Gloy Y S and Gries T 2016 Analysis of the heat setting process *IOP Conf. Ser.: Mater. Sci. Eng.* **141** 012018
[3] Brenner F C, Chen C S and Britt D G 1966 The mechanical behavior of fabrics: Part IV: effects of finishing *Text. Res. J.* **36** 216
[4] Popper P and Backer S 1968 Instrument for measuring bending-moment curvature relationships in textile materials *Text. Res. J.* **38** 870
[5] Skelton J 1968 22—a theoretical and experimental investigation of the crease recovery of plain-weave fabrics woven from staple yarns *J. Text. Inst.* **59** 261
[6] Senthilkumar M and Anbumani N 2011 Dynamics of elastic knitted fabrics for sports wear *J. Ind. Text.* **41** 13
[7] Zorraquino M A, Althaus R L, Roca M and Molina M P 2011 Heat treatment effects on the antimicrobial activity of macrolide and lincosamide antibiotics in milk *J. Food Prot.* **74** 311
[8] Kamruzzaman M, Talukder M E, Pervez M N, Kamal A and Ying-Jie C. *A novel approach to rejoin the carbon fiber tow by fiber splicing equipment*. in *Proc. of the 3rd International Conference on Material Engineering and Application* 2016. China: Atlantis.
[9] Kan C-W and Lam Y-L 2013 Low stress mechanical properties of plasma-treated cotton fabric subjected to zinc oxide-anti-microbial treatment *Mater.* **6** 314
[10] Furvik N B, Bernskiöld A and Gralen N 1955 49—Heat-setting of nylon fabrics *J. Text. Inst. Trans.* **46** T662
[11] Zohdy M H, El Hossamy M B, El-Naggar A W M, Fathalla A I and Ali N M 2009 Novel UV-protective formulations for cotton, PET fabrics and their blend utilizing irradiation technique *Eur. Polym. J.* **45** 2926
[12] Oda H 2011 Development of UV absorbers for sun protective fabrics *Text. Res. J.* **81** 2139
[13] Yazhini K B and Prabu H G 2015 Study on flame-retardant and UV-protection properties of cotton fabric functionalized with ppy–ZnO–CNT nanocomposite *RSC Adv.* **5** 49062
[14] Shahidi S 2017 Ultraviolet protection properties of metal salts treated cotton fabrics *J. Nat. Fibers* **14** 166
[15] Andrady A, Torikai A, Redhwi H H, Pandey K K and Gies P 2015 Consequences of stratospheric ozone depletion and climate change on the use of materials *Photochem. Photobiol. Sci.* **14** 170
[16] Gies H P, Roy C R and Holmes G 2000 Ultraviolet radiation protection by clothing: comparison of in vivo and in vitro measurements *Radiat. Prot. Dosim.* **91** 247
[17] Dubrovski P D and Golob D 2009 Effects of woven fabric construction and color on ultraviolet protection *Text. Res. J.* **79** 351
[18] Hashem M, Ibrahim N A, El-Sayed W A, El-Husseiny S and El-Enany E 2009 Enhancing antimicrobial properties of dyed and finished cotton fabrics *Carbohydr. Polym.* **78** 502
[19] Holloway A C, Mueller - Harvey I, Gould S W J, Fielder M D, Naughton D P and Kelly A F 2015 Heat treatment enhances the antimicrobial activity of (+) - catechin when combined with copper sulphate *Lett. Appl. Microbiol.* **61** 381
[20] Zhou C-E and Kan C-W 2015 Optimizing rechargeable antimicrobial performance of cotton fabric coated with 5, 5-dimethylhydantoin (DMH) *Cellul.* **22** 879

Acknowledgement

This work was supported by the China National Textile and Apparel Council (2013 "Textile Vision" Applied Basic Research), a grant from the Hubei Province Science and Technology Support Program (Project 2013BAA043), and the Collaborative Innovation Plan of Hubei Province for Key Technology of Eco-Ramie Industry (E JIAO KE HAN 2014 No.8).

IOP Conf. Series: Materials Science and Engineering **284** (2017) 012011 doi:10.1088/1757-899X/284/1/012011

Studies on degradation performance of Mg-4.0Zn-1.5Sr alloy with coated of the laser surface processing combining alkaline treatment

CUI Tong[1], GUAN Ren-guo[1], Ma Xin-rui[1], QIN Hai-ming[2], SONG Fu-lin[2]

1 College of Materials Science and Engineering, Northeastern University, Shenyang, China;
2 Department of Pathology, Northern Hospital, Shenyang, China
E-mail:ct114928@163.com

Abstract. The surface modification of biomaterial Mg-4.0Zn-1.5Sr alloy has been done by means of laser surface processing combining alkaline treatment process, as well as the degradation performance of Mg-4.0Zn-1.5Sr alloy with and without coatings in Hank's solution has been analyzed comparatively. The results indicate that the optimal parameters of laser surface processing are that the power is 3 kW, the current 200 A, the width 1mm, the defocus amount 135 mm and the scanning speed 1mm/s. The optimal parameters of alkaline treatment are that the solution is NaOH, the concentration 0.5 mol/L, the temperature 80 ℃ and the time 12 h. There are only two phases of Mg $(OH)_2$ and magnesium matrix, and the surface generated most of Mg $(OH)_2$ which can improve the corrosion resistance of the alloy after laser combining alkaline treatment, as well as the corrosion rate is almost the stable, which is much smaller than both of uncoated and laser surface processing. The study of the electrochemical corrosion behavior shows that the corrosion potential of the alloy with coated of laser combining alkaline treatment is improved 0.1277 V than that of laser treatment, and the corrosion current is decreased 470.2 muA than laser treatment. The corrosion resistant ability of Mg-4.0Zn-1.5Sr alloy is greatly improved by means of laser combining alkaline treatment.

1. Introduction

In recent decades, as a novel biomedical planting material, the study of magnesium alloys is paid more attentions [1-3]. It is well known that magnesium (Mg) and magnesium alloys possess many properties superior to other metallic biomaterials. However, magnesium alloys studied at present are mostly focused on industrial area, usually containing some elements that are harmful to human healthy such as aluminium, manganese or rare earth elements. In addition, the rapid degradation rate and relatively low biological activity restrict its widespread use in clinical applications. Therefore, the research on magnesium-based implant materials which are consisted with all of the nourishment elements and with the lower degradation rate has become an important subject with growing interests [4-6].

The surface modification is one of the most effective ways to improve the alloy corrosion resistance, including the laser surface treatment can greatly improve the fatigue corrosion resistance, and alkali heat treatment can significantly improve the corrosion resistance. In this paper, In order to improve the corrosion resistance of the alloy, the Mg-4.0Zn-1.5Sr alloy alloys is designed, and laser surface processing combining alkaline treatment process is optimized to improve the corrosion resistant properties of the alloy, and the degradation performance of the alloy with and without coating has been studied.

2. Materials and experimental

In the experiment, pure magnesium with a purity of 99.9 % was adopted. The Mg-4.0Zn-1.5Sr (wt.%) alloy ingots was produced by melting with Ar gas protection. The magnesium and its alloy ingots were rolled into sheets in three sequential steps at 350 ℃. The final sheet size was 13×13×1 mm. By means of YAG laser device to make laser surface treatment, and hot type magnetic stirrer to make alkaline treatment. the optimal parameters of laser surface processing are that the power is 3 kW, the current 200 A, the width 1 mm, the defocus amount 135 mm and the scanning speed 1 mm/s. The optimal parameters of alkaline treatment are that the solution is NaOH, the concentration 0.5 mol/L, the temperature 80 ℃ and the time 12 h. The alloy samples with and without coating were immersed in SBF. To make the SBF, 700 mL of deionized water was poured into a 1L beaker in which the temperature was held at (36.5±1.5) ℃. 8.035 g NaCl, 0.355 g $NaHCO_3$, 0.225 g KCl, 0.231g $K_2HPO_4 \cdot 3H_2O$, 0.311 g $MgCl_2 \cdot 6H_2O$, 0.292 g $CaCl_2$, 0.072 g Na_2SO_4, 6.118 g $(HOCH_2)_3CNH_2$ and 5 ml of 1 mol/L HCl were added into the beaker sequentially. Deionized water was added until reaching a final volume of 1000 ml and pH was finally adjusted to 7.4. For degradation studies, the specimens were hung vertically by a fine line and fully immersed in SBF. The study of electrochemical behavior of the alloy by means of electrochemical equipment CHI600B at 37 ℃ in SBF. The surface morphology and composition of corrosion products were observed and analyzed by means of Scanning electron microscopy (SEM) and X-ray diffraction (XRD).

3. Results and discussion

3.1 The surface morphology and microstructure of Mg-4.0Zn-1.5Sr alloy, Mg-4.0Zn-1.5Sr alloy with coating of laser surface processing and with coating of laser surface processing combining alkaline treatment

The metallographical image of the cast structures of Mg-4.0Zn-1.5Sr alloy is shown in Figure 1. It can be seen that the alloy which grain is uniform and fine is mainly composed of a gray α-Mg matrix and a black second phase and the second phase is mainly concentrated near the grain boundaries.

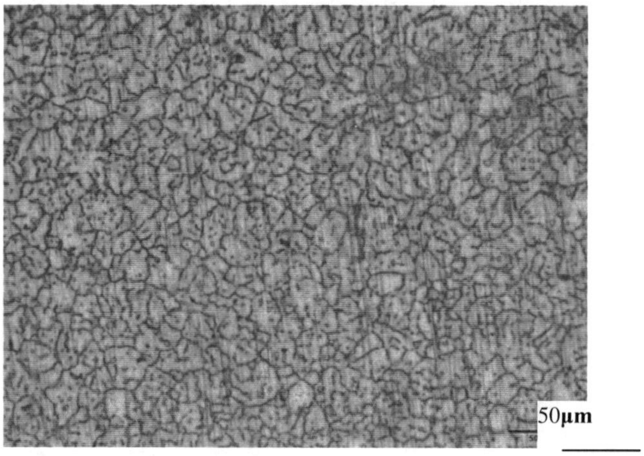

Figure 1. The metallographical image of the cast structures of Mg-4.0Zn-1.5Sr alloy

The SEM image of sample after laser surface processing is shown in Figure 2(a). It can be seen that there are many molten pools after laser scanning, and the surface composition is very close to the cast alloy. The metallographical image of Mg-4.0Zn-1.5Sr alloy after laser surface processing combining alkaline treatment is shown in Figure 2(b). It can be seen that after alkaline treatment the surface is changed smooth and compact compared with only laser surface processing. The SEM image and EDS analysis of sample after laser surface processing and alkaline treatment is shown in Figure 3 and Table 1. It can be seen that after alkali heat treatment, there are only two elements of Mg and O. It can be explained that the main composition is Mg $(OH)_2$, and the coating is uniform and density, which is separated with the matrix. The XRD analysis result of Mg-4.0Zn-1.5Sr after laser surface processing and alkaline treatment is shown in Figure 4. It can be seen that after alkaline treatment, there are only two phases of Mg $(OH)_2$ and magnesium matrix. In addition, the large amount of Mg $(OH)_2$ generated the surface can improve the corrosion resistance of the alloy.

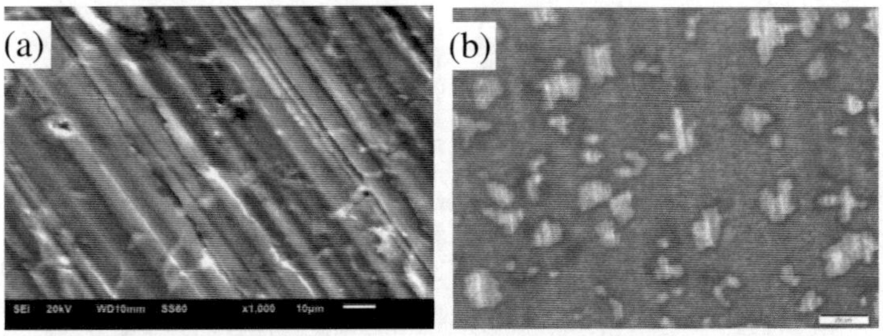

Figure 2. The metallographical image of Mg-4.0Zn-1.5Sr alloy after (a) laser surface processing ; (b) laser surface processing combining alkaline treatment

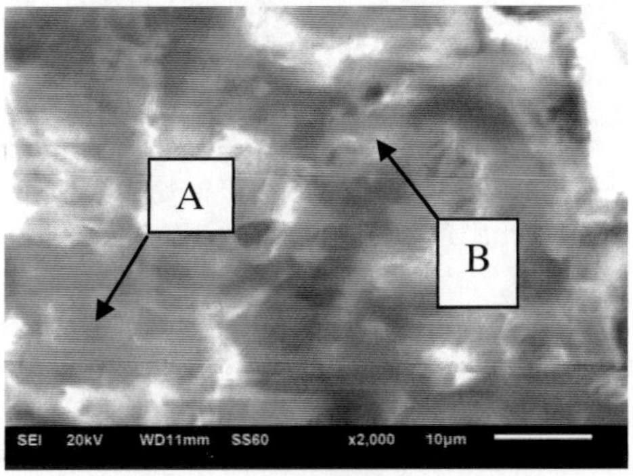

Figure 3. The SEM image and EDS analysis of sample after laser surface processing and alkaline treatment

Table 1. The percentage elements of place A to B by EDS analysis

The elements	Position A		Position B	
	Wt%	At%	Wt%	At%
O	54.02	64.41	50.54	59.75
Mg	45.98	35.59	49.47	40.25

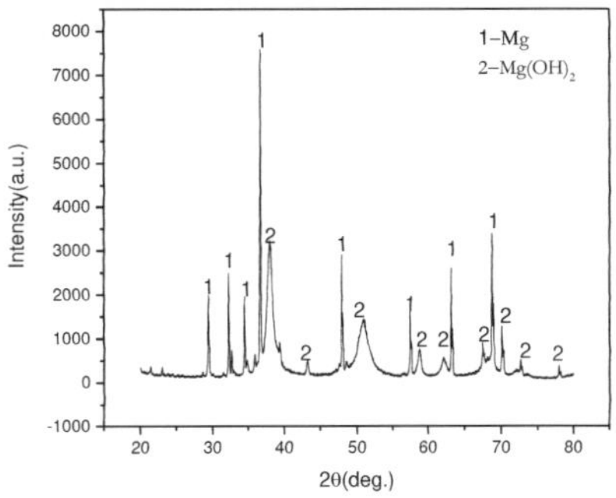

Figure 4. XRD analysis result of Mg-4.0Zn-1.5Sr alloy after laser surface processing and alkaline treatment

3.2 The test of Mg-4.0Zn-1.5Sr alloy with coating of laser surface processing combining alkaline treatment after immersed in SBF

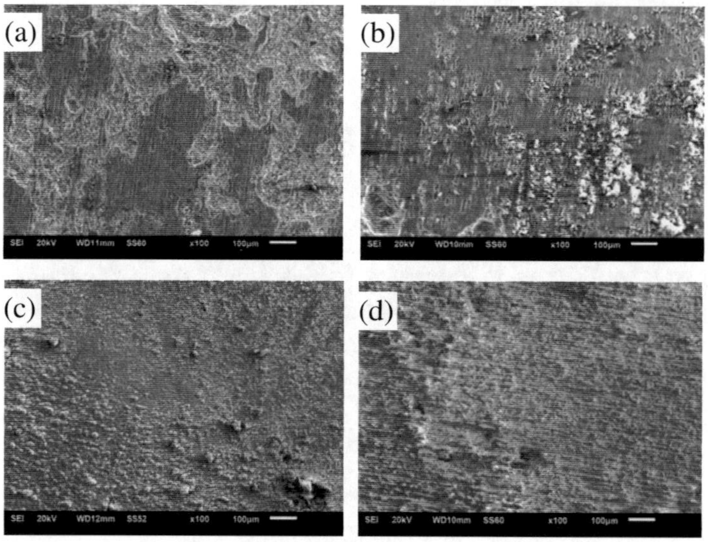

Figure 5. The surface metallographs of Mg-4.0Zn-1.5Sr with laser surface processing and alkaline treatment after immersion in SBF for (a) 1 d; (b) 7 ds; (c) 12ds (a) and (d): 15ds

The surface metallographs of Mg-4.0Zn-1.5Sr with laser surface processing and alkaline treatment after immersion in SBF for the different times is shown in Figure 5. It can be seen that only a small area is very slight damage happened in the surface after the sample immersed in SBF for 1 day shown in Figure 5(a). In Figure 5(b) immersed in SBF for 7 days, only local area is slight damage with shallow corrosion depth, and does not have the phenomenon of continuous, pitting corrosion and no lateral extension. In Figure 5(c) and Figure 5(d) immersed in SBF for 12 and 15 days, respectively. It can be seen that only local area is slightly destroyed with a shallow corrosion depth expressing that the corrosion is very slight. And corrosion coating remained very good adhesion with matrix, there is no large area falls off phenomenon happened for 15 days. The sample surface has a lot of small particles bump for 12 days, which is gone for 15 days. It is possible that corrosion products or coating fall off. The loss mass change compared of Mg-4.0Zn-1.5Sr alloy, Mg-4.0Zn-1.5Sr alloy with coating of laser surface processing and with coating of laser surface processing combining alkaline treatment immersed in SBF is shown Figure 6. The change trend of loss mass of uncoated alloy is similar with the laser surface treatment, which is increased gradually, but the loss mass ratio is different. As well as the both values of loss mass are closed. While, for the laser surface processing combining alkaline treatment, the loss mass ratio is almost stable during the immersed times with the slower increased at immersed times in SBF. As well as the value of loss mass is much lower than the alloy with uncoated and laser surface processing. The corrosion rate of Mg-4.0Zn-1.5Sr alloy with coated of laser surface processing and alkaline treatment in SBF is shown in Figure 7. For the alloy with uncoated and laser surface processing, the corrosion rate is increased before 7 days, then declined during 7 days to 15 days, but for the laser surface processing combining alkaline treatment, the corrosion rate is almost the stable, which is much smaller than both of uncoated and laser surface processing.

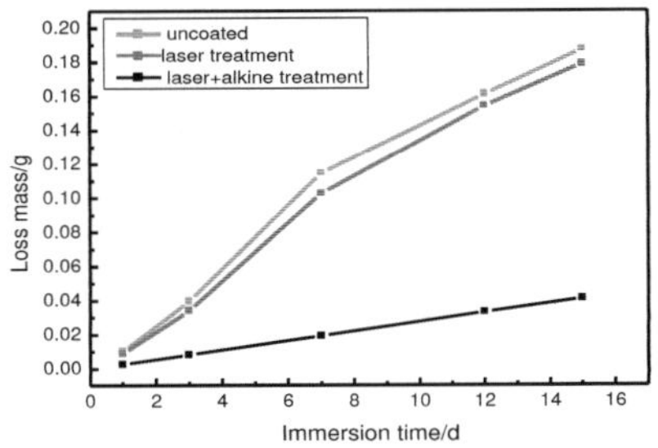

Figure 6. The loss mass change Mg-4.0Zn-1.5Sr alloy with coated of laser aalkaline treatment in SBF

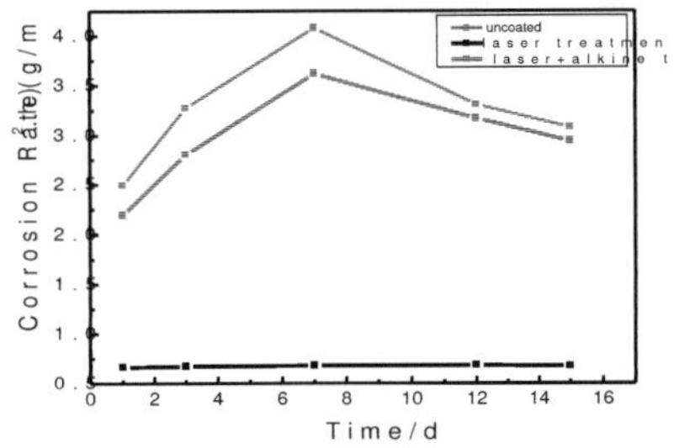

Figure 7. The corrosion rate of Mg-4.0Zn-1.5Sr alloy with coated of laser surface processing and alkaline treatment in SBF

3.3 The electrochemical corrosion behavior of Mg-4.0Zn-1.5Sr alloy with coated of laser combining alkaline treatment

Dynamic polarization test is used on Mg-4.0Zn-1.5Sr alloy with coated of laser combining alkaline treatment. The sample effective area is 1 cm^2, scanning speed is 0.01 V/s. Corrosive medium pH is 7.35. In general, the higher the corrosion potential, corrosion current density is smaller, the material of the corrosion rate is smaller, and the material has stronger corrosion resistance ability. Dynamic polarization test result on Mg-4.0Zn-1.5Sr alloy with coated of laser + alkaline treatment is shown in Figure 8, and the results are shown in table2. For Mg-4.0Zn-1.5Sr alloy with coated of laser combining alkaline treatment, the corrosion potential is improved 0.1277 V than that of laser treatment, and the corrosion current is decreased 470.2 muA than laser treatment. The result is consistent with the weight-loss method which the corrosion resistant is greatly improved by means of laser and alkaline treatment.

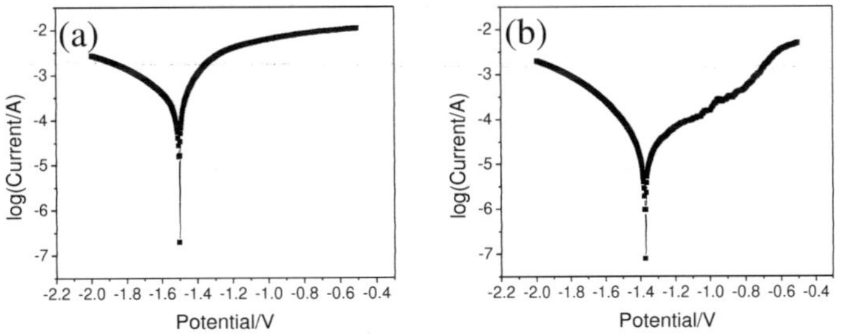

Figure 8 The dynamic polarization curves of Mg-4.0Zn-1.5Sr alloy after alkaline treatment (a): Laser surface treatment; (b): Laser combining alkaline treatment

Table 2. The electrochemical corrosion results of Mg-4.0Zn-1.5Sr alloy with coated of laser and alkaline treatment

The process	Effective area/cm^2	Corrosion potential /V	Corrosion current/uA	Corrosion current density /uA/cm^2
Uncoated	1	-1.5062	566	566
Laser treatment	1	-1.4995	479	479
Laser + alkaline treatment	1	-1.3718	8.8	8.8

4. Conclusion

The optimal parameters of laser surface processing are that the power is 3 kW, the current 200 A, the width 1mm, the defocus amount 135 mm and the scanning speed 1mm/s. The optimal parameters of alkaline treatment are that the solution is NaOH, the concentration 0.5 mol/L, the temperature 80 ℃

and the time 12 h. There are only two phases of Mg $(OH)_2$ and magnesium matrix, and the surface generated most of Mg $(OH)_2$ which can improve the corrosion resistance of the alloy after laser combining alkaline treatment, as well as the corrosion rate is almost the stable, which is much smaller than both of uncoated and laser surface processing. The study of the electrochemical corrosion behavior shows that the corrosion potential of the alloy with coated of laser combining alkaline treatment is improved 0.1277 V than that of laser treatment, and the corrosion current is decreased 470.2 muA than laser treatment. The corrosion resistant ability of Mg-4.0Zn-1.5Sr alloy is greatly improved by means of laser combining alkaline treatment.

5. References

[1] Witte F. The history of biodegradable magnesium implants: A review. Acta Biomater, 2010, 6(5): 1680-1692.

[2] Brar H S, Platt M O, Sarntinoranont M, et al. Magnesium as a biodegradable and bioabsorbable material for medical implants. JOM, 2009, 61(9): 31-34.

[3] Guan Ren-guo, Johnson Ian, Cui Tong, Zhao tong, Zhao Zhan-yong, Liu Hui-nan. Electrodeposition of hydroxyapatite coating on Mg-4.0Zn-1.0Ca-0.6Zr alloy and in vitro evaluation of degradation, hemolysis and cytotoxicity, Journal of Biomedical Materials Research, Part A, 100(4): 999-1015, 2012.

[4] Zhang E L, Yin D S, Xu L P, et al. Microstructure, mechanical and corrosion properties and biocompatibility of Mg-Zn-Mn alloys for biomedical application. Mater Sci Eng A, 2010, 527(13-14): 3195-3199.

[5] Tang Hui, Yu Dezhen, Luo Yan, et al. Preparation and characterization of HA microflowers coating on AZ31 magnesium alloy by micro-arc oxidation and a solution treatment. Applied Surface Science, 2013, 264 (1): 816-822.

[6] Zhao Quanming, Guo Xiong, Dang Xiaoqian, et al. Preparation and properties of composite MAO/ECD coatings on magnesium alloy. Colloids and Surfaces B: Biointerfaces, 2013, 102 (1): 321-326.

Acknowledgments:

This work was financially supported by National Natural Science Foundation of China (No.51374068).

Impact properties of geopolymeric concrete: a state-of-the-art review

Fang Chi, Xie Jianhe*, Zhang Bingxue, Yuan Bing and Wang Chonghao

School of Civil and Transportation Engineering, Guangdong University of Technology, Guangzhou, 510006, China
* E-mail: jhxie@gdut.edu.n

Abstract. The application of geopolymer instead of cement in construction engineering provides an effective way to achieve sustainable development. A large number of studies have shown that geopolymeric concrete (GC) has excellent static performance. However, there are relatively few studies on the dynamic performance of GC which is of dominate importance in cases such as explosion, earthquake and other impact loads. This paper presents a brief review of recent research on the impact properties of GC, in order to facilitate the research development in this field. The effects of strain rate, fiber type, alkali type, elevated temperature, water environment and flow state on the impact resistance performance of GC were analysed. Research findings revealed that GC exhibited better impact properties than ordinary Portland concrete (OPC).

1. Introduction

China is a large resource-consuming country, the economic growth of which is increasingly constrained by shortage of resources and environmental pollution. In 2015, the cement production over the whole world was 4 billion tons and China consumed about 2.5 billion tons, that is to say, 60% of all cement are poured on China's land. It should be mentioned that the production of each 1 ton of Portland cement requires about 1.5 tons of limestone, and consumes a lot of coal and electricity, and emits about 1 ton of CO_2. At the same time, the amount of solid wastes increases dramatically in China. For example, every 1 ton of steel production is accompanied by the output of dozens of tons of steel slag, slag and other wastes. The accumulation of these solid wastes not only occupies a large amount of land, but also causes serious environmental pollution. In order to significantly reduce CO_2 emissions and recycle industrial wastes to achieve sustainable development in China, using geopolymer instead of cement to develop green concrete is one of the efficient methods.

Geopolymer is considered as the third generation cement after lime and ordinary Portland cement, which is generically used to describe a amorphous alkali aluminosilicate. A variety of aluminosilicate materials such as kaolinite, feldspar and industrial solid residues such as fly ash, metallurgical slag, mining wastes etc. have been used as solid raw materials in the geopolymerization technology [1]. Geopolymeric concrete (GC) is a new kind of concrete material prepared by geopolymer used as cementing material [2]. The production process of GC not only consumes less energy and emits low CO_2, but also recycles solid wastes and conserves land resources. Thus, GC generates more environmental benefits than ordinary Portland concrete (OPC). Moreover, GC has been widely reported to exhibite high resistance to fire and acids, and does not produce the high evolution of reaction heat associated with OPC, reducing cost and potential cracking issues when the material is placed in large volumes [3].

Nowadays, numerous studies have been conducted on the mechanical properties of geopolymer concrete [1-4]. However, the majority of these studies are focused on geopolymer concrete under monotonic loading, with a limited number under dynamic loading, very limited information is available for the impact performance of GC which is of dominate importance in cases such as pile driving. Thus, To facilitate the research development in this field and promote the application of this new type of green concrete in civil engineering, this paper reviews the performance of GC under impact loading, and presents the recommendations for future research.

2. Impact Resistance of GC

Geopolymer has unique three-dimensional network structure of inorganic polycondensation, and its microstructure and physical-chemical properties are relatively better than traditional cement. Therefore, the bonding property between the geopolymer and aggregate is superior to that between the cement and aggregate. Many experimental results indicated that compared with the cement concrete, GC has less inherent defects and a denser internal structure resulting in a stronger impact resistance [1,4-6].

2.1. Effect of Strain Rate on Impact Performance

The strain rate range of Split Hopkinson Pressure Bar (SHPB) test is usually 10^2-10^4/s, which can represent the strain rate sensitivity of in-site engineering material. Thus, many researchers conducted SHPB test to investigate the effect of strain rate on the impact performance of GC [4-6], as shown in Table 1. The findings have reached a consensus that strain rate affects significantly the impact performance of GC. It can be indicated from these experiments [4-6] that the compressive strength, deformation properties and impact toughness of GC under impact loading increase with the increase of strain rate.

Table 1. SHPB test results of geopolymeric concrete

Data sources	Test number	Strain rate（s^{-1}）	Dynamic compressive strength(MPa)	Critical strain ($\times 10^{-3}$)	Specific energy absorption ($kJ \cdot m^{-3}$)
Fan et al. [4]	1	34.6	72.3	5.60	340
	2	54.8	87.1	6.77	690
	3	76.4	98.1	11.97	1070
Xu et al.[5]	1	38.5	62.2	8.20	174.4
	2	55.9	73.3	9.44	393.2
	3	73.0	86.5	9.96	658.1
Lu and Zheng [6]	1	39.3	62.2	8.20	250.9
	2	54.7	72.4	10.23	548.3
	3	83.2	86.6	10.26	730.2

2.2. Effect of Fibre Type on Impact Performance

GC is a quasi-brittle material as same as OPC. Usually, adding fibre into concrete is an efficient method to enhance toughness, including basalt fibre and carbon fibre. In order to investigate the effect of basalt fiber on the impact resistance of GC, Li et al. [7] conducted the impact test of the GC with volume fraction of 0.3% basalt fibre by SHPB. This result revealed that fibre had no significant improvement in 28-day-old compressive strength, but it could obviously enhance the deformation and energy absorption capacity of GC. Xu et al. [8,9] studied the deformation characteristics of the GC with 0.1%, 0.2% and 0.3% basalt fibre and carbon fibre under impact loading, and then found that the optimal addition of fibre is 0.2%. Similarly, Zhu et al. [10] conducted a series of impact tests of the GC with different carbon fibre content, and then pointed out that the optimal addition of fibre is 0.3%. It can be seen that the addition of the appropriate amount of basalt fibre and carbon fibre can effectively improve the impact resistance of GC, and its optimal addition is 0.2%-0.3%.
To compare the effects of basalt fibre and carbon fibre on the impact resistance of GC, Luo et al. [11] found that the early impact properties of 3-day-old and 7-day-old GC with basalt fibre is better than

that with carbon fibre, while Xu et al. [12] poited out that of 28-day-old GC is just on the contrary. The role of the fibre in the impact mechanism and its quantitative effect are complex and require further investigation.

2.3. Effect of Alkali Type on Impact Performance

The use of alkali activator is one of the key procedure in the preparation of GC. Undoubtedly, the type of alkali activator is critical to the performance of GC. Luo et al. [13,14] conducted some impact tests on slag and fly ash based GC to compare the effect of NS activator (NaOH and sodium silicate) and NN activator (NaOH and Na2CO3) on the dynamic performance. The results indicated that NN is more beneficial than NS to enhance the compressive strength of GC, while NS is more beneficial than NN to improve the deformation capacity. The reason may be that different alkali could change the hydration-producing substance and microstructures of GC. Therefore, suitable alkali-activator should be chosen to use in practical application according to the specific engineering requirements.

2.4. Effect of Water Environment on Impact Performance

Liu et al. [15] and Ren et al. [16] conducted a series of tests to investigate the influence of long-time water environment on the dynamic mechanical properties of GC. These experimental results shown that the dynamic compressive strength of GC increased after exposure to water environment (180 days), but the quasi-static strength of the specimens decreased. It can be seen that water environment enhance the strain rate sensitivity of the dynamic compressive strength of GC, which is the same as OPC in water environment [17]. Moreover, the higher dynamic elastic modulus indicated that water environment has a stiffening effect on GC. It should be noted that the influence of water environment on the impact toughness of GC is relatively small. Thus, it can be concluded that GC has a good impact resistance in water environment. Further research is required to investige others environments such as saltwater erosion and freezing-thawin on the impact resistance of GC.

2.5. Effect of Temperature on Impact Performance

To investigate the effect of elevated temperature on the impact resistance of GC, Su et al. [18] designed high temperature SHPB system to conduct a series of impact test (200℃-800℃). As shown in Figure 1 [18], the dynamic compressive strength of GC grows higher at 200℃ than at room temperature, but suffers a dramatic drop at 800℃. However, at 400℃ and 600℃, it is close to that at room temperature. It indicates that GC could keep its integrity when exposed to temperatures from 200℃ to 600℃, resulting in a good machenical properties. For GC, the elevated temperature may promote the dissolution and polycondensation of aluminosilicate that has not reacted previously, leading to more compact structure and higher strength at 200℃ [18]. With the temperature increases, free water vaporizes and the polycondensation terminates, so the strength decreases dramatically after 800℃. A similar observation was also reported for GC static test [19]. For OPC, calcium hydroxide decomposes after 400℃ and most strength is lost after 600℃. Thus, it can be seen the impact energy absorption ability of GC is better than that of OPC subjected to elevated temperature.

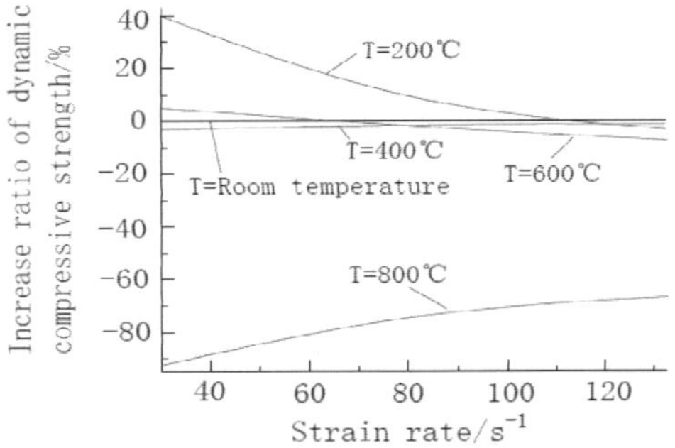

Figure 1. The increase ratio of dynamic compressive strength compared to room temperature [18].

2.6. Impact Performance of Highly-fluidized GC

GC usually is lowly-fluidized because the setting and hardening of geopolymer is fast. To popularize the application of GC, highly-fluidized GC (HFGC) is receiving increasing attentions [20,21]. Xin et al. [20,21] conducted an experimental study on the impact properties of HFGC. The results revealed that the dynamic strength of HFGC exhibits significant strain rate correlation, which is similar to lowly-fluidized GC and OPC. Compared with lowly-fluidized GC, the strain rate sensitivity threshold of HFGC is higher. However, limited information is available for HFGC. Therefore, further research is needed to investigate of impact resistance of HFGC leading to improved design methodologies.

3. Summary

Geopolymeric concrete, as a kind of green concrete, produces significant environmental benefits, which can not only effectively recycle solid waste and save a lot of resources, but also can reduce CO_2 emissions. In addition, it exhibits good static and impact resistance performance. The dynamic compressive mechanical properties of GC has strong strain rate dependency. Compared with OPC, the impact resistance of GC is better subject to elevated temperature or in water environment. To improve the impact resistance of GC, adding basalt fibre or carbon fibre into it is an efficient method. Therefore, it can be concluded that GC has unique advantages and broad application prospect in cases such as heat-resistance structure, hydraulic structure and rapid-repair of pavement and so on. However, the impact resistance of GC is not clear yet. Further research is required to investigate the effects of alkali dosage, curing conditions, various environments, etc. In addition, the application of recycled aggregates from construction and demolition waste currently has become a hot focus in civil engineering. Consequently, the impact resistance of GC mixed with recycled aggregates should be further investigated.

4. References

[1] Singh B, Ishwarya G and Gupta M 2015 Geopolymer concrete: A review of some recent developments *J. Construction and Building Materials* **85** 78-90

[2] Hardjito D, Wallach S E and Rangan B V 2004 On the development of fly ash-based geopolymer concrete *J. Aci Materials Journal* **06** 467-72

[3] Deventer J S J V, Provis J L and Duxson P 2012 Technical and commercial progress in the adoption of geopolymer cement *J. Minerals Engineering* **03** 89-104

[4] Fan F L, Xu J Y, Li W M, Yang J Y and Zhai Y 2009 Impact mechanical properties of slag- and fly ash-based geopolymer concrete *Explosion & Shock Waves*

[5] Xu J Y, Li W M, Fan F L and Bai E L 2009 Study on mechanical properties of geopolymeric concrete under impact loading *Journal of Vibration & Shock* **28** 46-50

[6] Lu Q and Zheng R 2008 Impact Compression Experiment of Basalt Fiber Reinforced Geopolymeric Concrete *Construction Technology*

[7] Li W and Xu J 2009 Mechanical properties of basalt fiber reinforced geopolymeric concrete under impact loading *Materials Science & Engineering A* **505** 178-86

[8] Xu J, Li W, Huang X and Li P 2009 Deformation property of basalt fiber reinforced geopolymeric concrete under impact loading *Journal of the Chinese Ceramic Society* **37** 1137-41

[9] Xu J Y, Li W M, Fan F L and Bai E L 2010 Experimental study on impact properties of carbon fiber reinforced geopolymeric concrete using a SHPB *Journal of Building Materials* **13** 435-40

[10] Zhu J, Jinyu X U, Xin L, Weimin L I, Bai E and Gao Z 2014 Comparative Study on the Toughness Evaluation Index of Carbon Fiber Reinforced Geopolymeric Concrete *Journal of Building Materials* **17** 303-8

[11] Luo X, Xu J Y and Li W M 2009 Early impact properties of fiber reinforced geopolymeric concretes *Journal of Vibration & Shock* **28** 163-8

[12] Xu J and Li 2010 Dynamic properties of fiber reinforced geopolymeric concrete *China Civil Engineering Journal* **43** 127-32

[13] Luo X, Xu J Y, Bai E L and Li W M 2014 Comparative study of the effect of the type of alkali on the strain rate effect of geopolymer concrete Explosion & Shock Waves 34 340-6
[14] Luo X, Xu J Y, Li W and Bai E 2014 Effect of alkali-activator types on the dynamic compres sive deformation behavior of geopolymer concrete Materials Letters 124 310-2
[15] Liu F D, Liu J Y and Ren W 2015 Mechanical Properties of Geopolymer Concrete in Water Environment under Impact Loading *Bulletin of the Chinese Ceramic Society* **11** 3289-302
[16] Ren W, Xu J, Liu J and Su H 2015 Dynamic mechanical properties of geopolymer concrete after water immersion *Ceramics International* **41** 11852-60
[17] Rossi P, Mier J G M V, Boulay C and Maou F L 1992 The dynamic behaviour of concrete: influence of free water *Materials & Structures* **25** 509-14
[18] Su H, Xu J and Ren W 2016 Mechanical properties of geopolymer concrete exposed to dynamic compression under elevated temperatures *Ceramics International* **42** 3888-98
[19] Kong D L Y and Sanjayan J G 2010 Effect of elevated temperatures on geopolymer paste, mortar and concrete *Cement & Concrete Research* **40** 334-9
[20] Xin L, Jinyu X U, Haoyang S U, Bai E and Weimin L I 2014 Strength Properties of Highly Fluidized Geopolymer Concrete under Impact Loading *Journal of Building Materials* **17** 72-7
[21] Xin L, Xu J Y, Su H Y, Li W M and Bai E L 2014 Deformation behaviors of highly-fluidized geopolymer concrete during dynamic compression at high strain rates *Explosion & Shock Waves* **34** 216-22

Acknowledgments
The authors gratefully acknowledge the financial support provided by the Science and Technology Planning Project of Guangdong Province (No. 2014A010105044), the Science and Technology Planning Project of Guangzhou City (No. 201704030057 and No. 201707010364).

IOP Conf. Series: Materials Science and Engineering **284** (2017) 012013 doi:10.1088/1757-899X/284/1/012013

First principles study of Ti_5Si_3 intermetallic compounds with Cu additions: Elastic properties and electronic structure

P F Zhang, Y X Li and P K Bai

College of Material Science and Engineering, North University of China, Taiyuan 030051, China
E-mail: liyuxin326@hotmail.com

Abstract. The structural stabilities, elastic and electronic properties of Ti_5Si_3 intermetallic compounds with Cu additions have been studied by density functional theory. The calculated lattice constants (a, c) and the equilibrium volume are found to decrease with the increasing of Cu atomic number.Then the independent elastic constants are calculated, the results shows that all the elastic constants of compounds satisfy the mechanical stability criteria, which indicates all these hexagonal structures are mechanically stable, furthermore, the elastic constants (C11, C44 and C66) decrease with the increasing of Cu atomic number. The calculated bulk modulus B, shear modulus G and Young' s modulus E decrease, while Poisson' s ratio σ and Cauchy pressure increase with the increasing of Cu atomic number, and the ductility is slightly improved. The calculated DOS showed that the density of states (DOSs) of the Feimi level increases and the stability of compounds decreases with the increasing of Cu atomic number.

1. Introduction

Titanium silicide (Ti_5Si_3) has attracted considerable attention due to its low density (4.32 g/cm^3), high melting temperature (2130℃), good strength at elevated temperatures, high oxidation resistance, and excellent creep resistance [1-2]. Nevertheless, the defect of Ti_5Si_3, such as low fracture toughness (2.5 Mpa/m^2) below the ductile-brittle transition temperature [3], severely restricted its practical applications.

In fact, by adding Cu to the structure, Ti_5Si_3 becomes a considerably more hopeful material for engineering applications by improving mechanical property and electronic structure [4]. This study is to combine experimentally determined structural data with first-principle calculations to fully understand bonding changes that occur with the addition of Cu to Ti_5Si_3. The main purpose of the this study is to investigate the lattice constants, formation enthalpy, density of states (DOS) and elastic constants of Ti_5Si_3 intermetallic compounds with Cu additions.

2. Calculation methods

The structural, electronic and elastic properties of Ti_5Si_3 intermetallic compounds with Cu additions was calculated by First-principles of Cambridge Serial Total Energy Package (CASTEP). The density functional theory (DFT) was used by this software package that employed the plane wave pseudo-potential method as a basically theory [5]. The generalized gradient approximation of Perdew Burke Ernzerhof (GGA-PBE) parameterized by PERDEW described to the exchange and correlation terms [6]. The electron-ion interaction was described by Ultrasoft pseudo-potential with the cutoff energy 400 eV. The calculations of total energy and electronic are followed by cell optimization with self-consistant field below 5.0×10^{-6} eV/atom, the maximum force on the atom was below 0.01eV/ Å, the maximum displacement between cycles was below 5.0×10^{-4} and the maximum stress was below 0.02 GPa.

3. Results and discussion

3.1. Structure and stability

Ti-Si-Cu system intermetallics possess the $D8_8$ crystal structure of Ti_5Si_3 prototype with the space group $P6_3/mcm$ (No. 193). As shown in Fig. 1, the unit cell of Ti_5Si_3 contains 10 Ti atoms and 6 Si atoms in which Ti atoms occupy at 4d (0.33, 0.67, 0), 6g (0.23, 0, 0.25) sites and Si atoms are at 6g (0.60, 0, 0.25). Cu is incorporated in the intermetallic phase of Ti_5Si_3 by occupation at sites 4d (0.33, 0.67, 0) or 6 g (0.23, 0, 0.25). This is gave rise to $Ti_{10}Si_5Cu_1$, $Ti_9Si_6Cu_{1\text{-}4d}$ and $Ti_9Si_6Cu_{1\text{-}6g}$ intermetallic phases. Fig. 2 given is the crystal structures of these compounds. Considering the phase stability, formation enthalpy of intermetallic phases are calculated as follows:

$$\Delta H = \frac{1}{a+b+c}\left(E_{tot} - aE_{solid}^{Ti} - bE_{solid}^{Si} - cE_{solid}^{Cu}\right) \quad (1)$$

where E_{tot} is the total energy of the unit cell, E_{solid}^{Ti}, E_{solid}^{Cu} and E_{solid}^{Si} are the total electronic energies per atom of Ti and Cu, and Si, and a, b and c are to the numbers of Ti, Si and Cu atoms in unit cell. Table 1 gives the calculation results of formation enthalpy($\triangle$H). It can be seen that all the formation enthalpies of the compounds are negative, which indicates these intermetallic compounds are stable. Except $Ti_{10}Si_6$ phase, $Ti_9Si_6Cu_{1\text{-}4d}$ phase possess the lowest formation enthalpy when Cu occupied 4d (0.33, 0.67, 0) or 6 g (0.23, 0, 0.25) sites, which indicates that the $Ti_9Si_6Cu_{1\text{-}4d}$ phase possess the highest structure stability, and then followed by$Ti_9Si_6Cu_{1\text{-}6g}$ and $Ti_{10}Si_5Cu_1$. Therefore, Cu preferentially occupies 4d-Ti site of Ti_5Si_3.

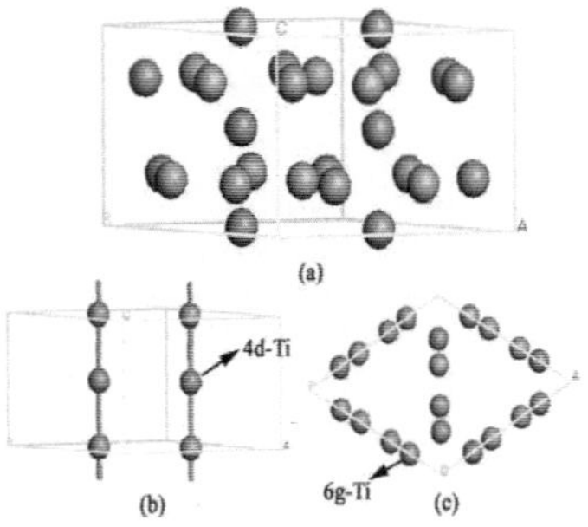

Fig.1 (a) $D8_8$ crystal structure of Ti_5Si_3; (b) Schematic diagram of 4d-Ti arrangement; (c) Schematic diagram of 6g-Ti arrangement.

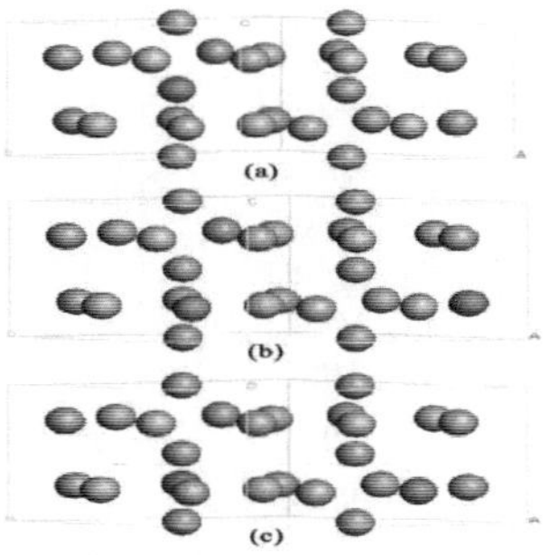

Fig.2 (a) Cu doped 4d-Ti sites of Ti_5Si_3; (b) Cu doped 6g-Ti sites of Ti_5Si_3; (c) Cu doped Si sites of Ti_5Si_3.

Table 1 Calculated formation enthalpy (Δ H) of Cu atom to dope Ti sites of Ti_5Si_3

Compound	$Ti_{10}Si_6$	$Ti_{10}Si_5Cu_1$	$Ti_9Si_6Cu_{1\text{-}4d}$	$Ti_9Si_6Cu_{1\text{-}6g}$
Δ H (eV/atom)	-6.377	-5.203	-5.462	-5.341

Table 2 gives the calculation value of lattice parameters and equilibrium volume by different Cu atomic number occupying of Ti_5Si_3 stable sites, and the lattice parameters (a c) decrease. Because of the atomic radius of Ti is bigger than that of Cu, which induces lattice distortion of compound. At the same time, the equilibrium volume also decreases.

Table 2 Calculated lattice parameter of different Cu atom to dope 4d-Ti sites of Ti_5Si_3.

Compound	a/(Å)	b/(Å)	c/(Å)	$V/(Å)^3$
$Ti_{10}Si_6$	7.4716	7.4716	5.132	248.119
$Ti_9Si_6Cu_{1\text{-}4d}$	7.4004	7.4004	5.124	243.79
$Ti_8Si_6Cu_{2\text{-}4d}$	7.2783	7.2783	5.114	239.681
$Ti_7Si_6Cu_{3\text{-}4d}$	7.2623	7.2623	5.014	232.681
$Ti_6Si_6Cu_{4\text{-}4d}$	7.2592	7.2592	5.012	228.704

3.2. Elastic properties

Table 3 gives the calculated elastic constants Cij in Cu-Ti-Si intermetallic compounds. In Table 3, the values of C11, C44, C66 decrease with the increasing of Cu atomic number. The traditional mechanical stability criteria for hexagonal crystal systems are given as [7]:

$$C_{11} > 0,\ C_{11} - C_{12} > 0,$$
$$C_{44} > 0,\ (C_{11} + C_{12})C_{33} - 2C_{13}^2 > 0 \qquad (2)$$

Table 3 Calculated elastic constants of different Cu atom to dope 4d-Ti sites of Ti_5Si_3.

Compound	C_{11}	C_{12}	C_{13}	C_{33}	C_{44}	$C_{66} = \frac{1}{2}(C_{11} - C_{12})$
$Ti_{10}Si_6$	282.62	107.91	54.00	271.63	94.17	87.36
$Ti_9Si_6Cu_{1\text{-}4d}$	272.89	110.65	45.03	281.85	73.34	80.93
$Ti_8Si_6Cu_{2\text{-}4d}$	264.72	111.97	49.83	284.05	59.63	76.38
$Ti_7Si_6Cu_{3\text{-}4d}$	258.20	112.45	53.33	269.69	49.32	72.88
$Ti_6Si_6Cu_{4\text{-}4d}$	243.60	113.33	46.08	251.63	14.6	65.13

In Table 3, all the elastic constants of phase satisfy the mechanical stability criteria of Eq. (2), and all these hexagonal structures are excellent mechanical stability. Shear modulus (BV, BR, GV, GR) in terms of the elastic constants Cij' s are given by[7]. And the final values can be taken as:

$$B = (B_V + B_R)/2 \qquad (3)$$

$$G = (G_V + G_R)/2 \qquad (4)$$

$$E = \frac{9BG}{3B + G} \qquad (5)$$

$$\sigma = \frac{1}{2}\left(1 - \frac{E}{3B}\right) \qquad (6)$$

According to expression (3-6), Shear modulus (BV, BR, GV, GR) can be calculated in Table 4. The calculated bulk modulus and the shear modulus decrease with the increasing of Cu atomic number in Table 4, which indicates the resistance of volume change by applied pressure and the resistance of reversible deformations upon shear stress is decreased, respectively.

The ratio of the bulk to B/G represents the ductility of materials. When The B/G is higher than 1.75, suggesting that the material shows a ductile behavior. In Table 4, the value of $Ti_{10}X_6$ and $Ti_9Si_6Cu_{1\text{-}4d}$ are 1.46 and 1.60, respectively, which indicates $Ti_{10}X_6$ and $Ti_9Si_6Cu_{1\text{-}4d}$ are brittle material. While the values of the other phases are higher than 1.75, which indicates that the others are ductile materials.

Table 4 Calculated bulk modulus B, shear modulus G, Young' s modulus E and Poisson' s ratio σ of different Cu atom to dope 4d-Ti of Ti_5Si_3.

Compound	B	G	E	σ	B/G
$Ti_{10}Si_6$	140.32	95.82	237.16	0.36	1.46
$Ti_9Si_6Cu_{1-4d}$	136.21	85.37	225.25	0.39	1.60
$Ti_8Si_6Cu_{2-4d}$	132.68	67.29	209.18	0.41	1.97
$Ti_7Si_6Cu_{3-4d}$	131.47	59.35	198.74	0.42	2.21
$Ti_6Si_6Cu_{4-4d}$	127.19	41.37	188.37	0.45	3.07

From Table 4, it can be also seen that Young' s modulus (B) of $Ti_{10}Si_6$ is largest, this indicates $Ti_{10}Si_6$ is the stiffest, and then followed by $Ti_9Si_6Cu_{1-4d}$, $Ti_8Si_6Cu_{2-4d}$, $Ti_7Si_6Cu_{3-4d}$ and $Ti_6Si_6Cu_{4-4d}$. For another, Poisson' s ratio σ increased with the increasing of Cu atomic number, which indicates the ductility of materials increases with the increasing of Cu atomic number.

Besides B/G, Young' s modulus E and Poisson' s ratio σ, Cauchy pressure (C12-C44) is another important parameter for the mechanical properties of materials [8]. The higher values of Cauchy pressure (C12-C44) correspond to better toughness of materials. Fig. 3 shows the effect of the number of Cu atom on Cauchy pressure (C12-C44). It can be seen that the toughness of materials increases with the increasing of Cu atomic number, which the same with above calculation results.

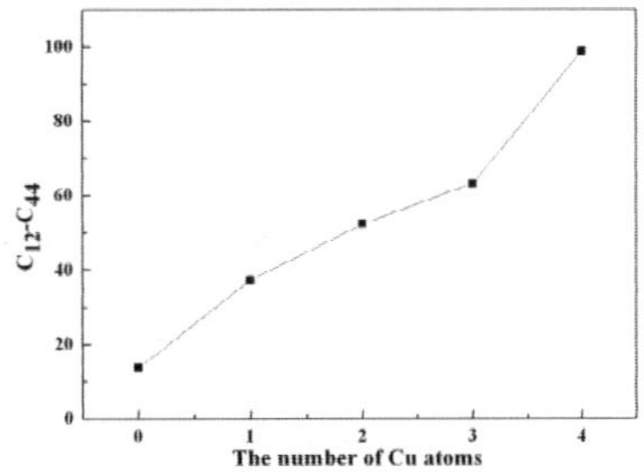

Fig. 3 Cauchy pressure (C12 - C44).

3.3. Electronic structure

Fig. 4 gives the total and partial electronic density of states (DOSs) for Ti_5Si_3. It can be seen that the values of total DOS at Fermi level (Ef) are all higher than 0, which implying that Ti_5Si_3 exhibits metallic behavior. From Fig. 4, it can be also seen that the main bonding peaks between -10 eV and -5 eV mainly originate from the contribution of valence electron numbers of Si(3s) orbits, the main bonding peaks between -5 eV and -2 eV are the result of the bonding Ti(3d), Ti(4s) and Si(3s) hybridized. The main bonding peaks from -2 to 2 eV are the contribution of Ti(3d) with little Si(3p) and Ti(4s). Furthermore, from the total density of states (DOS) for Ti_5Si_3, it can be also seen that the pseudogap is wider and deeper, and the Feimi level is close proximity to the valley bottom. The pseudogap at the Fermi level was always associates with stronger stability, which indicates Ti_5Si_3 is stable.

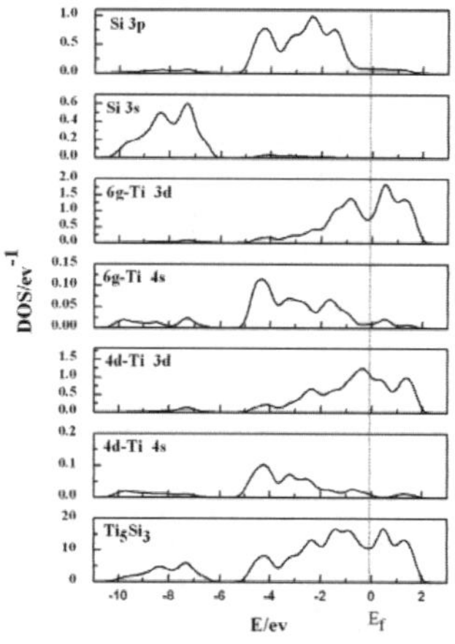

Fig. 4 The total and partial density of states (DOS) of Ti_5Si_3.

Fig. 5 shows the total density of states (DOSs) of different Cu atom to dope 4d-Ti of Ti_5Si_3. It is noted that the Feimi level of Ti_5Si_3 is the lowest, the Feimi level of $Ti_6Si_6Cu_{4\text{-}4d}$ is the highest, which indicates that the density of states (DOSs) of the Feimi level increases and the stability of compounds decreases with the increasing of Cu atomic number. Furthermore, the Feimi level moves toward the high pseudogap side, which indicates the valence electron occupies the anti-bonding states of d(Ti-4d)- d(Ti-4d).

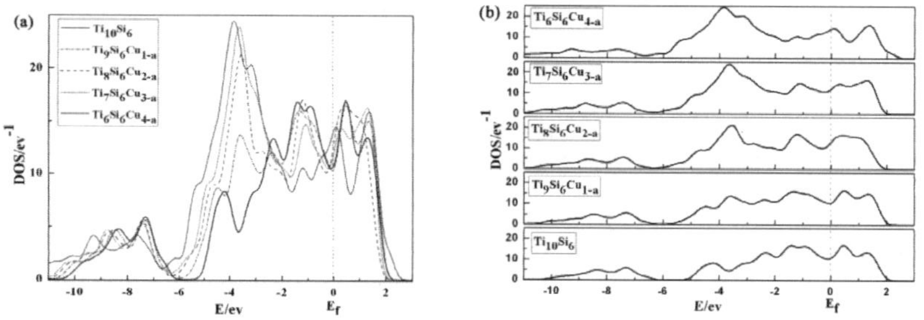

Fig. 5 The total density of states (DOS) of different Cu atom to dope 4d-Ti of Ti_5Si_3.

4. Conclusions

Structural stability, elastic and electronic properties of Cu-Ti-Si system intermetallic compounds with $D8_8$ hexagonal structure were investigated.

The calculated lattice constants (a, c) and elastic constants (C11, C44 and C66) decrease with the increasing of Cu atom.

The polycrystalline elastic modulus were obtained by the VRH method, and the obtained B, G and E decreased and Poisson' s ratio σ_{xy} and Cauchy pressure (C12-C44) increased with the increasing of Cu atomic number, whereas the ductility was improved.

The calculated DOS showed that the density of states (DOSs) of the Feimi level increases and the stability of compounds decreases with the increasing of Cu atomic number.

References

[1] Stoloff N S 1999 *J. Mater. Sci. Eng. A.* **261** 169.
[2] Meschter P J and Schwartz D S 1989 *J. JOM* **41** 52.
[3] Yeh C L, Chen W H and Hsu C C 2007 *J J. Alloys Compd.* **432** 90.
[4] Pettifor D G, 1992 *J Mater.Sci. Tech.* **8** 345.
[5] Hohenberg P and W. Kohn 1964 *J Phys. Rev. B*, **136** 384.
[6] Perdew J P and Levy M 1985 *J Phys Rev B*, **31** 6264.
[7] Wu M M, Li W, Bi Y T, Li M P and Wen J D 2010 *J J. Alloys Compd.* **506** 412.
[8] Pugh S F 1954 *Philos Mag.* **45** 823.

Functionalization of silica-gel with polyamidoamine and 2-mercaptobenzothiazole and its adsorption property for lead

X Z Wu[1], Y Liu[1], L L Luo[1] and Z Y Chen[1]

1 Department of Chemistry and Bioengineering, Guilin University of Technology, Guilin 541004, P.R. China
E-mail: 2645642603@qq.com

Abstract. Adsorbents PAMAM-n.0MBTSG (n=1-4) have been prepared by immobilizing polyamidoamine (PAMAM) and 2-mercaptobenzothiazole (MBT) on silica-gel. Characterized with FTIR, SEM, TGA, the preconcentration of Pb^{2+} with PAMAM-n.0MBTSG(n=1-4) has been investigated by graphite furnace atomic absorption spectroscopy (GFAAS). SEM showed that the surface morphology of adsorbent changed with the generation increase of PAMAM. Adsorption capacity of PAMAM-n.0MBTSG for Pb^{2+} (n=1-4) reached 16.22, 19.84, 22.92 and 27.56 mg g^{-1} respectively. Pb^{2+} (1.0 ng mL^{-1}) in 2000 mL solution could be quantitatively absorbed with PAMAM-4.0MBTSG and eluted to obtain a preconcentration factor (PF) of 200. GFAAS method for analysis of Pb^{2+} with PAMAM-4.0MBTSG as adsorbent was proposed and successfully applied to analysis of Pb^{2+} of standard reference material, sea water and squid sample.

1. Introduction

In recent years, the analysis of lead in environmental samples has become an important research field. In order to overcome the interference of co-existing substances in the determination of metal ions, Solid-phase extraction adsorbents, such as nano-materials [1,2], activated carbon [3], chelating resin [4], functionalized silica-gel [5] have been already applied to enrichment of lead in aqueous solution, fruits or salt, etc. Because of their special structure, dendrimer-like polyamidoamine (PAMAM) has been applied to silica-gel modification and metal ions adsorption [6,7]. EDTA modified Gn-PAMAM-SBA-15 (n=1-4) inorganic–organic hybrid materials were successfully used to absorb Pb^{2+}[8].

In present work, dendrimer-like highly branched polyamidoamine (PAMAM) and 2-mercaptobenzothiazole (MBT) have been immobilized on silica-gel surface to prepare adsorbents PAMAM-n.0MBTSG (n=1-4). After characterized, PAMAM-n.0MBTSG (n=1-4) was used to preconcentrate Pb^{2+}. With PAMAM-4.0MBTSG as adsorbent, micro-column enrichment-GFAAS method for determination of Pb^{2+} was proposed and applied to analysis of Pb^{2+} in standard reference material, sea water and squid sample.

2. Materials and methods

2.1 Reagents and instruments

Silica-gel (60-100 meshes, surface area: 300-600 m^2 g^{-1}) was bought from Qingdao Shuoyuan Chemical Co. Ltd., China. γ-aminopropyltriethoxysilane was bought from Beijing Shenda Fine Chemical Co. Ltd. and distilled at 123℃ under 10 mm Hg. Standard reference material (GSB 04-1767-2004) was bought from General Research Institute for Nonferrous Metals, Beijing, China. 2-

mercaptobenzothiazole (MBT) was bought from Aladdin Industrial Corporation. The other reagents used in preconcentration were analytical grade. Suitable amount of $Pb(NO_3)_2$ was dissolved in HNO_3 solution (pH=2) to obtain 1000 mg L^{-1} Pb^{2+} stock solution. Pb^{2+} stock solution was diluted step by step to obtain Pb^{2+} standard solutions.
Infrared spectra were obtained at room temperature with Thermo Nexus 470 FTIR spectrometer. The surface morphology of APSG and PAMAM-n.0MBTSG (n=1-4) were obtained by scanning electron microscopy (SEM, Hitachi S-4800). Thermal analyzer (SDT-Q600, Waters) was used to obtain thermogravimetric analysis (TA) curves of sample. Temperature range of TG analysis was from room temperature to 600 ℃ with heating rate of 5 ℃ min^{-1} and N_2 atmosphere.

2.2 Preparation of adsorbents

PAMAM modified silica-gel PAMAM-n.0SG (n=1-4) were synthesized according to reference [7]. According to acid-base titration method described in reference [6], $-NH_2$ content of γ-aminopropyltriethoxysilane modified silica-gel (APSG) and PAMAM-n.0SG (n=1-4) was 0.60, 0.86, 1.30, 1.70 and 1.97 mmol g^{-1} respectively. PAMAM-n.0MBTSG (n=1-4) was prepared as below according to reference [9]. 5.85 g (about 35.0 mmol) 2-mercaptobenzothiazole was dissolved with 60 mL 95% ethanol in a 250 mL flask. 4.5mL formaldehyde solution (37%) and 20.0 g PAMAM-1.0SG were added to the flask. Then the flask was placed in 50 ℃ oil bath and the mixture was stirred for 8 h to prepare PAMAM-1.0MBTSG.
PAMAM-n.0MBTSG (n=2,3,4) were prepared as PAMAM-1.0MBTSG. The above solid products PAMAM-n.0MBTSG (n=1-4) were filtrated and washed with 95% ethanol for 3 times. After extracted with 95% ethanol for 5 hours in Soxhlets, solid products were dried at 50 ℃ under vacuum and yellow adsorbents PAMAM-n.0MBTSG (n=1-4) were obtained respectively. The ideal synthetic route of adsorbents was illustrated in Scheme 1.

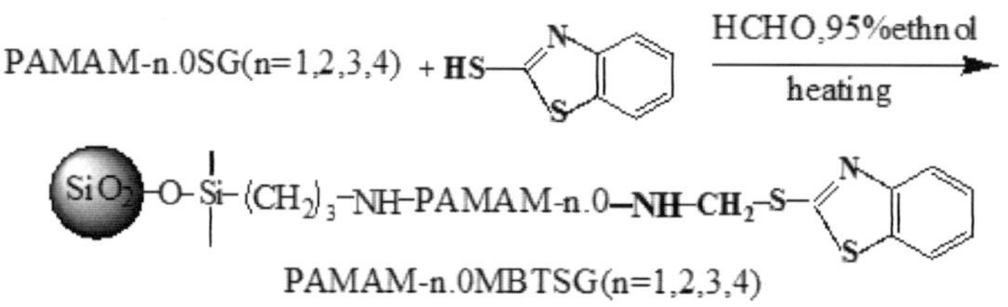

Scheme 1. The ideal synthetic route to PAMAM-n.0 MBTSG (n=1,2,3,4)

2.3 Preconcentration experiment

Preconcentration of Pb^{2+} was completed in a self-made micro-column (7.0 cm × 0.5 mm i.d., packed with about 0.6 g adsorbent) and peristaltic pump (BT01S-YZ1515) (Yi Kang Xin Da Technology Co., Ltd, China). PTFE tubing (0.8 mm i.d.) was used for connection.
The preconcentration experiment was carried out as below. (1) micro-column was cleaned with blank solution. (2) Pb^{2+} solution was driven through micro-column and the outflow solution was collected. (3) Blank solution was driven through micro-column for 30 s to remove the matrix ions. (4) Pb^{2+} was desorbed with eluent in the reverse direction and effluent solution was collected. Pb^{2+} in outflow solution and effluent solution was determined by atomic absorption spectrometer (TAS-990, Beijing Purkinje General Instrument Co., Ltd.) to obtain adsorption percentage and desorption percentage of Pb^{2+}.

3. Results and discussion

3.1 Characterization

In FTIR spectra of PAMAM-n.0MBTSG (n=1-4), the characteristic absorption of stretching vibration of C=O in amide groups and N-H scissorvibration appeared at 1658 cm^{-1} and 1558 cm^{-1} respectively, which suggested that PAMAM was introduced onto the surface of silica-gel. Absorption at 1508 cm^{-1}

and 1452cm^{-1}, attributed to characteristic vibration of aromatic rings, indicated that 2-mercaptobenzothiazole was linked to the surface of PAMAM-n.0SG.

Figure 1 presents SEM images of APSG and PAMAM-n.0MBTSG (n=1-4). As the increase of PAMAM grafting generation, many surface pores became smaller and the surface became smoother gradually, representing that some of micropore could be closed by PAMAM. The phenomenon indicated that PAMAM has successfully grafted onto silica-gel surface and played a dominant role in the changing of the surface morphology.

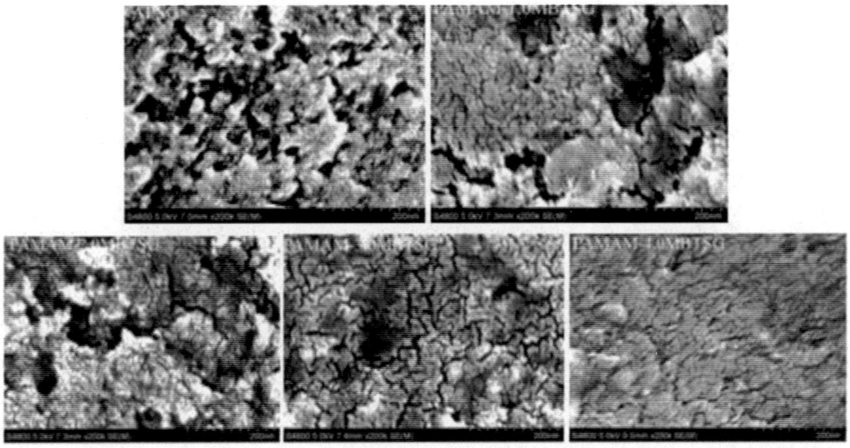

Figure 1. The SEM images of APSG and PAMAM-n.0 MBTSG (n=1,2,3,4)

TG curves in Figure 2 shows the weight loss of APSG and PAMAM-n.0MBTSG (n=1-4). The weight loss of 2.09-3.90% at about 110 ℃ was due to evaporation of physically adsorbed water. While the temperature reached 600 ℃, decomposition of organic molecular layer and condensation of the existing silanol groups [10] brought an obvious total weight loss of 11.69%, 19.54%, 23.76%, 27.84% and 29.18% for APSG and PAMAM-n.0MBTSG (n=1-4) respectively.

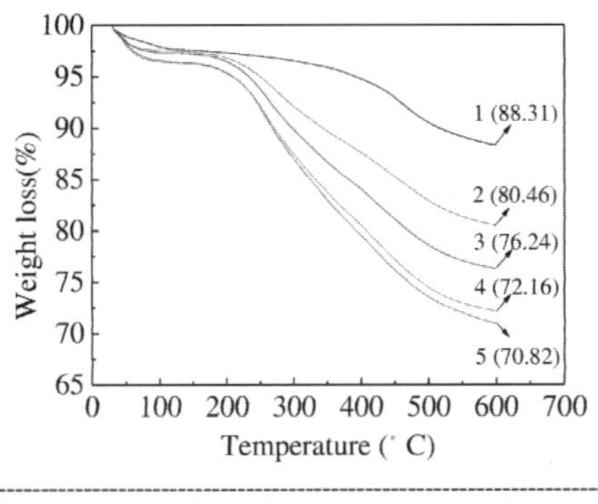

Figure 2. TG curves of APSG and PAMAM-n.0 MBTSG (n=1,2,3,4)
1-APSG, 2-PAMAM-1.0MBTSG, 3-PAMAM-2.0MBTSG, 4-PAMAM-3.0MBTSG, 5-PAMAM-4.0MBTSG.

3.2 Effect of pH

0.2 μg mL^{-1} Pb^{2+} solutions with pH in the range of 3-9 were used to investigate the adsorption of Pb^{2+} on PAMAM-n.0MBTSG. While with silica-gel or APSG as adsorbent, adsorption percentage of Pb^{2+} were all less than 70.0% in pH range of 3-9, which indicated silica-gel or APSG was not suitable for quantitative enrichment of Pb^{2+} in the system. With PAMAM-1.0MBTSG as adsorbent, adsorption percentage of Pb^{2+} gradually increased with increase of pH from 3.0 to 6.0 and then was kept about 100% while solution pH from 6.0 to 9.0. With PAMAM-n.0MBTSG (n=2-4) as adsorbents, adsorption percentage of Pb^{2+} gradually increased with increase of pH from 3.0 to 5.0 and then was kept about 100% while solution pH from 5.0 to 9.0. The experiment results indicated that the surface modification of silica-gel with PAMAM and 2-mercaptobenzothiazole could effectively improve silica-gel adsorption property for Pb^{2+}. Based on the above facts, only the adsorption of Pb^{2+} with PAMAM-n.0MBTSG (n=1-4) as adsorbent was studied and pH 6.0 medium solution was selected for Pb^{2+} adsorption in further experiments.

3.3 Effect of eluent concentration

The effect of concentration of HCl solution on desorption of Pb^{2+} from PAMAM-n.0MBTSG was investigated in the range of 0.05-4.0 mol L^{-}. HCl solution with concentration of 0.05-1.0 mol L^{-1} was sufficient for elution of Pb^{2+}. While HCl solution was more than 1.0 mol L^{-1}, the existing HCl solution in micro-column could not completely eluted with blank solution and would decreased adsorption of Pb^{2+} in the next enrichment cycle. In order to guarantee elution of Pb^{2+} and adsorption of Pb^{2+} in the next cycle, 0.5 mol L^{-1} HCl solution of was used as eluent for the elution of Pb^{2+} in further experiments.

3.4 Effect of flow rate

In the range of 0.6-6.0 mL min^{-1}, effect of flow rate on adsorption and desorption of Pb^{2+} was investigated when 10 ml Pb^{2+} solution (0.2 μg mL^{-1}) was used. The adsorption of Pb^{2+} was complete and then kept over 95% until the sampling flow rate increased to 3.0 mL min^{-1}. While flow rate was more than 3.0 mL min^{-1}, adsorption percentage of Pb^{2+} on PAMAM-4.0MBTSG was obviously higher than that on PAMAM-n.0MBTSG (n=1-3). In the following experiments, 2.4 mL min^{-1} was chose as flow rate for quantitative enrichment of Pb^{2+}.

Desorption percentage of Pb^{2+} reached more than 95% respectively with flow rate not more than 3.6 mL min^{-1}. Desorption percentage of Pb^{2+} with PAMAM-4.0MBTSG as adsorbent was higher than that with PAMAM-n.0MBTSG (n=1,2,3) as adsorbent if the same flow rate was used. Therefore, 3.0 mL min^{-1} was selected as flow rate for desorption of Pb^{2+}.

3.5 Effect of solution volume

With 0.2 μg mL^{-1} Pb^{2+} solution as sample solution, the effect of solution volume (5.0, 10.0, 15.0, 20.0, 30.0 mL respectively) on adsorption of Pb^{2+} was studied. When Pb^{2+} solution volume was not more than 10.0 mL, adsorption percentage of Pb^{2+} on PAMAM-n.0MBTSG (n=1-4) were all more than 95% respectively. In further experiments 10.0 mL solution was applied to preconcentration of Pb^{2+}.

The effect of eluent volume on recovery of Pb^{2+} was studied with 0.5 mol L^{-1} HCl solution (5.0, 10.0, 15.0, 20.0, 25.0 mL respectively) as eluent. When the volume of eluent was more than 10.0 mL, the recoveries of Pb^{2+} were all more than 95%. 10.0 mL HCl solution (0.5 mol L^{-1}) was used for desorption of Pb^{2+} in the further experiments.

3.6 Effect of co-existing substance

Anions such as Cl^{-}, NO_3^{-}, SO_4^{2-}, PO_4^{3-} (added as sodium salt) and ions such as Na^{+}, K^{+}, Ca^{2+}, Mg^{2+}, Zn^{2+}, Mn^{2+}, Al^{3+}, Cu^{2+}, Co^{2+}, Cr^{3+}, Cd^{2+}, Ni^{2+}, Ba^{2+}, Sr^{2+}, Fe^{3+}, Ag^{+}, NH_4^{+} (added as nitrate) were added respectively when 10 mL solution of 0.2 μg mL^{-1} Pb^{2+} was enriched. Interference from Fe^{3+} of 5.0 μg mL^{-1} could be masked by triethanolamine. The tolerance of ions and anions could reaches no less than 5 μg mL^{-1} or 500 μg mL^{-1} respectively with relative error not more than 5.0%. The results indicated that it was hopeful to enrich Pb^{2+} of trace level selectively with PAMAM-n.0MBTSG (n=1-4) as adsorbent.

3.7 Adsorption capacity

Under the selected experimental conditions, 40.0 μg mL^{-1} Pb^{2+} solution with hydro-chloric acid as medium (pH=6.0) was driven through micro-column with a flow rate of 2.4 mL min^{-1}. The outflow solution through micro-column was collected every 5 minutes and Pb^{2+} concentration was determined until Pb^{2+}concentration in outflow solution was not changing. The adsorption capacity of PAMAM-n.0MBTSG (n=1-4) were 16.22, 19.84, 22.92 and 27.56 mg g^{-1} respectively. After used for no less than 100 cycles of adsorption and desorption, the adsorption performance of PAMAM-n.0MBTSG has not changed yet.

3.8 Low concentration enrichment

According to enrichment procedure, solutions (containing 2.0 μg Pb^{2+}) of different volume (10 mL, 100 mL, 1000 mL or 2000 mL) flowed through micro-column, and then Pb^{2+} adsorbed on PAMAM-4.0MBTSG was eluted and determined. 2.0 μg Pb^{2+} in solutions (solution of volume than 2000 mL was not tested) could be quantitatively absorbed on PAMAM-4.0MBTSG and eluted to obtain a preconcentration factor (PF) of 200, which makes it possible for analysis of ng mL^{-1} Pb^{2+} with GFAAS and higher sample solution volume. In order to decrease enrichment time, 10 mL Pb^{2+} solution was still selected as sample solution.

3.9 Analytical performances

GFAAS method for determination of Pb^{2+} combining microcolumn preconcentration system with PAMAM-n.0MBTSG (n=1-4) as adsorbent was proposed. The limit of detection (LOD) for Pb^{2+} was 4.4 ng mL^{-1} with 10 mL sample solution and a loading flow rate of 2.4 mL min^{-1}. RSD was 2.1% by 11 times parallel preconcentration and determination of Pb^{2+} (0.2 μg mL^{-1}). The calibration gragh using the proposed method was linear in the range of 0.05-0.25μg mL^{-1} with the regression equation of $A=2.6783\times C+0.0177$ (where C is the concentration of Pb^{2+} in μg mL^{-1}) with a correlation coefficient (R) of 0.9993. If a sample solution of 2000 mL was used, the analytical concentration range of Pb^{2+} was lower to 0.25-1.25 ng mL^{-1}, obtained by dividing the linear concentration (0.05-0.25 μg mL^{-1}) by preconcentration factor of 200. Sample frequency reached 10 samples h^{-1}.

3.10 Sample analysis

With the proposed method, standard reference sample (GSB 04-1767-2004), sea water and squid sample were analyzed. 0.10μg mL^{-1} solution A (pH=6.0) (pH was adjusted with 0.01mol L^{-1} NaOH and HCl solution) was obtained by dilution of standard reference sample solution step by step. Pb^{2+} in 10.0 ml solution A was enriched and determined. The analysis resultsshown in Table 1 was in good agreement with certified value.

Table 1. Analysis of Pb^{2+} in standard reference sample

Sample	Certified Value / (μg)	Found[a] / (μg)	RSD / (%)	Reletive error / (%)
Solution A	1.0	0.98	0.3	-2.00

[a] Mean (n=6)

Pb^{2+} in 100.0 ml sea water (pH = 6.0) was enriched and determined. A suitable amount squid sample was digested to obtain sample solution (pH was adjusted to 6.0), Pb^{2+} in 100 ml squid sample solution was enriched and determined. The analysis results are shown in Table 2. The above results of analysis indicate that the proposed method is reliable for the preconcentration and detection of Pb^{2+} in environmental sample.

Table 2. Analysis of Pb^{2+} in sea water and addition-Recovery test

Sample	Added /(μg)	Determined[a] /(μg)	RSD /(%)	Recovery / (%)
Sea water	0	0.38	2.21	--
	0.2	0.58	1.72	101.0
	0.5	0.80	1.12	102.4
	0.8	1.17	0.47	99.8

[a] Mean (n=6)

4. References

[1] Mahmoud M E, Abdou A E H, Nabil G M 2015 *J. Ind. Eng. Chem.* **32** 365
[2] Sheshmani S, Nematzadeh MA and Shokrollahzadeh S 2015 *Int. J. Biol. Macromol.* **80** 475
[3] Feist B, Mikula B 2014 *Food Chem.* 147 302
[4] Kagaya S, Kajiwara T, Gemmei-Ide M, Kamichatani W and Inoue Y 2016 *Talanta* **147** 342
[5] Cui Y, Chang X, Zhu X, Luo H, Hu Z, Zou X and He Q 2007 *Microchem. J.* **87** 20
[6] Qu R, Niu Y, Sun C, Ji C, Wang C and Cheng G 2006 *Microporous Mesoporous Mater.* **97** 58
[7] Wu X Z, Liu P, Pu Q S, Sun Q Y and Su Z X 2004 *Talanta* **62** 918
[8] Jiang Y, Gao Q, Yu H, Chen Y and Deng F 2007 *Microporous Mesoporous Mater* **103** 316
[9] Pu Q, Su Z, Hu Z, Chang X and Yang M 1998 *J. Anal. At. Spectrom.* **13** 249
[10] Arakaki L N H and Airoldi C 2000 *Polyhedron* **19** 367

Acknowledgments

This work was supported by the National Natural Science Foundation of China (No. 51262005), Guangxi Colleges and Universities Key Laboratory of Food Safety and Detection, and the project of high level innovation team and outstanding scholar in Guangxi colleges and universities.

IOP Conf. Series: Materials Science and Engineering **284** (2017) 012015 doi:10.1088/1757-899X/284/1/012015

Theoretical study on the edge-functionalization effect on nanoporous carbons for adsorption capacity and selectivity of CO_2 over N_2

S X Wei [1], Z H Wu [1], X Q Lu [1] *

1 College of Science, China University of Petroleum, Qingdao, Shandong 266580, P. R. China
E-mail: luxq@upc.edu.cn

Abstract. The adsorption and separation of CO_2/N_2 mixture in the edge-functionalized nanoporous carbons (NPCs) were investigated by grand canonical Monte Carlo simulation. Results demonstrated that edge-functionalization effectively improved the pore structure characteristics of NPCs. The edge-functionalization can enhance the uptake of CO_2/N_2 and the selectivity of CO_2 over N_2 in NPCs. The isosteric heat and radial distribution functions confirmed that temperature had a negative effect on the single-component adsorption of CO_2/N_2 whereas pressure had a positive effect before adsorption reaches a stable equilibrium state. Overall, this work highlighted the effects of edge-functionalization on the adsorption and separation of CO_2/N_2 in NPCs, and provided an effective strategy for designing and selecting solid adsorbents for carbon capture and storage.

1. Introduction

Carbon capture and storage (CCS) has received widespread concerns as a promising strategy to alleviate excessive emissions of CO_2 into the atmosphere [1]. One of the crucial issues for CCS technology is to screen adsorbent materials with excellent CO_2 adsorption capacity and selectivity, low regeneration energy, and good tolerance in equipment corrosion [2]. Nanoporous carbon materials exhibit remarkable advantages because of their lightweight property, high surface area and porosity, easy regeneration [3]. Apart from intrinsic advantages of carbon materials, chemical functionalization is extensively used to enhance carbon materials of CO_2 capture and selectivity [4].

The separation of CO_2 from N_2 is research focus because post-combustion processes in fuel-based power plants often release CO_2/N_2 mixture gas. Xiang et al. [5] successfully incorporated –OH, –NH_2, and –SO_2Cl functional groups into one phase for porous organic materials and observed that functionalized covalent organic polymers exhibited considerably higher selectivity for CO_2 over N_2 than initial structures. Babarao et al. [6] indicated that the introduction of O- and N-containing functional groups on porous aromatic frameworks could enhance the adsorption capacity of CO_2 and selectivity of CO_2 over N_2. Despite these findings, the in-depth theoretical investigations on the effects of edge-functionalization on the adsorption and separation of CO_2/N_2 in NPCs materials have not been convincingly elucidated. In this work, the density functional theory (DFT) and grand canonical Monte Carlo (GCMC) was performed to explore the adsorption and separation behaviour of CO_2/N_2 mixtures in the edge-functionalized NPCs

2. Model and computing methodology

Coronene-shaped graphene build non-functionalized NPC. Four functional groups, –H, –OH–NH_2, and–COOH, are used to passivate the edge of NPC to form the edge-functionalized NPCs, as shown in Figure 1. Atomic partial charge was calculated after the geometry optimization by using the B3LYP functional and incorporating the 6-31+g(d, p) basis set in Gaussian 09 package [7]. The LJ potential parameters for both molecules were from the TraPPE model [8]. For NPC frameworks, the universal force field (UFF) was adopted to set atomic LJ potential parameters. More informations can be found in previous works [9]. Calculations on the adsorption uptake and selectivity of CO_2 over N_2 were performed by GCMC in MuSiC [10].

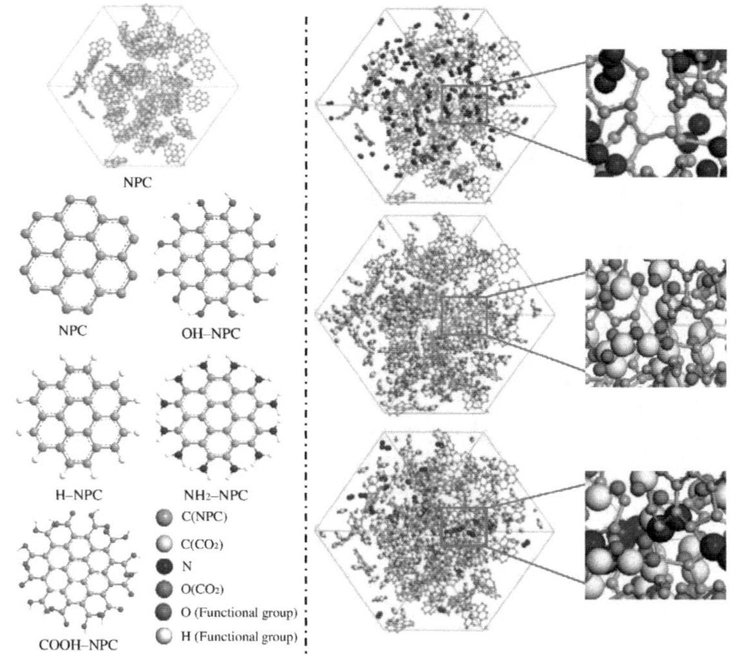

Figure 1. Initial configurations of NPCs with different functional groups, and snapshots of CO_2/N_2 adsorption

3. Results and discussions

3.1. Pore topology and morphology

Pore physical characteristics are critical factors for the gas adsorption and separation. Table 1 lists the pore topology and morphology of the frameworks evaluated by Sarkisov [11] and Duren [10] methods and Poreblazer v3.0 procedure [8]. The Φ of the edge-functionalized NPCs ranges from 68.48% to 76.50% relative to 66.58% for the unmodified NPC. The porosity of the edge-functionalized NPCs is smaller than that of some PAFs (84.46%-91.22%) [12] But larger than that of HKUST-1 (72%) [13]. The accessible surface area of the edge-functionalized NPCs fluctuates from 3006.17 to 3548.36 m2 g-1, and these values are larger than those of microporous MOFs (600-2858 m2 g-1) [14]. Figure 2 displays the geometric pore size distributions (PSDs). Miniscule ultramicropores (< 7.00 Å) and numerous micropores (7.00-20.00 Å; 84.50%-96.80% of the total) are observed in the edge-functionalized NPCs in accordance with the IUPAC classification [15]. In summary, edge-functionalization improves the effective accessible surface area and increases the pore spaces and diameters, and facilitates the creation of a favorable environment for gas adsorption in gas-framework systems.

IOP Conf. Series: Materials Science and Engineering **284** (2017) 012015 doi:10.1088/1757-899X/284/1/012015

Table 1. Physical characteristics and isosteric heat of NPCs with different functional groups. (Gas probe molecule = He).

Parameters	None	H–	OH–	NH_2–	COOH–
V_P (cm^3g^{-1})	1.30	1.26	1.41	1.33	1.36
D_L (Å)	7.06	8.33	11.90	8.19	8.52
D_m (Å)	12.87	11.26	17.99	12.44	16.97
Porosity, α (%)	66.58	68.48	76.54	71.78	71.53
Microporosity (%)	84.9	84.5	96.8	88.1	95.0
Surface area (m^2 g^{-1})	3191.	3091.31	3006.17	3492.48	3548.30
Qst of CO_2 (kJ mol^{-1})	24.63	27.276	30.339	34.252	30.704
Qst of N_2 (kJ mol^{-1})	8.388	12.176	14.180	13.129	12.996

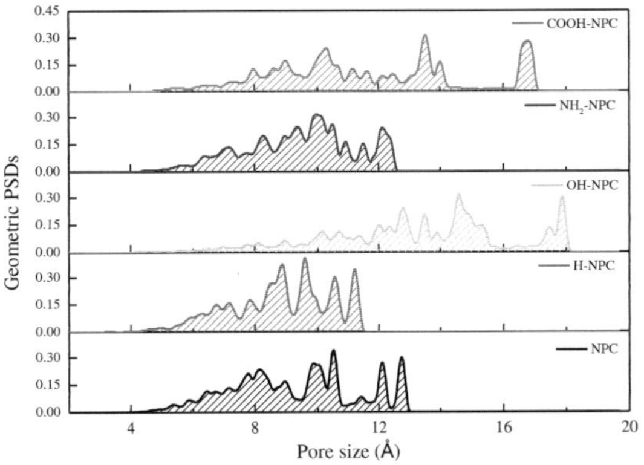

Figure 2. Geometric pore size distributions (PSDs) of NPCs (different functional groups).

3.2. Single-component adsorption of CO_2/N_2

Figure 3a presents the absolute adsorption isotherms of CO_2 at 298 K. The uptake of CO_2 in the edge-functionalized NPCs are significantly higher than that of the unmodified NPC. At low pressures (0-2 MPa), the uptake of CO_2 follows the sequence of NH_2–NPC > COOH–NPC > OH–NPC > H–NPC > NPC. That is, a functional group with largeelectronegativity increases the adsorption at low pressures. At high pressures (2-10 MPa), the saturated uptake of CO_2 follow the sequence of OH–NPC > NH_2–NPC > COOH–NPC > H–NPC > NPC. This finding is attributed to the combined effect of pore size and porosity [16]. Figure 3d delineates the absolute adsorption isotherms of N_2 in the edge-functionalized NPCs at 298 K. Edge-functionalization significantly enhances the uptake of N_2. The uptake of N_2 follows the sequence of OH–NPC > NH_2–NPC > H–NPC > COOH–NPC > NPC. The effects of temperature on the adsorption of CO_2/N_2 are presented in Figure 3a–c and 3d–f. The results show that the uptake of CO_2/N_2 in the edge-functionalized NPCs decreases with the increase of temperature due to the exothermic nature of the adsorption process. Overall, the adsorption of CO_2/N_2 in the edge-functionalized NPCs exhibits type–I Langmuir adsorption behavior [17]. This result indicates that edge-functionalization can improve the ability of NPCs to separate CO_2 from N_2.

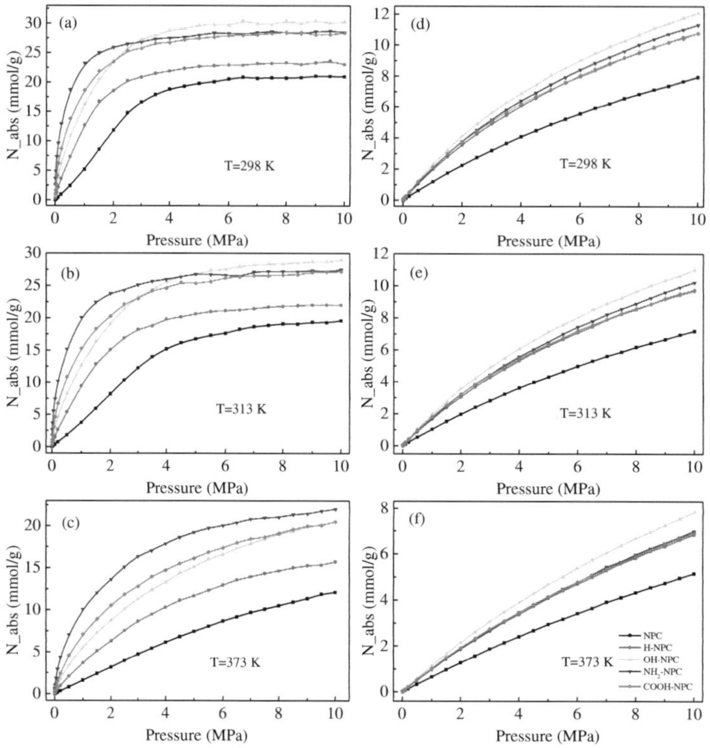

Figure 3. Absolute adsorption isotherms of CO_2 (a-c) and N_2 (d-f) in the edge-functionalized NPCs at 298 K, 313 K, and 373 K

3.3. Effect of functional groups on adsorption

We calculated the average isosteric heats (Qst) and radial distribution functions (RDFs, g(r)) to obtain the gas adsorption strength and distribution, as shown in Table 1 and Figure 4. Isosteric heat is a good indicator exhibiting the affinity of gas molecules toward porous adsorbents [18]. The Qst of CO_2 remains basically constant (24.630 kJ mol^{-1}) in NPC, whereas the Qst of CO_2 increases to 27.276-34.252 kJ mol-1 in the edge-functionalized NPCs due to the introduction of functional groups. The isosteric heat of CO_2 in NPCs is in good consistence with that of CO_2 uptake at low pressure. The sequence of Qst of N_2 agrees well with the sequence of N_2 uptake. For both CO_2 and N_2, the isosteric heat analyses clearly show that the adsorption capacity of gases depends on the strength of the gas-framework interaction. Comparatively, the isosteric heat of CO_2 is larger than the corresponding value of N_2 in NPCs, and this result is attributed to the stronger interactions in CO_2-frameworks.

To estimate the influence of functional groups on gas distribution, we calculated the RDFs of CO_2/N_2 around the heteroatoms in functional groups. As shown in Figure 4a, the strongest peak occurs at r = ~3.75 Å due to the strong interaction between the CO_2 and electronegative N atoms in the framework. For COOH–NPC, the g(r) presents two evident peaks. The first peak represents the adsorbed phase, and the second peak represents the equilibration phase. The RDFs of N_2 are shown in Figure 4b. The g(r) of N_2 in the edge-functionalized NPCs presents weak first peaks, indicating that the functional groups have minimal effects on N_2 adsorption. This is ascribed to the weak quadrupole moment of N_2 and the resulting weak interaction between N_2 and heteroatoms in the edge-functionalized NPCs. These CO_2/N_2 radial distribution functions are consistent with the corresponding isosteric heat values. Therefore, both isosteric heat and RDFs analyses can explain why the functional groups exert a better effect on the adsorption of CO_2 than that of N_2.

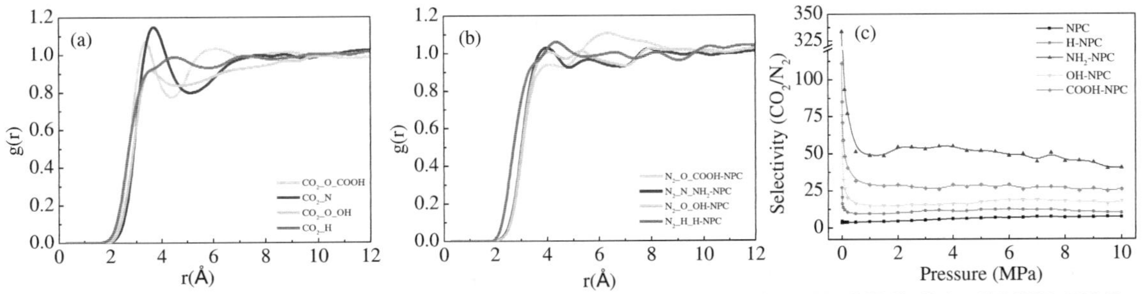

Figure 4. Radial distribution functions of CO_2 (a) N_2 (b) around N in NH_2–NPC, O in COOH–NPC, O in OH–NPC, and H in H–NPC for CO_2/N_2 mixture at 10 MPa. (c) Selectivity of CO_2 over N_2 in the edge-functionalized NPCs at 298 K with the equal mixture ratio.

3.4. Selectivity of CO_2 over N_2 with equal molar fraction

Selectivity of CO_2 over N_2 $S_{CO_2/N_2}=(x_{CO_2}/x_{N_2})/(y_{CO_2}/y_{N_2})$, is an excellent indicator of the ability of certain species to separate from its mixed gases.Figure 4c shows the selectivity of CO_2 over N_2 with equal molar fraction in the edge-functionalized NPCs at 298 K. The selectivity of CO_2 over N_2 declines sharply initially, and then flattens out to a constant value with the increase in pressure. Functional groups improve the selectivity of CO_2 over N_2, especially at low pressures. Consistent with the results of isosteric heat and RDFs analyses, the selectivities demonstrate the positive effect of functional groups. At low pressures, the adsorption of CO_2 is prior to that of N_2 due to the strong attractive interactions between CO_2 and the framework, leaving N_2 to fill the void space and thus forms a high selectivity of CO_2 over N_2. With the increase in pressure, the increasing repulsion between the gas molecules weakens the attraction between the gas molecules and the framework and then renders adsorption reaching an equilibrium state. Thus, the selectivity of CO_2 over N_2 tends to be a constant value. This finding further highlights the positive effect of functional groups on the separation of CO_2 from N_2.

4. Summary

The edge-functionalization effects on adsorption and selectivity of CO_2/N_2 gas in NPCs were investigated by DFT and GCMC simulations. Edge-functionalization creates a favorable environment for the adsorption of CO_2/N_2. The results of isosteric heat and radial distribution function confirm that the uptake of CO_2 is more significantly enhanced by edge functionalization than that of N_2, indicating the enhanced selectivity of CO_2 over N_2. Temperature has a negative effect on the uptake of CO_2/N_2 in the edge-functionalized NPCs. This work highlights edge-functionalization effects on the adsorption and separation of CO_2/N_2 in NPCs, and provides an effective strategy for designing and selecting solid adsorbents for carbon capture and storage.

5. References

[1] Boot-Handford M E, Abanades J C, Anthony E J, Blunt M J, Brandani S, Dowell N M, Fernández J R, Ferrari M C, Gross R, Hallett J P, et al. 2014 *Energ. Environ. Sci.* **7** 130-89.
[2] Lu X, Jin D, Wei S, Wang Z, An C and Guo W 2015 *J. Mater. Chem. A* **3** 12118-32.
[3] Samanta A, Zhao A, Shimizu G K H, Sarkar P and Gupta R 2012 *Ind. Eng. Chem. Res.* **51** 1438-63.
[4] Dasgupta T, Punnathanam S N and Ayappa K G 2015 *Chem. Eng. Sci.* **121** 279-91.
[5] Xiang Z, Mercado R, Huck J M, Wang H, Guo Z, Wang W, Cao D, Haranczyk M and Smit B 2015 *J. Am. Chem. Soc.* **137** 13301-7.
[6] Babarao R, Dai S and Jiang D e 2011 *Langmuir* **27** 3451-60.
[7] Jursic B S 1998 *J. Mol. Struct.* THEOCHEM **452** 145-52.
[8] Potoff J J and Siepmann J I 2001 AIChE J. **47** 1676-82.
[9] Lu X, Jin D, Wei S,Zhang M, Zhu Q, Shi X, Deng Z, Guo W and Shen W 2015 *Nanoscale* **7** 1002-12.

[10] Düren T, Millange F, Férey G, Walton K S and Snurr R Q 2007 *J. Phys. Chem. C* **111** 15350-6.
[11] Sarkisov L and Harrison A 2011 *Mol. Simul.* **37** 1248-57.
[12] Yang Z, Peng X and Cao D 2013 *J. Phys. Chem. C* **117** 8353-64.
[13] Wen H M, Wang H, Li B, Cui Y, Wang H, Qian G and Chen B 2016 *Inorg. Chem.* **55** 7214-8.
[14] Liu Y and Wilcox J 2012 *Environ. Sci. Technol.* **46** 1940-7.
[15] Thommes M, Kaneko K, Neimark A V, Olivier J P, Rodriguez-Reinoso F, Rouquerol J and Sing K S W 2015 *Pure Appl. Chem.* **87** 1051–69.
[16] Yi H, Li Y, Tang X, Li F, Li K, Yuan Q and Sun X 2015 *J. Chem. Eng. Data* **60** 1388-95.
[17] Balbuena P B and Gubbins K E 1993 *Langmuir* **9** 1801-14.
[18] Guillerm V, Weselinski L J, Alkordi M, Mohideen M I H, Belmabkhout Y, Cairns A J and Eddaoudi M 2014 *Chem. Commun.* **50** 1937-40.

IOP Conf. Series: Materials Science and Engineering **284** (2017) 012016 doi:10.1088/1757-899X/284/1/012016

Research progress of nano self - cleaning anti-fouling coatings

Y Liu [1,2], Y J Zhao[1,2], J L Teng[1,2], J H Wang[1,2], L S Wu[1,2], Y L Zheng[1,2]

1Nari group corporation (state grid electric power search stitute), Naijing, 211106, China
2Beijing Guodian Futong science and technology development corporation, Ltd, Beijing, 100070, China
E-mail: caser825@163.com

Abstract. There are many methods of evaluating the performance of nano self-cleaning anti-fouling coatings ,such as carbon blacking method, coating reflection coefficient method, glass microbead method, film method, contact angle and rolling angle method, organic degradation method, and the application of performance evaluation method in self-cleaning antifouling coating. For the more, the types of nano self-cleaning anti-fouling coatings based on aqueous media was described, such as photocatalytic self-cleaning coatings, silicone coatings, organic fluorine coatings, fluorosilicone coatings, fluorocarbon coatings, polysilazane self-cleaning coatings. The research and application of different kinds of nano self-cleaning antifouling coatings are anlysised, and the latest research results are summed .

1. Introduction

It is an effective way to ensure the normal operation of the equipment in harsh environments by preparing nano high self-cleaning coating materials with special functions on the surface of the equipment shell and parts. In the ceramic, glass, metal and other equipment shell surface coated self-cleaning paint, the establishment of super-hydrophobic or sparse surface, thus changing the shell material material surface function. The small water droplets, dust, grease, icing and other dirt adhered on the surface of the shell can fall off automatically so as to achieve the purpose of self-cleaning.

For the study of nano self-cleaning antifouling coatings, it has been widely concerned by the domestic and international coatings industry for a long time. In this paper, the performance evaluation methods of nano self-cleaning antifouling coatings and the research progress of self-cleaning antifouling coatings are mainly discussed.

2. Study on evaluation method of nano self - cleaning coatings

There are many measures to evaluate the performance of nano self-cleaning coatings. It is very important to select suitable self-cleaning performance evaluation methods for different kinds of coatings.

2.1. Carbon black collection method [1]

Carbon black collection method is a simple and relatively close to the actual method of pollution assessment of natural environment.

The process of carbon black collection is to spread the carbon black powder evenly on the surface of the nano self-cleaning anti-fouling coating; Then, a certain amount of water is loaded into the injector and drops on the surface of the coating. At the same time, in order to allow the water droplets to roll

away from the surface of the coating, the test device must be tilted to a certain Angle. Schematic diagram of the principle of Carbon black collection method . The state of the water droplets on the surface of the coating reflects the self-cleaning ability of the nano self-cleaning anti-fouling coating. Song Zhenyu et al [1] using carbon black collection method to characterize the surface coating agent on the PVC substrate before and after treatment in the self-cleaning performance. The results showed that PVC substrate after nano self-cleaning coating, can effectively improve the surface of the substrate self-cleaning effect.
The carbon black collection method can only characterize the self-cleaning of the coating, and can not quantify the self-cleaning of the paint.

2.2. Coating reflection coefficient method [2-3]
The coating reflection coefficient method is suitable for the evaluation of nano self - cleaning performance of solvent - based exterior wall paint.
Reflective coefficient test, the fly ash mixed with water as a source of pollution, brushing in a paint on the model. After the fly ash layer coated on the template is dried, the reflection coefficient of the surface of the paint sample is measured after the rinsing of the water flow for a specified number of times. The decrease rate of reflection coefficient is calculated to investigate the self-cleaning performance of the coating surface.The specific operation is as follows: The first step, mixed; The second step, measuring the reflection coefficient. The entire testing process needs to be repeated five times.
The coating reflection coefficient method is close to the natural pollution situation and can be quantitatively analyzed, but because of its complex operation and large working cycle of the limitations, it can not quickly determine the coating coating self-cleaning level.

2.3. Glass microbeads method [4]
The glass microbeads method is a common method used to characterize the self-cleaning properties of the anti-fouling of silicone rubber used for insulators.
The self-cleaning of the coating is measured by glass microbeads method, which can be tested by special test bed or special self-cleaning coefficient tester. Weigh 3g glass beads as a source of pollution. Remove the sample and collect the glass beads that naturally fall from the surface of the sample and weigh it, and weigh the mass as m (unit: g).Calculate the self-cleaning coefficient η:

$$\eta=m/0.3\times100\%. \tag{1}$$

The advantages of glass microbeads are simple test operation, short work cycle, and can be used to analyze the self-cleaning properties of coatings. However, this method is different from the natural fouling situation, which can not explain the self-cleaning performance of coatings in natural fouling environment.

2.4. Film method [4]
The film method is also a method for characterizing the self-cleaning properties of RTV silicone rubber antifouling coatings. Unlike the glass bead method, the film method is a qualitative test.

2.5. Contact angle and rolling angle method [5-7]
At the level of solid on the surface of a drop of water, when the gas liquid solid three-phase solid surface contact balance, tangential gas liquid interface and the solid-liquid interface formed by the tangent to the folder in which the liquid phase angle is called contact angle [5-6]. According to the size of contact angle formed by water and interface on solid surface, the wetting degree can be divided into 2 kinds: hydrophilic material and hydrophobic material.
The rolling angle is the critical angle formed by an inclined surface and a horizontal surface when the droplet is just rolling on the inclined surface. Zhang Zhiqiu and other[7] people in the study repeated 7 times the sample surface rolling angle and its average value, to verify this conclusion. When the static

contact angle value is greater than 150 degrees and the rolling angle value is less than 5 degrees, the coating can be considered as a surface with super hydrophobic properties.

2.6. Organic degradation method [8-9]

The organic degradation method is generally used for the characterization of photocatalytic self-cleaning coatings. The self-cleaning properties of the coatings are evaluated according to the amount of photodegradation organic compounds. The organic compounds in degradation include alizarin red , methylene blue, formaldehyde and benzene compounds, etc..

Organic matter degradation method is simple but only applicable to the performance characterization of photocatalytic self-cleaning coatings. In fact, the pollution in the actual environment is not only single organic matter, but also solid dust, inorganic pollutants and so on. Therefore, the structure of the degradation material and the actual pollution of the composition are different, and it is not sufficient to prove the self-cleaning performance of the coating in practical application.

3. Research progress of nano self - cleaning antifouling coating

3.1. Nano photocatalytic self-cleaning coating [10-12]

The nano photocatalytic self-cleaning coating can be applied to the outer walls of equipment and facilities, and can effectively degrade organic pollutants such as adhesion, pollution and corrosion that are adhered to the outer surface of the equipment.

Photocatalytic self-cleaning coatings are currently widely used in glass, ceramic and some metal surface. It is widely used in building exterior anti-fouling, can be effective anti-bacterial mold, to prevent the surface caused by microbial degradation of paint and other phenomena occur.

3.2. Silicone self-cleaning antifouling coating [13]

Lee et al. used sol-gel method to prepare the compact structure and can be applied to the silicone insulating coating of electronic devices. However, the curing temperature of silicone coating is high, which is suitable for factory prefabrication and has limitations in site construction.

3.3. Organic fluorine self-cleaning antifouling coating [14-16]

Organic fluorine coatings have excellent corrosion resistance, wear resistance, insulation and pollution resistance. They are widely used in various fields. They are excellent coatings with many advantages. Organic fluorine coating has the name of "coating king". Its film-forming substances are mainly organic fluorine resin and organic fluorine modified resin / modified organic fluorine resin .

3.4. Fluorine silicon self-cleaning antifouling coating [17]

The physical modified fluorine silicon self-cleaning antifouling coating mainly uses the modified fluorosilicone resin itself as water repellent, oil repellent and other excellent characteristics. Zhang Jinghai uses tetrafluoroethylene, vinyl acetate, allyl alcohol and eleven acid four monomers to obtain organic fluorine resin, then after copolymerization of organic fluorine resin and silicone resin copolymer, adding fillers and additives mixing, self-cleaning and anti fouling coatings for silicon fluoride. It is found that the paint has excellent hydrophobic, hydrophobic and aging resistance up to 20 years, and can be used in marine antifouling coatings.

3.5. Fluorocarbon self-cleaning antifouling coating [18]

C-F-containing paint collectively referred to as fluorocarbon coating, fluorocarbon coating material for the fluorocarbon resin. Fluorocarbon materials are widely used in construction, petrochemical, railway and road traffic, and military and many other fields. Fluorocarbon coating as a durable protective coating, can withstand blooming, cracking, corrosion and air pollution, chemical erosion, etc[18]. Commonly used fluorocarbon materials, such as polytetrafluoroethylene, polyvinylidene fluoride and other melt-type fluorine resin, with the same silicone coating similar to the need to film at high temperatures, it is only applicable to the factory coating.

3.6. Self cleaning antifouling coating of Poly [19-20]

The dry film thickness of varnish developed from poly (urethane) material is usually between 10 nm~2μm. Compared with the alkyd alkyd varnish system and the two-component polyurethane varnish system, two types of varnish show better gloss and weather resistance. In comparison with the silicone modified polyester varnish system, the performance of silicone varnish is superior to that of . If the polysilazane based self-cleaning coatings by certain means for power system, its high hardness will greatly increase the insulator surface coatings, such as the use of electrical equipment for outer wall coatings such as paint the length of time, reduce maintenance costs, is an important direction for the development of electric power industry coatings.

4. Conclusion

The evaluation method of nano self-cleaning antifouling coating has been continuously improved with development for many years. The antifouling effect of nano self-cleaning coating has been evaluated more accurately by using of various methods. With the progress of science and technology, the type of self-cleaning costing increases and the effect is gradually increasing at the same time. Different types of coatings also have a variety of problems, such as higher curing temperature, high cost, poor environmental adaptability and so on. Choosing proper self-cleaning antifouling paint is the key to practical application. I believe that with the continuous optimization of technical means, the development of a comprehensive performance of self-cleaning anti fouling paint is just around the corner.

5. Reference

[1] Song Z Y, Huo R T, Zhou J X, Qi C Z, Hou L L, Wang X Q. Modified SiO2 and Its Applicationin Anti-Dirty and Self-Cleaning SurfaceTreatment. 2009 *J.PANT&COATINGS Industry,* **39(3)** 6-9.

[2] GB/T 9759-2001 《*Solvent - based exterior coatings - Appendix A Test method for stain resistance of coatings*》.

[3] Yu L Y, Li X Y, Luo H. Preparation of Nano - TiO2 Modified Silicone Adhesive Coatings. 2010 *J,CHINA Adhesives,* **19(3)** 37-40.

[4] DL/T 627-2012 《*Insulator for curing at room temperature Silicone rubber anti-fouling flash coating Appendix B (normative) Self-cleaning test method*》.

[5] Chen H Z, Geng T, Zhang X, Zhang P Y. Research Progress of Superhydrophobic Surface. 2013 *J.Chemical Research*, **24(4)** 434-440.

[6] Liu K, Yao X, Jiang L. Recent developments in bio-inspired special wettability. 2010 *J.Chem.Soc.Rev,* **39** 3240-3255.

[7] Xu Q F , Bikash M, Alan M. Lyons.Fabricating Superhydrophobic Polymer Surfaces with Excellent Abrasion Resistance by a Simple Lamination Templating Method. 2011 *J. Appl.Mater.Interfaces*, **3** 3508-3514.

[8] Tu H J, Liu A H. Preparation of Photocatalysis Self-Clean Coatings and its Photocatalytic Activity . 2011 *J. Guangdong Chemical Industry*, **39(7)** 93-108.

[9] Zhou W Y, Gao Q Z, Liu W A, Fu S M, Yang Z H. Synthesis and Application of Nano-Photocatalysts Emulsion. 2009 *J. PANT&COATINGS Industry*, **39(6)** 19-21.

[10] Qian L C, Shao L S, Wang C. Research and Progress of Stickiness-Proof and Self-Cleaning on Surface Materials. 2008 *J. Journal of Kunming University of Science and Technology (Science and Technology)*. **33(3)** 30-34.

[11] Zhou S X, Li C D, Li Y G. Research Progress of Nano-TiO2 Photocatalytic Self-cleaning Technology . 2012 *J. Journal of Hefei Normal University*, **30(6)** 83-87.

[12] Xu M J, Wang T Z. Development and Application of Glass in the 21st Century in Environmental Protection - Development of Surface Modified Glass in Environmental Protection. 2009 *J. Glass*, **7** 1-8.

[13] Wang Y, Xiao X Y, Wan C X. Research and application of self-cleaning exterior wall coatings . 2008 *J. New Chemical Materials*, **36(3)** 12-15.

[14] Lee L H, Chen W C, Liu W C. Structural control of oligomeric methyl silsesquioxane precursors and their thin film properties. 2002 *J. Journal of polymer Science Part A:Polymer Chemistry*, **40** 1560-1571.

[15] Tang F,Zhang X Y, Fu H Q,Huang H, Chen H Q. New trend of study on organofluorine coatings . 2004 *J. Electroplating & Finishing*, **23(2)** 39-42.

[16] Tian J, Zhou Z F, Xu J F. *P.Non-toxic marine antifouling paint*. CN1097447, (1993-07-15).

[17] Qiu J Y, Cheng Y L, Guo X L, Xiong S D, Yi C F. Study on Organic Fluorine-Silicon Modified Acrylate Copolymer Emulsion. 2008 *J. PANT&COATINGS Industry*, **3(3)** 16-18.

[18] Yang Y. Development of fluorocarbon coatings. 2006 *J. Coating and plating*, **4** 7-10.

[19] Liu H L, Zhao X L, Tian C Y, Han L, Li J Qiu B H, Wan Q. Study on Carbon Anti Oxidation Composite Coating Prepared with the PSZ and Si-Mo Si2 Powders. 2013 *J. Journal of Synthetic Crystals*, **42(9)** 1960-1964.

[20] Zhao J B. Preparation of Polysilazane and Its Application in Coatings. 2015 *J. Shanghai Coatings,* **53(4)** 24-28.

IOP Conf. Series: Materials Science and Engineering **284** (2017) 012017 doi:10.1088/1757-899X/284/1/012017

Effect of synthesized ZnO nanoparticles on thermal conductivity and mechanical properties of natural rubber

R Suntako[1]

1 Department of Physics, Faculty of Liberal Arts and Science, Kasetsart University Kamphaeng Saen Campus, Nakhon Pathom 73140, Thailand
E-mail: rudeerat_p@hotmail.com

Abstract. Zinc oxide (ZnO) is widely used in rubber industry as a cure activator for rubber vulcanization. In this work, comparison of cure characteristic, mechanical properties, thermal conductivity and volume swell testing in oil no.1 and oil no.3 between natural rubber (NR) filled synthesized ZnO nanoparticles (sZnO) by precipitation method and NR filled conventional ZnO (cZnO). The particle size of sZnO is 41.50 nm and specific area of 27.92 m^2/g, the particle size of cZnO is 312.92 nm and specific surface area of 1.35 m^2/g. It has been found that NR filled sZnO not only improves rubber mechanical properties, volume swell testing but also improves thermal conductivity and better than NR filled cZnO. Thermal conductivity of NR filled sZnO increases by 10.34%, 12.90% and 20.00%, respectively when compared with NR filled cZnO in same loading content (various concentrations of ZnO at 5, 8 and 10 parts per hundred parts of rubber). This is due to small particle size and large specific surface area of sZnO which lead to an increase in crosslinking in rubber chain and enhance heat transfer performance.

1. Introduction

Natural rubber (NR) is an unsaturated hydrocarbon or called cis-1,4 polyisoprene. The mechanical properties of vulcanized natural rubber are dependent on several factors such as degree and type of crosslinking, type and amount of fillers, degree of filler dispersion [1]. The important properties of NR are high elasticity and abrasion resistance, good tensile strength, good resilience, low heat build-up and excellent dynamic property, insoluble in water, alcohol and acetone. Due to the presence of a double bond in each repeat unit, ozone in the air will attack double bond in rubber chain and affect to degradation. It is always used in many applications such as automotive part, tires, footwear and engineering products. In addition, NR has a low thermal conductivity 0.135-0.150 W/m·K and then fine electrical insulating property. This is weakness to completely vulcanized of the surface and inner of the thick rubber product due to poor heat transfer. In fact the increment of the thermal conductivity leads to an increase in heat transfer performance. Zinc oxide (ZnO) is a type of material that used as an activator in vulcanizing with accelerators, reinforcement, aiding in heat dissipation [2] and other fillers such as aluminum nitride [3] and graphite [4] also enhance thermal conductivity. Generally, the conventional rubber grade uses ZnO particle size in range of 0.1-0.4 μm, the specific surface area around 10-20 m^2/g and thermal conductivity of 34 W/m·K [5]. The reduction of particle size and increment of specific surface area may have important contribution to improve thermal conductivity [6]. The objective of this research is studied the effect of synthesized ZnO nanoparticles (sZnO) by precipitation method comparing with conventional ZnO (cZnO) which is usually used in rubber industry on thermal conductivity. Additionally, cure characteristic, volume swell testing and mechanical properties are also carried out.

2. Materials and Methods

2.1. Synthesis of sZnO

$Zn(NO_3)_2.6H_2O$ and NaOH are used as precursors in precipitation method. Polyacrylic acid (PAA) is used as a polymeric additive. The process of synthesis sZnO is followed by the previous studied [7]. The morphology and primary size of sZnO and conventional ZnO (cZnO) are observed by a transmission electron microscope (TEM). The specific surface areas are investigated by a Brunuer-Emmett-Teller (BET) technique.

2.2. Preparation of rubber compound

The formulations of rubber compound are showed in table 1. Compound is prepared in an internal mixer 3 liter with rotor speed of 30 rpm. STR 20 is first masticate for 2 min, N-330, N-550, aromatic oil, 6PPD, stearic acid and ZnO are then added and mixed continued up to 5 min. The accelerator (CBS and DPG) and sulphur are added on a two roll mill at room temperature (35°C) and sheeted off a thickness at 8 mm.

2.3. The cure characteristics and mooney viscosity testing

The cure characteristics of rubber compound are measured at 170°C for 10 min with a Moving Die Rheometer (MDR2000, Alpha Technologies). The differential torque (M_H-M_L), cure time (T_{10}) and optimum cure time (T_{90}) are determined. Mooney viscosity (ML 1+4) at 125°C for 10 min are measured by Mooney Viscometer (MV2000, Alpha Technologies).

2.4. The mechanical properties and volume swell testing

The mechanical properties such as hardness, tensile strength, elongation at break and modulus are determined. The hardness test is performed using a Hardness tester (Teclock), according to ASTM D2240. Tensile strength, elongation at break and modulus are performed at 23±2°C with a crosshead speed of 500 mm/min using an Autograph (AG-IS, Shimadzu), according to ASTM D412.

2.5. The thermal conductivity testing

The specimen is prepared by compression molding at 165°C for 8 min, into rectangular sheets (100 mm × 100 mm × 10 mm). Thermal conductivity of the rubber is measured by Thermal Constants Analyser (TPS 2500S), according to ISO 22007-2.

Table 1. The formulation of rubber compound.

Materials	Chemical name	phr (part per hundred of rubber)					
		cZnO 5 phr	cZnO 8 phr	cZnO 10 phr	sZnO 5 phr	sZnO 8 phr	sZnO 10 phr
STR 20	Natural rubber	100	100	100	100	100	100
N-330	Carbon black	35	35	35	35	35	35
N-550	Carbon black	50	50	50	50	50	50
Aromatic oil	Aromatic oil	5	5	5	5	5	5
6PPD	N'phenyl-p-phenylenediamine	1	1	1	1	1	1
Stearic acid	Stearic acid	1	1	1	1	1	1
cZnO	Conventional ZnO	5	8	10	-	-	-
sZnO	Synthesized ZnO nanoparticles	-	-	-	5	8	10
CBS	N-cyclohexyl-2-benzothaiazole sulfenamide	1	1	1	1	1	1
DPG	Diphenyl guanidine	1	1	1	1	1	1
Sulphur	Sulphur	2	2	2	2	2	2

3. Results and discussion

3.1. The morphology of ZnO particles
TEM images of sZnO and cZnO are showed in figure 1. The sZnO exhibits a spherical shape with the average primary size of 41.50 nm and specific surface area of 27.92 m^2/g. The particle size of cZnO is 312.92 nm and specific surface area of 1.35 m^2/g.

3.2. The cure characteristics and mooney viscosity
The comparison of cure characteristics and mooney viscosity of NR filled sZnO and cZnO are presented in table 2. Mooney viscosity (ML1+4), the differential torque (M_H-M_L), cure time (T_{10}) and optimum cure time (T_{90}) are increased with increasing ZnO content. It is indicated that increasing degree of crosslink density or stiffness of the rubber compound. This phenomenon is occurred in both types of ZnO. In the case of NR filled sZnO, the experimental results showed that the M_H-M_L is higher than NR filled cZnO in all loading content. This may be attributed to small particle size (nanoparticles) and large specific surface area of sZnO [8]. Furthermore, the ML1+4 of NR filled sZnO higher than filled cZnO due to fast curing because of a lower T_{10} and T_{90} are observed.

3.3. The mechanical properties
The mechanical properties of the NR filled with the various types and loading content of ZnO are showed in table 3. Results indicated that hardness, tensile strength and 100% modulus are gradually increased with increasing ZnO content [9]. The increment in these properties is due to enhancing degree of crosslink density in rubber chain. This phenomenone is confirm by M_H-M_L result which is a measure of shear modulus and degree of crosslinking. Whereas the elongation at break show a decreasing trend when the content of ZnO increases. This is due to the increasing crosslink density and it is indicated that the rubber exhibits hardly extension. At similar loading content, the mechanical properties of NR filled sZnO are higher than NR filled cZnO. It is clearly demonstrated that small grain size and grain boundaries of sZnO improve rubber mechanical properties more than cZnO.

3.4. The thermal conductivity
The effect of NR filled with the various types and ZnO loading on thermal conductivity are showed in figure 2. It can be seen that the thermal conductivity for both types of ZnO are increased with increasing ZnO content. This is possibly due to increase metal oxide content and attributed to the enhancement of heat transfer performance. In addition the fixed ZnO loading, NR filled sZnO shows a higher thermal conductivity than NR filled cZnO. While the thermal conductivity of NR filled sZnO increases by 10.34%, 12.90% and 20.00%, respectively when compared with NR filled cZnO in same loading content. The higher thermal conductivity of NR filled sZnO is due to large specific surface area and enhance the heat transfer ability, thus the reduction of particle size leads to an increase in thermal conductivity.

3.5. The mechanical properties
The mechanical properties of the NR filled with the various types and loading content of ZnO are showed in table 3. Results indicated that hardness, tensile strength and 100% modulus are gradually increased with increasing ZnO content [9]. The increment in these properties is due to enhancing degree of crosslink density in rubber chain. This phenomenone is confirm by M_H-M_L result which is a measure of shear modulus and degree of crosslinking. Whereas the elongation at break show a decreasing trend when the content of ZnO increases. This is due to the increasing crosslink density and it is indicated that the rubber exhibits hardly extension. At similar loading content, the mechanical properties of NR filled sZnO are higher than NR filled cZnO. It is clearly demonstrated that small grain size and grain boundaries of sZnO improve rubber mechanical properties more than cZnO.

3.6. The volume swell testing
Figure 3 summarizes the volume swell value in oil no.1 and oil no.3. It can be seen that the volume swell value is decreased with increasing ZnO content in both oil types. The decreasing of volume

swell value with higher ZnO content is associated mainly with the degree of crosslinking in rubber chain. This result is correlated with the higher differential torque (M_H-M_L); as previously reported in table 2. In addition, the effect of nanoparticles size of sZnO on volume swell is also investigated. It is showed that the volume swell value of NR filled sZnO in oil no.1 and oil no.3 lower than NR filled cZnO. Because of nanoparticles size significantly enhance crosslink density higher than microparticles size which leads to a decrease in the volume swell.

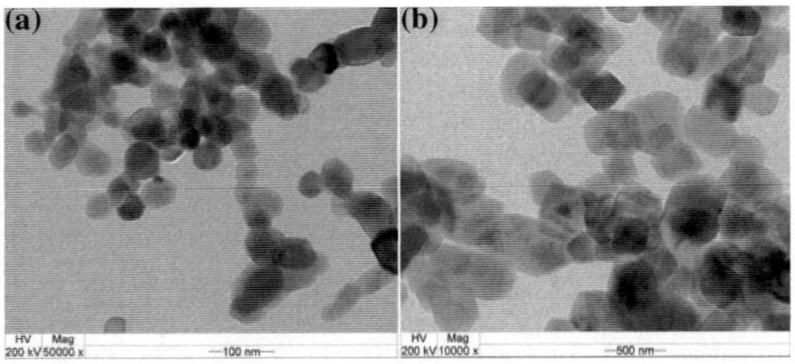

Figure 1. TEM images of (a) sZnO and (b) cZnO.

Table 2. The cure characteristics and mooney viscosity.

Properties	cZnO 5 phr	cZnO 8 phr	cZnO 10 phr	sZnO 5 phr	sZnO 8 phr	sZnO 10 phr
ML1+4 (MU)	65.78	72.71	76.02	67.33	74.83	79.20
M_H-M_L (lb-in)	20.70	21.96	22.02	21.40	22.65	23.52
T_{10} (min)	1.71	1.80	1.81	1.65	1.74	1.78
T_{90} (min)	4.15	4.28	4.36	4.06	4.11	4.21

Table 3. The mechanical properties of the NR filled with various ZnO types.

Properties	cZnO 5 phr	cZnO 8 phr	cZnO 10 phr	sZnO 5 phr	sZnO 8 phr	sZnO 10 phr
Hardness (Shore A)	78	79	80	79	80	82
Tensile strength (MPa)	19.70	20.40	22.10	21.70	22.70	23.20
Elongation (%)	266	254	231	275	254	247
100% modulus (MPa)	7.19	8.43	8.88	7.82	8.62	9.12

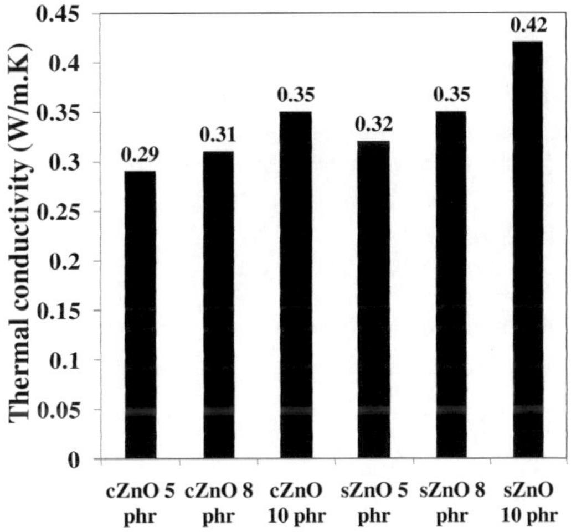

Figure 2. Thermal conductivity of NR filled with the various types and content of ZnO.

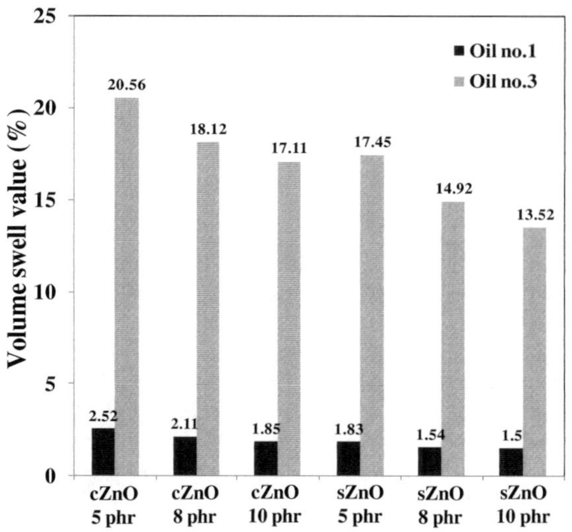

Figure 3. Volume swell resistance of NR filled with the various types and content of ZnO.

4. Summary

This work is studied the effect of sZnO which is synthesized by precipitation method and compared with cZnO on cure characteristics, mechanical properties, thermal conductivity and volume swell in oil no.1 and oil no.3. The test results found greatly improve all rubber properties such as higher tensile strength and modulus, reduce volume swell in oil no.1 and oil no.3. This is due to nanoparticles size enhances the degree of crosslinking in rubber chain which confirm by the differential torque (M_H-M_L) result. Furthermore, the small particle size and large specific surface area of sZnO also improve heat transfer performance or thermal conductivity.

5. Acknowledgments

Author gratefully acknowledge the financial support from Kasetsart University Research and Developmemt Institute (KURDI) and Research Promotion and Technology Transfer Center (RPTTC), Faculty of Liberal Arts and Science, Kasetsart University Kamphaeng Saen Campus.

6. References

[1] Ismail H and , Freakley P K 1997 *Polym-Plast Technol.* **36** 873
[2] Sim L C, Ramanan S R, Ismail H, Seetharamu K N and Goh T J 2005 *Thermochim. Acta.* **430** 155
[3] Namitha L K, Ananthakumar S and Sebastian M T 2015 *J. Mater. Sci: Mater. Electron.* **26** 891
[4] Song S H, Jeong H K, Kang Y G and Cho C T 2010 *Korean J. Chem. Eng.* **27** 1296
[5] Lee B and Dai G 2009 *J. Mater. Sci.* **44** 4848
[6] Jeong J, Li C, Kwan Y, Lee J, Kim S H and Yun R 2013 *Int. J. Refrig.* **36** 2233
[7] Suntako R 2017 *Key. Eng. Mat.* **748** 79
[8] Kim I J, Kim W S, Lee D H, Kim W and Bae J W 2010 *J. Appl. Polym. Sci.* **117** 1535
[9] Suntako R 2014 *Adv. Mater. Res.* **1044-1045** 23

IOP Conf. Series: Materials Science and Engineering **284** (2017) 012018 doi:10.1088/1757-899X/284/1/012018

Preparation and characterization of tourmaline tetradecen-ylsuccinate-vinyl acetate copolymer

Y M Hu and M C Li

School of Materials Science and Technology, China University of Geosciences, Beijing 100083
E-mail: huyingmo@cugb.edu.cn

Abstract. This In this paper, the tetradecenylsuccinic anhydride was used to modify tourmaline to prepare a polymerizable organic tourmaline tetradecenylsuccinate, and then copolymerized with vinyl acetate to prepare tourmaline-containing functional copolymer. The structures and morphologies of modified tourmaline and copolymer were characterized by means of IR, SEM and EDX analyses. The experimental results showed that the tourmaline was embedded into copolymer by the process of surface modification and copolymerization and illustrated well dispersion stability in copolymer, and meanwhile the tourmaline-containing functional copolymer had benign far infrared radiation and negative ions releasing properties, and a good mechanical performance.

1. Introduction

The tourmaline was a kind of cricoids silicate mineral belonging to trigonal system [1], the chemical formula was written as $XY_3Z_6(B0_3)_3Si_6O_{18}(OH)_4$, and here X could be K^+, Na^+ and Ca^+; Y could be Fe^{2+}, Mg^{2+}, Al^{3+}, Li^+ and Fe^{3+}; Z could be Al^{3+}, Fe^{3+}, and Cr^{3+} [2,3]. The tourmaline was paid a close attention recently for its specific performances, such as piezoelectric and pyroelectric properties, negative ions release and far infrared radiation et al [4-7]. So the tourmaline was widely used in many fields like water treatment, industrial, agriculture and daily life care, etc [8-11].

But tourmaline could be not well dispersed in a polymer matrix for its strong surface polarity [12], and led to the poor performances of tourmaline/polymer composites [13]. So, the surface modification of tourmaline was studied and combined with polymers to get the high-performance tourmaline/polymer composites [14-16]. If the tourmaline powder could directly be introduced into the polymer chain to prepare tourmaline containing functional polymer, it could be of paramount importance for preparing tourmaline containing functional polymer and value-added applications of tourmaline.

In this work, the surface modification of tourmaline was carried with tetradecenylsuccinic anhydride to get polymerizable organic tourmaline, and then copolymerized with vinyl acetate to synthesize tourmaline containing functional copolymer. The structures and morphologies of modified tourmaline and copolymer were characterized via IR, SEM and EDX analyses, and the synthetic process was studied, its physical and mechanical properties were tested quantificationally.

2. Experimental procedures

2.1. Materials and instruments

Tourmaline powder with diameters of d_{50} 1.75 mm and d_{97} 5.23 mm (purity: 98%) was purchased from Yanxin Mineral Co. Ltd., Hebei, China; N,N-dimethylformamide (DMF, AR grade) was purchased from West Long Chemical Co.; and liquid paraffin was purchased from Tianjin Fuchen Chemical

Reagent Factory; Tetradecenylsuccinic Anhydride from Shanghai Jianglai Industrial Co., Ltd; Acetic acid vinyl ester from Beijing chemical plant.

Fourier transform infrared spectroscopy, Spectrum100, Bose PerkinElmer Instruments (Shanghai) Co., Ltd.; X-ray powder diffraction (XD-2), Beijing Purkinje General Instrument Corporation; Scanning electron microscopy, S-450, Japan's Hitachi Ltd.; Contact angle tester (JC2000C), Shanghai Zhongchen Digital Technology Co., Ltd.; Turbidity (SGZ-2), Shanghai Yue Feng Instrument Co., Ltd; Negative ion tester, AIC-2M, Haina Environmental Technology Development (Shenzhen) Co. Ltd.; Microcomputer control electronic universal testing machine. CMT4304, XieQiang Instrument Manufacturing (ShangHai) Co., Ltd.; Far infrared emissivity measuring instrument, JPC-5X, PARAGON Co.,Ltd.

2.2. The preparation of polymerizable organic tourmaline

The super-fine tourmaline powder (10 g) and DMF (40 mL) were added to 250 ml three-neck flask, and heated to predetermined temperature with stirring constantly, then, amount of tetradecenylsuccinic anhydride was added to flask, and reacted 3 hrs at constant temperature. The tourmaline tetradecenylsuccinate (TTDS) were separated by filtration under vacuum, and washed three times with anhydrous ethanol, finally dried and grinded to be used.

2.3. The synthesis of tourmaline tetradecenylsuccinate-vinyl acetate copolyme

Vinyl acetate (50 mL, 0.54 mol) and modified tourmaline powder was poured into a 250 mL three-neck flask and heated to 70 ^{o}C. The 8.4mL methanol solution, in which dissolved azobisisobutyronitrile (0.187 g, 1 mmol), were added into the flask with funnel, amount of DMF was added into the flask after 2 hours to adjust the viscosity of copolymer and continued to stir for 8 h at constant temperature to get the tourmaline tetradecenylsuccinate-vinyl acetate copolymer (p(TTDS-VA) copolymer).

2.4. The measurement of the contact angle

The powered samples were pressed into a wafer and then the distilled water was dripped onto the surface of the wafer for the contact angle measurements with contact angle tester (JC2000C). Every experiments were repeated three times to get the average results [17].

3. Results and discussion

3.1. The modified tourmaline

3.1.1 IR spectra

The IR spectra of modified and unmodified tourmaline were showed in Figure 1. The spectrum of unmodified tourmaline (Figure 1a) exhibited its typical absorption bands at 3562 cm^{-1} (OH groups),

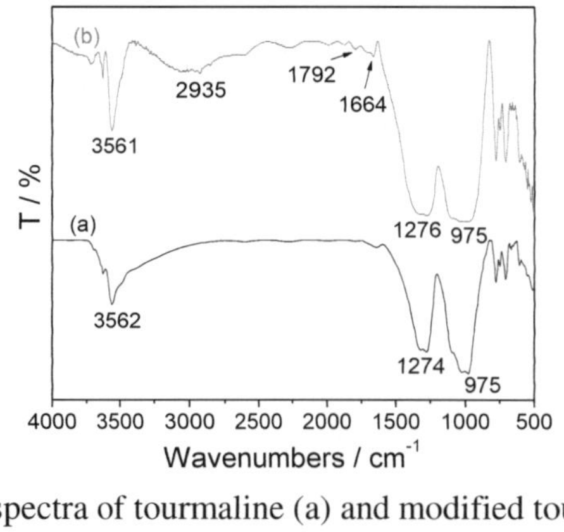

Figure 1. IR spectra of tourmaline (a) and modified tourmaline (b)

975 cm^{-1} (Si-O bonds) and 1276 cm^{-1} (B-O bonds) [18]. In comparison with Figure 1a, the IR spectrum of modified tourmaline (Figure 1b) showed new absorption bands corresponding to methyl at 2935 cm^{-1}, the absorption peak of carbonyl group at 1792 cm^{-1} and the olefinic bond peak at 1664 cm^{-1}. Besides, the other typical absorption peaks of tourmaline were almost unchanged, except the shift of several wavenumbers [19] for the electronic effect of tetradecenylsuccinate group. Those results indicated that tetradecenylsuccinate group was attached onto the surface of tourmaline by modification of tourmaline with tetradecenylsuccinic anhydride to produce polymerizable organic tourmaline tetradecenylsuccinate (TTDS).

3.1.2 SEM images
The SEM images of the unmodified and modified tourmaline were showed in Figure 2. There is no agglomerates were observed in the modified tourmaline particles (Figure 2a), and the conglomeration in the unmodified tourmaline particles was apparent (Figure 2b). This was because the organicmodification reduced the tourmaline's surface polarity and surface energy, its agglomeration phenomenon was improved obviously.

Figure 2. SEM images of modified tourmaline (a) and tourmaline (b)

3.1.3 The hydrophobicity test
In order to characterize the hydrophobicity of modified tourmaline quantificationally, the contact angle of tourmaline and modified tourmaline were tested and listed in Table 1. The contact angle of modified tourmaline was up to 119°, which indicates an excellent hydrophobicity compared to that of unmodified tourmaline. This result also indicated that organic group was introduced onto the surface of tourmaline after modification, and reduced the tourmaline's surface polarity and improved its surface hydrophobic properties.

Table 1. The contact angle of tourmaline and modified tourmaline

Sample	contact angle / ^{o}
Tourmaline	37
Modified Tourmaline	119

3.1.4 The negative ions releasing
The negative ions releasing amount of air, tourmaline and modified tourmaline were measured and ranked in Table 2. It showed that tourmaline have rather better negative ions releasing performance than air, and the negative ions releasing amount of tourmaline was increased by 3 times after surface organic modification. Those were because the organic modification reduced the surface energy of tourmaline, and improved its dispersion and surface areas, accordance with the result of SEM images.

3.1.5 Far-infrared radiation studies
The far infrared emissivities before and after modified tourmaline were showed in Table 3, which stated clearly that the far infrared radiation rate of modified tourmaline was higher than that of

unmodified tourmaline. This was ascribed that the organic modification improved its dispersion and surface areas, same with the result of negative ions releasing.

Table 2. Negative ions released of tourmaline and modified tourmaline

Sample	Negative ion release amount (Number/cm^{-1})
Air	180
Tourmaline	380
Modified Tourmaline	1150

Table 3. Far-infrared radiation rate of tourmaline and modified tourmaline

Sample	Far infrared radiation rate
Tourmaline	0.86
Modified Tourmaline	0.96

3.2 *Tourmaline tetradecenylsuccinate-vinyl acetate copolymer (p(TTDS- VA) copolymer)*

3.2.1 *IR spectra analysis*

The IR spectra of p(TTDS-VA) copolymer were shown in Figure 3. The absorption peaks of OH groups at 3459 cm^{-1}, B–O bonds at 1266cm^{-1} and Si–O bonds at 944 cm^{-1} were assigned to the typical bands of tourmaline, and the other absorption peaks of methyl and methylene groups appear at 2932 cm^{-1} to 2862 cm^{-1}, and carbonyl group occur at 1756 cm^{-1}, C-C bonds at 1438-1125cm^{-1}, and no absorption of double bond was found. Those results were in accordance with the characteristics of p(TTDS-VA) copolymer. This indicated that polymerizable organic tourmaline tetradecenylsuccinate (TTDS) had copolymerized with vinyl acetate to generate the p(TTDS-VA) copolymer.

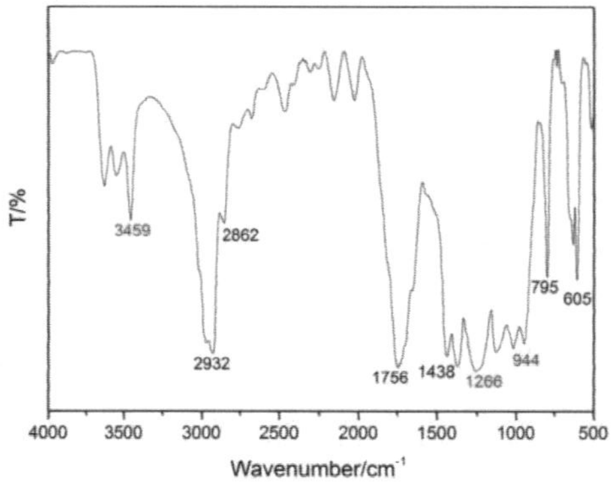

Figure 3. IR spectra of p(TTDS-VA) copolymer

3.2.2 *SEM and EDX analyses*

The SEM images and EDX analysis of the p(TTDS-VA) copolymer were showed in Figure 4. The SEM images of polyvinyl acetate and unmodified tourmaline (Figure 4b) that got with the same synthetic procedure of the p(TTDS-VA) copolymer had showed a serious tourmaline agglomeration, and the p(TTDS-VA) copolymer displayed a well distribution of tourmaline (Figure 4a). This was because that tourmaline was introduced into copolymer via the surface organic modification and copolymerization. The EDX analysis of p(TTDS-VA) copolymer was carried and showed in Figure 4c, the copolymer contained Na, Si, K and Ca elements, which are the typical elements of tourmaline and proved that tourmaline was introduced into copolymer.

3.2.3 *The measurement of negative ions*

The negative ions releasing amounts of p(TTDS-VA) copolymer with different tourmaline dosage were measured and showed in Figure 5. We could found that the amounts of negative ions releasing of p(TTDS-VA) copolymer were almost linear increased with the increasing of modified tourmaline

IOP Conf. Series: Materials Science and Engineering **284** (2017) 012018 doi:10.1088/1757-899X/284/1/012018

content. This further certified that modified tourmaline could be uniformly dispersed in p(TTDS-VA) copolymer [20].

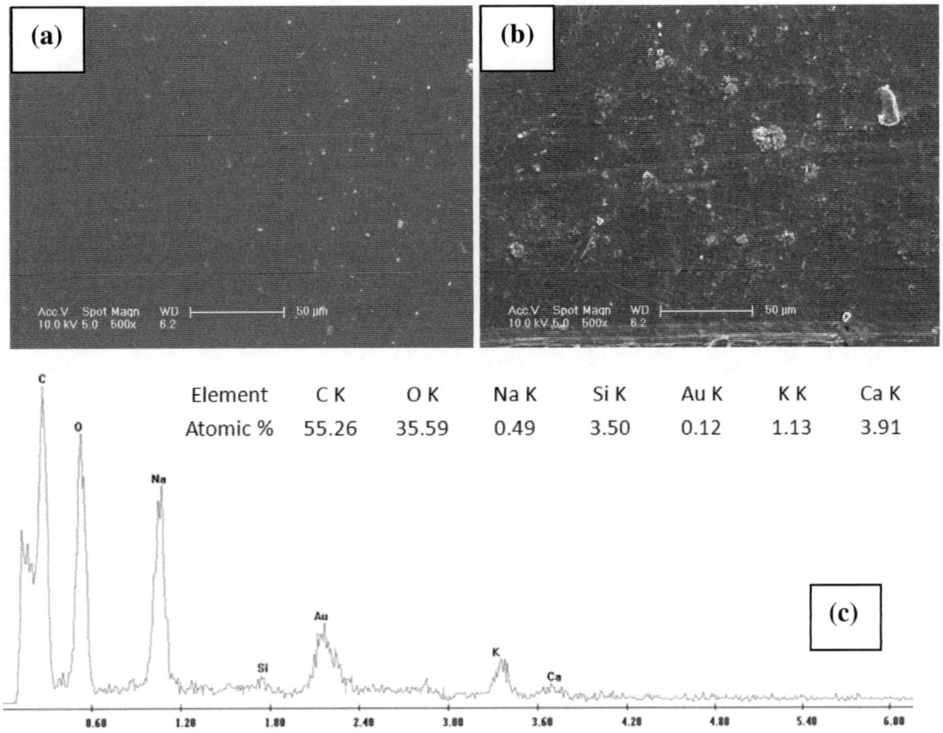

Figure 4. SEM and EDX of tourmaline tetradecenylsuccinic-vinyl acetate copolymer, (a) tourmaline tetradecenylsuccinic-vinyl acetate copolymer, (b) adding unmodified tourmaline, (c) EDS of tourmaline tetradecenylsuccinic-vinyl acetate copolymer

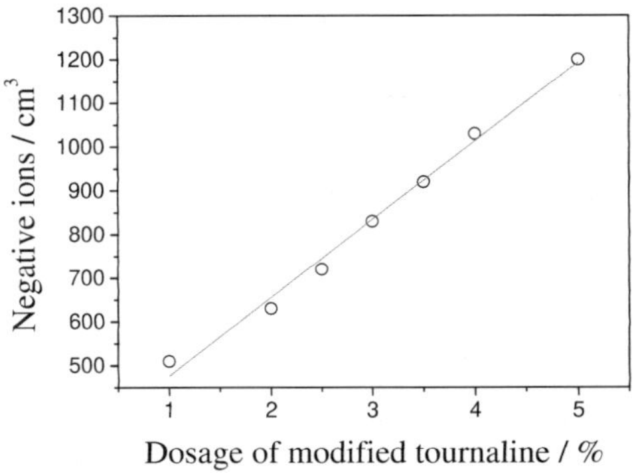

Figure 5. Negative ions releasing of p(TTDS- VA) copolymer

3.2.4 *Far infrared radiation studies*

The far infrared radiation of p(TTDS-VA) copolymer at the wavelength of 8-25µm was measured and showed in Figure 6. The data displayed that the far infrared radiation rates of p(TTDS-VA) copolymer with different tourmaline dosages were all high (about 0.95), and have an increased trend with the increasing of additive amounts of modified tourmaline, which was in keeping with the excellent dispersity of modified tourmaline.

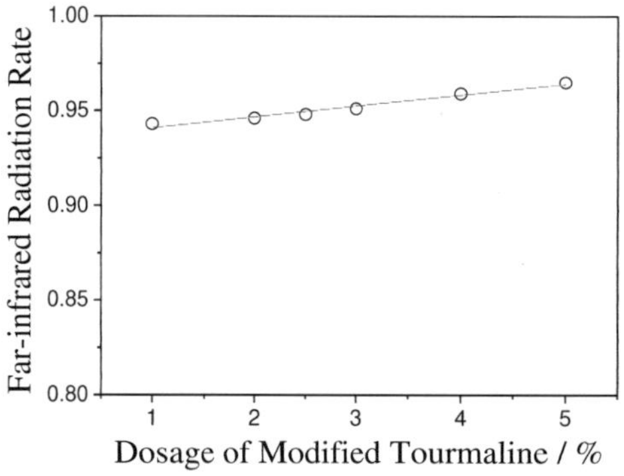

Figure 6. Far-infrared radiation of p(TTDS- VA) copolymer

3.2.5 The Mechanical property of the p(TTDS-VA) copolymer film

The p(TTDS-VA) copolymer with different additive amounts of modified tourmaline were pressed into specimen to test its elastic modulus and tensile strength According to the standard GB 13022-1991, and the results were shown in Figure 7. The Figure 7 exhibited that the elastic modulus and tensile strength of the p(TTDS-VA) copolymer film demonstrate the same trends and reached the maximum when the dosage of modified tourmaline was 3.5%. This was because that tourmaline was introduced into the polymer chains to enhance the force between polymer, and improved its mechanical properties, but excessive tourmaline could lead to the internal defects, so its mechanical properties decreased [21].

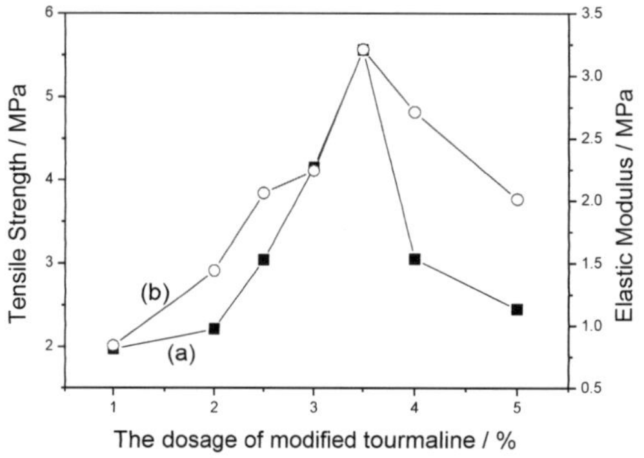

Figure 7. Mechanical property of the p(TTDS-VA) copolymer film, (a) Tensile strength, (b) Elasticity modulus

4. Conclusion

Tourmaline powder was modified with tetradecenylsuccinic anhydride as organic modifier in this work. The experimental results certified that tetradecenylsuccinate group was attached onto the surface of tourmaline to get polymerizable organic tourmaline tetradecenylsuccinate (TTDS), and then the tourmaline tetradecenylsuccinate-vinyl acetate copolymer was synthesized with the copolymerization of TTDS with vinyl acetate in DMF solution.

The results showed that the negative ions releasing quantities of modified tourmaline were increased by 3 times that of unmodified tourmaline, and the far infrared radiation rate of modified tourmaline was increased clearly after modification.

The tourmaline tetradecenylsuccinate-vinyl acetate copolymer displayed an excellent stability, and the negative ions releasing amount and far infrared radiation rate were all increased with the increasing of modified tourmaline content, and the good mechanical properties foreshadowed its potential prospects for development of novel functional materials and functional devices.

5. References

[1] Viti C, Brogi A and Liotta D 2016 J. Struct. Geo. **86** 1
[2] Dyar M D, Taylor M E and Lutz T M 1998 Amer. Min. **83** 848
[3] Rajajee V, Fletcher J and Rochlen L R 1999 Eur. J. Min. **11** 201
[4] Jiang S Y, Radvanec M and Nakamura E 2008 Lithos **106** 1
[5] Pal D C, Trumbull R B and Wiedenbeck M 2010 Chem. Geo. **277** 245
[6] Tokumura M, Znad H T and Kawase Y 2006 Chem. Eng. Sci. **61** 6361
[7] Machi Y 2001 J.Int. Soc. Life Inform. Sci. **19** 69
[8] Jie Z, Bing-Hui G E and Peng W 2009 J. Funct. Mat. **40** 556
[9] Garda G M, Trumbull R B and Beljavskis P 2009 Chem. Geo. **264** 207
[10] Hinsberg V J V and Marschall H R 2007 Chem. Geo. **238** 141
[11] Henry D J and Brodtkorb M K D 2009 J. South Amer. Ear. Sci. **28** 132
[12] Hu Y M, Bian J and Wu Z M 2014 China Non-Metal. Min. Ind. **108** 26
[13] Li M, Hu Y M and Li Y 2015 J. Surf. Sci. Nanotech. **13** 422
[14] Jen-Taut Y, Han-Hsing H and Wei W 2011 Internat. J. App. Math. Comp. Sci. **21** 431
[15] Zheng Y and Wang A 2010 Chem. Eng. J. **162** 186
[16] Tijing L D, Amarjargal A and Zhe J 2013 Current App. Phy. **13** 205
[17] Garces N Y, Stevens K T and Halliburton L E 2001 J. Crys. Growth **225** 435
[18] Jiang B J, Frost R L and Martens W N 2007 Vib. Spectr. **44** 42
[19] Zhang K Y, Cheng X H and Hong-Lu Q U 2004 Min. Metal. **13** 97
[20] Li C, Thostenson E T and Chou T W 2008 Comp. Sci. Techn. **68** 1445
[21] Tjong S C 2006 Mat. Sci. Eng. Rep. **53** 73

Acknowledgments
This work is supported by the National Natural Science Foundation of China (No.51372233).

IOP Conf. Series: Materials Science and Engineering **284** (2017) 012019 doi:10.1088/1757-899X/284/1/012019

The development of 3D food printer for printing fibrous meat materials

C Liu[1*], C Ho[2] and J Wang[3]

1 Graduate Institute of Manufacturing Technology, National Taipei University of Technology, Taipei. 10608, Taiwan
2 Graduate Institute of Manufacturing Technology, National Taipei University of Technology, Taipei. 10608, Taiwan
3 Institute of Mechanical Engineering, National Taipei University of Technology Taipei, Taiwan
E-mail: cnliu4357@yahoo.com.tw

Abstract. In this study, 3-D food printer was developed by integrating 3D printing technology with fibrous meat materials. With the help of computer-aided design and computer animation modeling software, users can model a desired pattern or shape, and then divide the model into layer-based sections. As the 3D food printer reads the design profile, food materials are extruded gradually through the nozzle to form the desired shape layer by layer. With the design of multiple nozzles, a wide variety of meat materials can be printed on the same product without the mixing of flavors. The technology can also extract the nutrients from the meat material to the food surface, allowing the freshness and sweetness of food to be tasted immediately upon eating it. This will also help the elderly's eating experience since they often have bad teeth and poor taste sensing problems. Here, meat protein energy-type printing is used to solve the problem of currently available powder slurry calorie-type starch printing. The results show the novel technology development which uses pressurized tank with soft piping for material transport will improve the solid-liquid separation problem of fibrous meat material. In addition, the technology also allows amino acids from meat proteins as well as ketone body molecular substances from fatty acids to be substantially released, making ketogenic diet to be easier to accomplish. Moreover, time and volume controlled material feeding is made available by peristaltic pump to produce different food patterns and shapes with food materials of different viscosities, allowing food to be more eye-catching.

1. Introduction

The development of 3D printing has grown significantly. The technology has been applied in various fields such as medicine, aviation, automobile and architecture modeling. There is no doubt that 3D printing will bring revolutionary change to every industry and this trend has reached the field of food safety. Currently, the 3D printing foods available on the market are based on starch-type, sugar-type, butter-type and gel-type powders that are processed into single, slurry-form printing material. The material is printed by using pressurized syringe with needle-type valve to control the material flow rate or piston type material container to squeeze out the power slurry material to form desired shape. So far, there is no 3D printer using fibrous meat or related studies. The reason is that it is difficult to extrude meat material by a 3D printer due to the fibrous tissue and the viscosity of the meat material. This study made improvements to the syringe-piston material container applied by most food material processors. The 3D meat printer developed in this study is equipped with peristaltic pump and pressurized tank to help the transport of food material. Compared to other food modeling machines available in the market, the 3D meat printer designed in this study will overcome the limitations of rough extrusion and container capacity, and thus, significantly enhance production efficiency and volume.

The 3D food printer is developed based on the concept of smart home appliance. In this study, the 3D food printer uses fibrous mud-like protein and fat from meat. This allows a wide variety of products to be manufactured, and the printed products are not only for display. The technology developed will improve household's willingness to use 3D printer and minimize its limitations, allowing the printer to be mass produced in the factory.

According to the Food & Health study conducted by IFIC Foundation in 2015, 69% of the adults feel excited about 3D printing technology that can produce any types of food. The Y generation (born in 1991~2000) is more passionate about this technology (79% of the population) because this generation is more willing to use software and machine for cooking.

In the past, 3D printing also known as Additive Manufacturing or Rapid Prototyping, which adopts the stacking method to add targeted materials layer by layer in forming the desired pattern rather than using the traditional production method via mold. With the help of software, the 3D CAD drawing will be converted into several cross-sectional thin slices. The benefit of this technique is that there is no processing shape limitation, any complex shape can be formed, and no supporting fixation is needed during the printing process. In 2011, 3D printing technology has been applied to the field of food processing. Researchers in the UK have developed the first 3D chocolate printer, which became the prototype of 3D food printer. Later, a Spanish company developed a 3D food printer named Foodini, which began the era of 3D food printer. The rapid development of 3D food printing technology has revolutionized the food and catering industries.

2. Literature and review

The basic concept of 3D printing, in fact, is similar to regular printer. In regular printing, the printer only need to print one layer of ink on a flat 2D (XY axis) printing paper, however, the image is only limited to two dimensional display. On the other hand, in 3D printing, one extra dimension (Z axis) printing (multiple layer stack printing) is utilized with special printing ink (food material) to add the printing material layer by layer in forming the desired shape [1]. With regards to the shape forming technique, 3D printing overcomes the limitation of traditional shape formation method. By integrating the fast & automatic shape forming system with computer simulated model, a wide variety of complex shape products can be manufactured with the needs of traditional molds and processing machines, which greatly shortens the manufacturing cycle of the product and significantly reduces the manufacturing cost.

2.1. Additive manufacturing technology classification

Additive manufacturing has evolved from the Rapid Prototyping (RP) in the early period to the Rapid Manufacturing (RM). The name was finalized as Additive Manufacturing (AM) by American Society for Testing and Materials (ASTM) in 2009 and related standards were stipulated [2]. ASTM has divided additive manufacturing into 7 categories, including:

I. VP (Vat Photopolymerization) uses a vat of liquid photopolymer resin, out of which the model is constructed layer by layer. An ultraviolet (UV) light is used to cure or harden the resin where required, whilst a platform moves the object being made downwards after each new layer is cured [3].

II. MJ (Material Jetting) uses wax and photopolymer resin as the main material. Sometimes metals are also used. The hardened wax model can be directly removed to proceed with shape formation. The well-known technology that uses this technique is objet.

III. BJ (Binder Jetting), the computer-controlled nozzle moves according to the cross-sectional information of the designed object and deposits alternating layers of the powder-based building material and the binding material. The binder acts as an adhesive between powder layers. After each layer, the object being printed is lowered to proceed with the next layer [4].

IOP Conf. Series: Materials Science and Engineering **284** (2017) 012019 doi:10.1088/1757-899X/284/1/012019

The current commercialized technology is 3D Printing and Gluing (3DPG) such as Chef Jet and Chef Jet Pro.

IV. ME (Material Extrusion) first uses heat source such as electricity to melt filament material then a 3D control system is adopted to move the melted filament material and deposit it layer by layer to form the desired 3D object. The printing material used (usually low melting point plastic such as ABS) first is made into filament and delivered into the nozzle via filament delivery system. The filament is then melted by heat inside the nozzle, and extruded along the desired cross-sectional profile and track controlled by a computer. The extruded material quickly hardens and bind with the nearby material. The process continues to form the desired shape layer by layer. The surface may be rough as the product finishes, therefore, post-polishing process is needed [5].

V. Currently, the commercialized technology is Fused Deposition Modeling (FDM) and Fused Filament Fabrication (FFM). Most of the food printers on the market use these two technologies; however, the products produced are not as good (in terms of printing accuracy) as those manufactured by 3DPG in which requirements for this technology is low and it is applicable to a wide variety of food materials [6].

VI. PBF (Powder Bed Fusion), a computer-controlled high energy laser beam then scans across the powder layer selectively according to the CAD information of each cross section of the desired model to fuse the powder. The scanned powder due to melting will bind together while the un-scanned powder will still be loosen and remains in position. The process repeats until the entire model is created [7].

VII. SL (Sheet Lamination) uses laser or knife to cut thin papers, plastics, metals or ceramics and then adopts thermal pressing or other technique to bind each layers together, obtaining the 3D object. During the process, laser or knife is controlled by computer to cut down a thin layer of the desired model. The unwanted area is cut into many small squires so that they can be removed later easily. The cut down layer material is bonded in place, over the previous layer, using heat pressing or other methods [8].

VIII. Directed Energy Deposition or Laser Metal Deposition (LMD) uses high power laser energy to melt the material on the processing surface. Metal powder is then injected into the melted material to complete the deposition process. The process is conducted with the help of positioning platform or robotic arm, which moves according to the desired 2D cross-sectional profile to build the 3D object [9].

2.2. Food 3D printing technology

The working principle of 3D food printer is similar to that of printing technology. However, the printing material used consists of one type or many types of liquid or mud-like foods that can be extruded rather than plastic or other industrial materials. The printer is composed of control computer, food material extruder, and transport device, etc. The food model and recipe is stored in the computer in advance. Food materials and other ingredients are placed first in the storage container. Once the user selected a preferred model and pressed the start button, the nozzle on the printer head will begin to move and extrude the food materials layer by layer to form the desired pattern, realizing 3D food printing.

2.2.1. Fused deposition modeling. The printing material is heated up to a certain temperature to form semi-melted material. The material is extruded on the flat supporting table where it quickly hardens. The process is repeated until the desired 3D object is formed [10].

2.2.2. Powder bed binder jetting. A thin layer of powder is spread first on the platform, and then glue is ejected by nozzle on the printer to bind the powder together. The process is repeated to spread another layer of powder on top of this bonded layer and glue is ejected again to bind them together. Finally, the un-bonded powder at each side is recovered for reutilization [10].

2.2.3. Selective laser sintering. This method uses computer to control the location of laser irradiation. Once the powder is irradiated by the laser beam, it will sinter to form aggregates. The process is repeated to add another layer of powder on top of this sintered layer followed by the sintering process until the final product is formed [11].

2.3. Summary of literature review

After reviewing the above literatures, it is found that currently available 3D food printers mainly use granular powder as the extruding material. However, 3D food printer that uses fibrous material as the extruding material will be more practical and useful, which will be the emphasis of this study. Moreover, most of the food printers on the market use piston-type mechanism or syringe container with flow rate controlling needle-head valve to extrude powder based food materials. One of the disadvantages of such design is the small storage container, which limits the possibility for mass printing and makes material replenishing more difficult. Another disadvantage is the difficulty in controlling the flow rate of solid-liquid separated fibrous meat material using the needle-head valve. As for the storage tank used in this study, it is large and independent, allowing more food material to be stored and making the change of food material to be more convenient. Such design is appropriate for household usage or factory mass production.

3. 3D Food Printer Developing Experiments

This section introduces the design framework of the food printer developed in this study, which includes process design, printer design concept, mechanical design (XYZ) of each component, and operation design (pressure tank and pump), etc.

3.1 System operation process flow diagram

In order to use the 3D food printer developed in this study, the 3D drawing file must be converted first to STL format. The converted file is then input into the printer and the desired food materials are placed inside the storage tank. After waiting for a while, the system can begin to operate. Figure 1 shows the system operation process flow diagram.

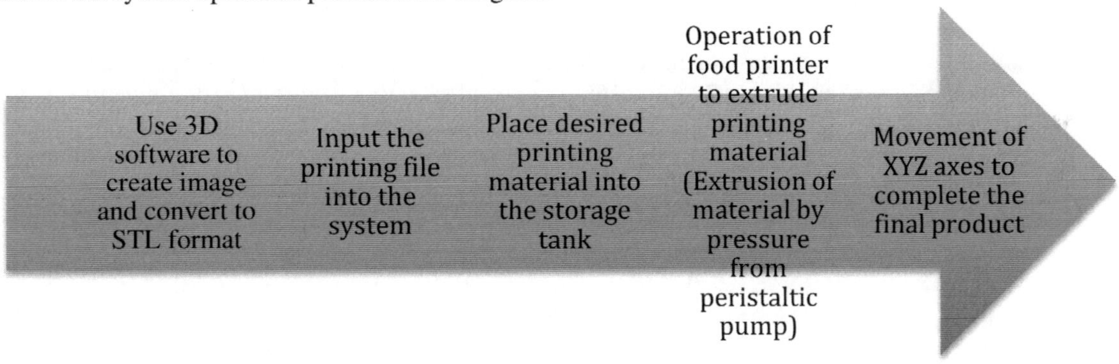

Figure 1. System operation process flow diagram

3.2 Novel extrusion technology of food printer

The extrusion method used by the food printer developed in this study is different from the traditional methods adopted by other food printers on the market. The traditional extrusion methods include pressurized syringe with needle-type valve to control the material flow rate or piston type material container to squeeze out the power slurry material. The system in this study uses pressure generated by a pump to squeeze the meat material out from the storage tank. By integrating peristaltic pump with multiple-nozzle device, the flow rate of meat material extruded can be easily controlled to form the desired pattern. Figure 2 shows the extrusion of technology of food printer.

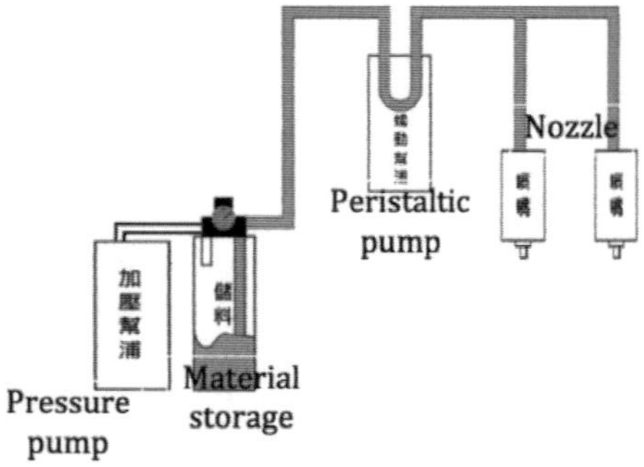

Figure 2. Extrusion technology of food printer

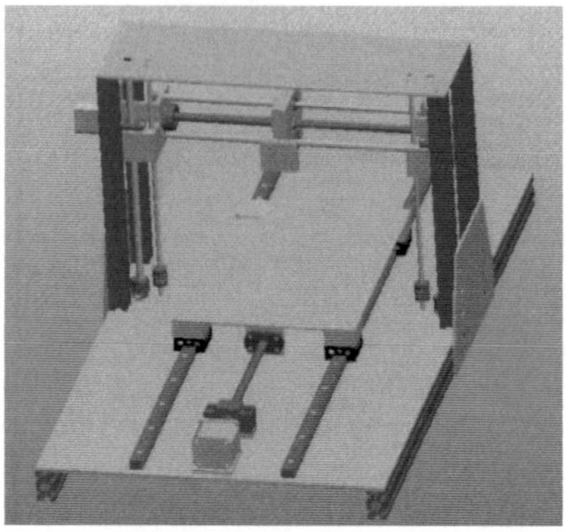

Figure 3. Mechanical design of food printer

3.3 Mechanical design

The purpose of the food printer designed in this study is to extrude food material in forming the desired pattern or shape. The area for the pattern formation is approximately 15cm x 30cm. However, since the mud-like food material is hard to stack up, the maximum height for the pattern formation is around 5cm.

In this study, four stepper motors were used to provide movement in the XYZ axis. The movement of Y axis is independent and carried out mainly by ball screw with the support of slide rail, which is responsible for moving the platform forward and backward. The movement of X and Z axis is interrelated and carried out mainly by ball screw with the support of optical axis. X axis is responsible for left and right movement while Z axis is responsible for up and down movement. This can prevent interfering product on the platform during operation, leading to distortion of the final product. Figure 3 shows the mechanical design of the food printer.

3.4 Material development

Besides creating a novel food processing method for elderly who has poor taste sensing and bad teeth, another important objective of this study is to offer a new and effective solution for dieting and cancer patients who require Ketogenic Diet to obtain Ketone body food elements.

The 3D food printing device developed in this study can provide significant amount of hypoxanthine (948 ppm of ketone group), 1170mg/100g of taurine and 35.41g/100g of unsaturated fatty acid, making ketogenic diet easier to carry out. In addition, it also lowers the diet of sugar and carbohydrates [12]. Evidences are shown in table 1, 2 and 3:

Table 1. Ketone group analysis results of original meat material and product.

Test method: After the samples were extracted and purified by solid phase extraction cartridge, the samples were analyzed by high performance liquid chromatography (HPLC-DAD)

Test result:

	Test limit (ppm)	Tested result of original meat material (ppm)	Tested result of product (ppm)
Guanline	50	58	133
Hypoxanthine	50	636	948
Adenine	50	50	98
Xanthine	50	54	61
Sum	--	798	1240

Table 2. Results of taurine analysis.

Test item	Tested result (mg/100 g)	Test method
Taurine	1170	Spectrophotometric method
Isoleucine	57.67	Determined by ethyl chloroformate derivatization and GC-MS determination
Leucine	137.04	
Branched Chain Amino Acid (BCAA)	303.20	
Valine	108.49	Determined by ethyl chloroformate derivatization and GC-MS determination

Table 3. Results of fatty acid analysis.

Test item	Test method	Test result	Test limit	unit
Saturated fat	The samples were extracted with oil, saponified and methylated, and the gas chromatograph (GC) was used. (Method of Test for Fatty Acids in Food)	47.9	0.1	g/100g
Trans fats		N/A	0.3	g/100g
Monounsaturated fatty acid		35.41	0.05	g/100g
Polyunsaturated fatty acid		1.26	0.05	g/100g

Before preparing the slurry meat material, its characteristics must first be studied. By selecting an appropriate slurry ratio, the actual printing material is prepared and used to test the extrusion system. Different from single powder material where the viscosity can be uniformly adjusted by controlling the amount solvent, fibrous meat material tends to have solid (meat)-liquid (juice) separation, making material extrusion to be more difficult. The researchers first add water into the material. However, after stirring, the granule size was too big. Then, fine gelatin powder with water was tested. It was found that 5:2 ratio of water to gelatin powder (100 g of water with 20 g of gelatin powder) is the most appropriate ratio for preparing the material. Figure 4, 5, 6, and table 4 show the result of material extrusion with different nozzle diameter. If the viscosity of the extruding material is too low, it may be difficult to form the desired shape. On the other hand, if the viscosity is too high, the extrusion of material may be difficult. Additional ingredients such as pork bone essence, pork liver essence and chicken essence can also be added to another pressure tanks.

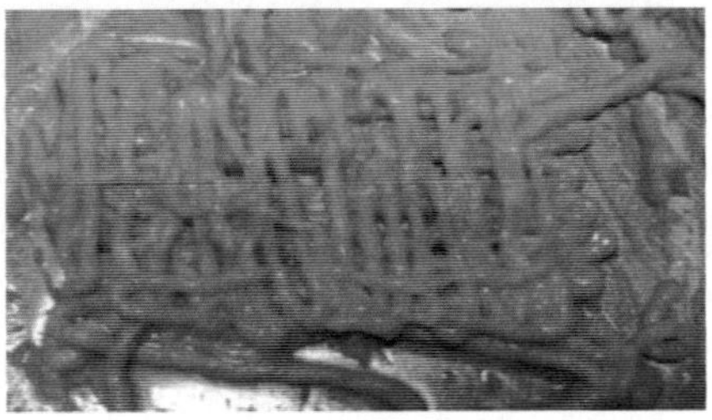

Figure 4. Diameter 2 mm

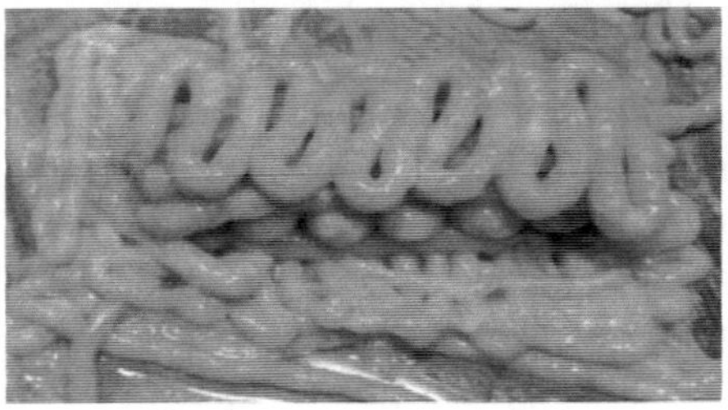

Figure 5. Diameter 4 mm

IOP Conf. Series: Materials Science and Engineering **284** (2017) 012019 doi:10.1088/1757-899X/284/1/012019

Figure 6. Diameter 5 mm

Table 4. Test parameters.

Slurry Solvent Ratio	Pressure Kg/cm^2	5mm	4mm	2mm
5:0	2.7	N.G	N.G	N.G
5:1	2.7	Not Smooth	N.G	N.G
5:2	2.7	Smooth	Smooth	Smooth

Meat materials (chicken, pork, and fish) can be cooked first to obtain the meat juice. The remaining meat can then be converted to slurry form by adding gelatin powder and meat juice with uniform stirring. The processed material is suitable for elderly who has bad teeth and poor taste sensing. In addition, after processing, the nutrients from the meat material are extracted to the food surface, allowing the freshness and sweetness of food to be tasted immediately upon eating it, greatly enhancing the eating experience. Therefore, the process developed in this study not only improves the sensing of taste but also makes chewing much easier.

3.5 Extrusion mode

One of the 3D food printers currently on the market uses chocolate as the printing material and extrude melted chocolate to form desired shape upon cooling. Another type of food printer uses granular and powder-based slurry as the printing material with piston or syringe-type container and needle-head valve to extrude the material. However, the blocking or overflow of printing material are commonly seen.

In this study, the extrusion method uses fibrous protein meat as the printing material which is completely different from that used by the commercial 3D food printers. In our system, pressurized air is used to push out the printing material. With the help of peristaltic pump, extrusion of material inside the soft silicon tube can be accurately controlled as shown in figure 7. The extrusion nozzle is made of teflon tube. All the above components are combined to come up with the novel 3D food printer.

Figure 7. Operation of peristaltic pump

3.6 Extrusion product

Different types of meat materials were placed separately in two different pressurized tanks. After tightening the pressurized tanks and switching on the pressure pump, food material will move slowly from soft silicon tube towards the nozzle. Once the air inside the front end of the nozzle is released, meat material will be extruded out from the soft silicon tube to form the desired pattern as shown in figure 8.

The system developed in this study uses two pressurized tanks and two extrusion nozzles. The first set of nozzle and pressurized tank can extrude fibrous meat material to form block shape with porous texture. The second set of nozzle and pressurized tank can inject liquid pork essence, chicken essence, etc. into the porous material to form juicy and moisturized food material, improving its appearance and test as shown in figure 9.

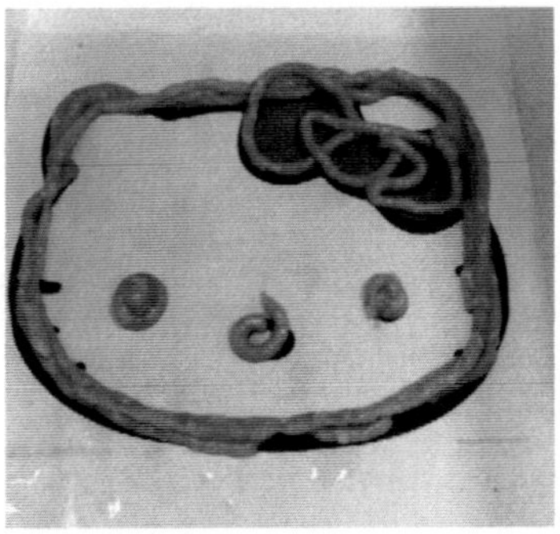

Figure 8. Extrusion to form desired pattern

Figure 9. Extrusion form block shape

4. . Results and discussion

4.1. Use of pressure tank to solve the limitation of syringe and piston type storage

Most of the food printers on the market use syringe and piston type material storage with needle-head valve to control the amount of food material extruded. These systems often encounter the problem of solid-liquid separation of the food material, leading to poor control of the extrusion volume, or small storage volume which requires frequent feeding of the printing material or is difficult to clean. In addition, overflow of printing material is often seen in system with needle-head valve. By using pressure tank developed in this study, the above mentioned limitations can be overcome.

4.2 Peristaltic pump control

The peristaltic pump and soft tube for material transport can be cleaned by simply using water while keeping peristaltic pump running. Most of the food printers on the market have the problems mentioned in Section 4.1. For instance, when the system is paused or prior to the start of the system, overflow of printing material may occur. In this study, peristaltic pump is used which can control the extrusion of printing material more accurately, making the printing process stable and smooth.

4.3 Three axes ball screw mechanism:

Most of the food printers on the market use belt to move XYZ axis, which may lead to unstable printing. As for the food printer developed in this study, ball screw was used for the movement of axis, which has the advantages of small vibration, accurate printing area, precise printing height and robust.

5. Conclusions

This study aims to develop a 3D food printer which has a printing dimension of 15cmx30cmx5cm. The food printer developed uses peristaltic pump to provide power for material extrusion, with the support of the driving board, the three axes can be driven separately. Comparing to the food printers currently on the market that only can use powder or granular food material to make slurry printing material, the novel food printer developed in this study can use fibrous meat as the printing material. The novel food printer improved the limitation of storage capacity, material property, and material type. It significantly enhance the operation efficiency and productivity. With the new food printer, a wide variety of food can be printed with improved chewing experience. The system uses ball screw as the main driving mechanism which has the advantages of large printing area, accurate printing height and fast printing speed.

6. References

[1] Jeffrey W Stansbury and Mike J. Idacavage 2016 *Dent Mater* **32.1** 54-64

[2] ASTM Standard F2792-12 Standard Terminology for Additive Manufacturing Technologies 2012 *ASTM International*

[3] Gregory I Peterson, Johanna J Schwartz, Di Z, Benjamin M. Weiss, Mark A. Ganter, Duane W. Storti and Andrew J. Boydston 2016 *ACS Appl. Mater. Interfaces* **8.42** 29037-43
[4] Gonzalez A, Mireles J, Lin Y and Wicker R 2016 *CERAM INT* **42.9** 10559-64
[5] Marlon W M Cunico, and Jonas d Carvalho 2013 *Rapid Prototyp. J.* **19.3** 144-52
[6] Jeffrey I Lipton, Meredith C, Franz N, Dan C and Hod L 2015 *TIFS* **43.1** 114-23
[7] Ryan T, Jason A Inzana, Edward M Schwarz, Stephen L Kates, Hani A Awad 2017 Ann. Biomed. Eng. **45.1** 23-44
[8] Miroslav P 2016 Anal. Lett. **49.18** 2865-82
[9] Kyle P, Clayson S and Johnson S *AM* 2016 **12** 121-31
[10] Jie S, Weibiao Z, Dejian H, Jerry Y H Fuh and Geok S Hong 2015 , FOOD BIOPROCESS TECH **8.8** 1605-15
[11] Fernanda C. Godoi, Sangeeta P and Bhesh R. Bhandari 2016 J Food Eng. **179** 44-54
[12] Robert G Levy, Paul N Cooper and Pratima G 2012 Cochrane Database Syst Rev **3** CD001903

Polarization-dependent plasmonic splitter based on low-loss polymer optical materials

Guang Qian[1], Xing-Chang Fu[2,3], Li-Jiang Zhang[2,3], Yi-Ran Liu[2,3], Ning Zhao[2,3] and Tong Zhang[1,2,3]

1 School of Instrument Science and Engineering, Key Laboratory of Micro-Inertial Instrument and Advanced Navigation Technology, Ministry of Education, Southeast University, Nanjing, China
2 Joint International Research Laboratory of Information Display and Visualization, School of Electronic Science and Engineering, Southeast University, Nanjing, China
3 Suzhou Key Laboratory of Metal Nano-Optoelectronic Technology, Suzhou Research Institute of Southeast University, Suzhou, China
E-mail: tzhang@seu.edu.cn

Abstract. A polarization-dependent optical beam splitter consisting of a straight long-range surface plasmon polariton (LRSPP) waveguide and an S-bend polymer waveguide was designed, fabricated and measured in this paper. At the splitting section, the two different waveguides are vertically coupled. The measurenment results show that the splitter operated in dual-channel mode at TM polarization, and single-channel mode at TE polarization. In addition, the polymer waveguide and LRSPP waveguide in the splitter exhibit low propagation loss of 0.51 dB/cm and 1.7 dB/cm, respectively. The hybrid beam splitter has wide potential applications in three dimensional (3D) multilayer photonic integrated circuits (PICs).

1. Introduction

Surface plasmon polaritons (SPPs) propagating electromagnetic waves along the metal-dielectric interfaces have been attracting a great deal of attention because of the ability to guide and manipulate optical signals at sub-wavelength scales that break the diffraction limit [1-3]. SPPs give effective ways to realize sub-wavelength optical devices for photonic integrated circuits (PICs). In addition, SPPs show special properties of transverse-magnetic (TM) polarization [4], simultaneous guiding of electrical and optical signals, and etc. Recently, applications of SPPs in polarization-dependent devices and tunable devices have been increasingly presented [5-8]. Particularly, LRSPP waveguides consisting of nanometer thin metal strips embedded in a homogeneous dielectric environment have been regarded as a promising alternative to the conventional integrated optical waveguides in some applications due to the distinct characteristics [9-11].

As a basic component, integrated optical beam splitter plays an important role in many optical applications involving optical sensors [12], integrated photonic quantum circuits [13], integrated microwave photonic circuits [14] and etc. LRSPP waveguide provide a compact device design and an effective way to realize the polarization-dependent and the tunable beam splitter to guide and manipulate optical signals [15-17].

In this paper, we proposed a polarization-dependent integrated optical beam splitter using hybrid vertical coupling between LRSPP waveguide and dielectric waveguide. The splitter was fabricated based on low-loss polymer optical materials and sliver. The measured results show that the device is dual-channel operating at TM polarization, and single-channel operating at TE polarization. The device supports the achievement of light splitting and switching in three dimensional (3D) multilayer PICs.

2. Device design

Figure 1 shows the schematic of the polarization-dependent integrated optical beam splitter. The device is based on Si wafer. The beam splitter consists of a straight LRSPP waveguide and an S-bend polymer-based dielectric waveguide. The LRSPP waveguide is made of a metal strip embedded in polymer cladding. The cladding materials for the LRSPP waveguide and the polymer waveguide are indentical. The polymer waveguide is located above the LRSPP waveguide with a spacing.

At the splitting section, the polymer waveguide is aligned vertically with the LRSPP waveguide to achieve the light coupling between the two waveguides. The working state of the device can be tuned by changing the polarization of input light. When the input light is TM-polarized, part of the light beam vertically couples into the LRSPP waveguide. Then the device is dual-channel operation. When the input light is TE-polarized, the LRSPP waveguide channel is blocked. In the device, low-loss polymer optical materials of photoactive UV curable fluorinated resins based on acrylate were used to make the dielectric waveguide core (n_{core}=1.46 at 1550 nm) and cladding ($n_{cladding}$=1.45 at 1550 nm), respectively. silver strip was used as the core of LRSPP waveguide.

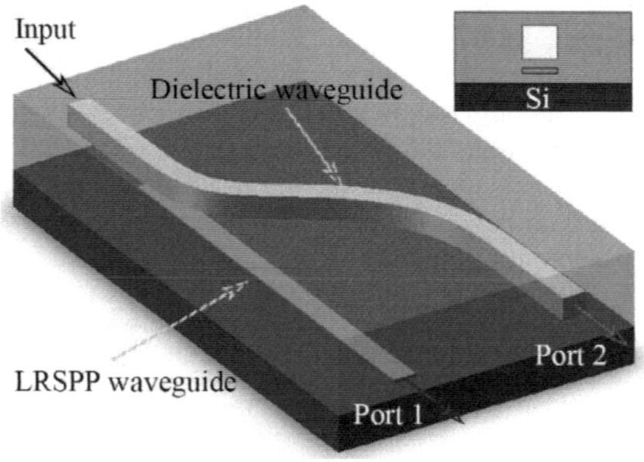

Figure 1. Schematic of the polarization-dependent integrated optical beam splitter using hybrid vertical coupling between LRSPP waveguide and polymer-based dielectric waveguide. Inset: cross-sectional sketch of the coupling section of the device.

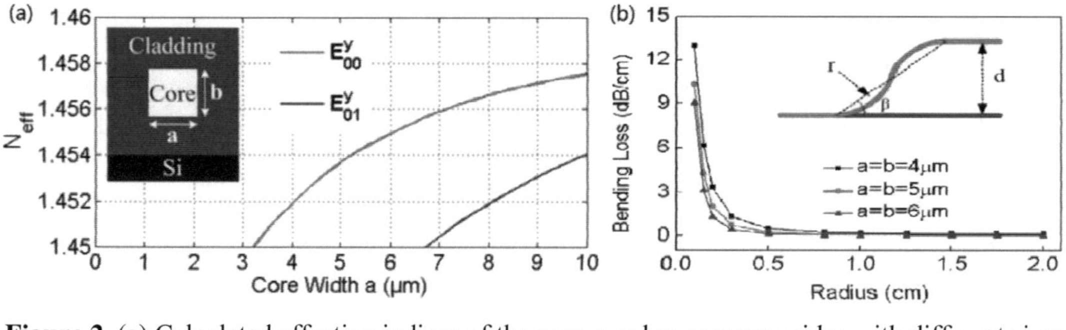

Figure 2. (a) Calculated effective indices of the square polymer waveguides with different sizes for E_{00}^{y} mode and E_{01}^{y} mode, respectively. (b) Calculated bending losses of waveguides with different bending radiuses at 1550 nm.

To ensure the single-mode operation of polymer waveguide, we calculated the effective refractive indices of square waveguide with different sides for E_{00}^{y} mode and E_{01}^{y} mode, as shown in figure 2(a).

It is known from figure 2(a) that the optimal core parameter for the square polymer waveguide only supporting E_{00}^{y} mode is about 3.2 μm < a < 6.8 μm. In addition, we calculated the bending losses of the polymer waveguides with different bending radiuses at 1550 nm by the beam propagation method (BPM), as shown in figure 2(b). When the bending radius is smaller than 0.5 cm, the bending loss of the polymer waveguide is sharply increased with the decreasing of the bending radius. In contrast, the bending radius has minor influence on the bending loss when the bending radius is greater than 0.5 cm. In addition, the wider-core waveguide has a lower bending loss.

To ensure the dual-channel operating of the device at TM polarization, the TM-polarized light in the polymer waveguide cannot couple into the LRSPP waveguide completely. For this, mode mismatch between the polymer waveguide and the LRSPP waveguide was required. Typically, on the mode match condition, the effective refractive indices of two different waveguides should be identical [18]. Therefore, we firstly calculated the effective refractive indices of LRSPP waveguides with different thicknesses and widths. Figure 3 shows the real parts of effective indices of LRSPP waveguides with different-size Ag strips.

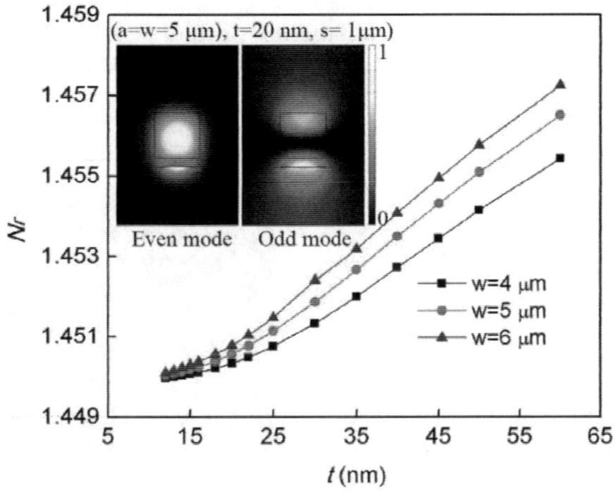

Figure 3. Real parts of effective indices of LRSPP waveguides with different-size Ag strips. Inset: normalized electric field distributions of the TM even mode and TM odd mode supported by the vertically coupled waveguides when t = 20 nm, a = w = 5 μm and the separation is 1 μm.

In order to achieve the low-loss splitter, the side of the polymer waveguide, the width and the thickness of the Ag strip were set to be 5 μm, 5 μm and 20 nm, respectively. The inset of figure 3 shows the normalized electric field distributions of the TM even mode and TM odd mode supported by the vertically coupled waveguides with a separation of 1 μm. In addition, the horizontal spacing of the two waveguides is 0.4 cm. The angle between the two waveguides β is 15° .

3. Experiments and Discussions

To verify the predicted polarization-dependent behavior, we have fabricated a hybrid beam splitter and measured its performance. Figure 4 (a) and (b) show the top view of sectional Ag strip and the cross-section view of the polymer waveguide in the hybrid beam splitter without upper cladding, respectively.

In order to test the polarization dependence of the beam splitter, a laser with wavelength of 1550 nm was used as the laser source. A polarization controller (PC) was used to turn the polarization of input light to the device. The output light spot was monitored by a near infrared CCD, and displayed by a displayer. Figure 5 shows the measured output light spots of the beam splitter at TM and TE polarization, respectively. The results show that the working state of the device is depended on the polarization of input light. The beam splitter is dual-channel operation at TM polarization, and single-channel operation at TE polarization. It is contributed by the TM-polarized property of LRSPP waveguide in the hybrid beam splitter. The polymer waveguide channel is conducting at both polarizations. It is due to the mode mismatch between the polymer waveguide and the LRSPP waveguide.

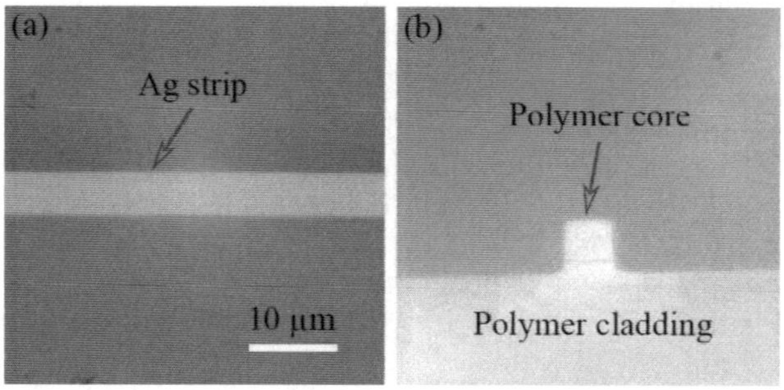

Figure 4. (a) Top view of sectional Ag strip in the hybrid beam splitter. (b) Cross-section view of the polymer waveguide in the hybrid beam splitter without upper cladding.

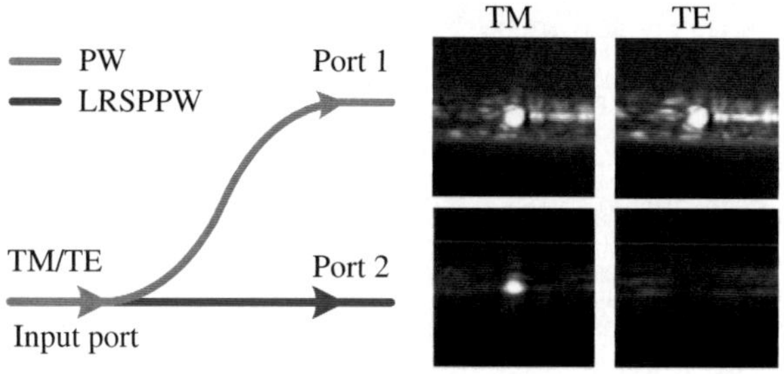

Figure 5. Measurement of the polarization dependence of the device.

By using the cut-back method, we measured the propagation loss of the polymer waveguide (PW) and the LRSPP waveguide (LRSPPW). The PW and the LRSPPW show low propagation loss of about 0.51 dB/cm and 1.73 dB/cm, respectively. At TM polarization, the fabricated splitter shows a splitting ratio of about 65%: 35% (Port 1 to Port 2).
In addition, according to the photoelectric multiplexing property of LRSPP waveguide, polarization-dependent integrated optical beam splitter capable of turning splitting ratio can be obtained based on the thermo-optic effect. In this tunable splitter, the silver strip should be used as the core of LRSPP waveguide and the heating electrode simultaneously. Similarly, tunable directional coupler and tunable Mach-Zehnder interferometer (MZI) can be obtained by using two identical hybrid beam splitters.

4. Summary

In this paper, we demonstrated a polarization-dependent hybrid beam splitter consisting of a straight LRSPP waveguide and an S-bend polymer waveguide for 3D multilayer PICs. We designed the structure of single-mode polymer waveguide and the coupling between the polymer waveguide and the LRSPP waveguide. Additionally, we fabricated a hybrid beam splitter utilizing polymer materials and measured its light transmission characteristic and polarization dependence. The results show that the working state of the splitter can be flexibly turned between single channel and dual channel by changing the polarization of input light. The hybrid beam splitter has great potential applications in 3D multilayer PICs.

5. References

[1] Bozhevolnyi S I, Volkov V S, Devaux E, Laluet J Y and Ebbesen T W 2006 *Nature* **440** 508
[2] Zhang X Y, Hu A, Wen J Z, Zhang T, Xue X J, Zhou Y and Duley W W 2010 *Opt. Express*, **18** 18945
[3] Berini P 2009 *Adv. Opt. Photonics* **1** 484
[4] Berini P, Charbonneau R and Lahoud N 2008 *IEEE J. Sel. Top. Quant.* **14** 1479
[5] Zhang X Y, Hu A, Zhang T, Xue X J, Wen J Z and Duley W W 2010 *Appl. Phys. Lett.* **96** 043109
[6] Volkov V S, Han Z, Nielsen M G, Leosson K, Keshmiri H, Gosciniak J, Albrektsen O and Bozhevolnyi S I 2011 *Opt. Lett.* **36** 4278
[7] Lee J, Song J, Sung G Y and Shin J H 2014 *Nano Lett.* **14** 5533
[8] Krupin O, Wei R W, Beland P and Adikan F R M 2016 *J. Lightwave Technol.* **34** 4673
[9] Zhang T, Qian G, Wang Y Y, Xue X J, Shan F, Li R Z, Wu J Y and Zhang X Y 2014 *Sci. Rep.* **4** 3855
[10] Fan H, Charbonneau R and Berini P 2014 *Opt. Express* **22** 4006
[11] Qian G, Fu X C, Zhang L J, Tang J, Liu Y R, Zhang X Y and Zhang T 2017 *Sci. Rep.* **7** 41146
[12] Lee S S, Ahn S W, Oh M C and. Shin S Y 1996 *IEEE Photon. Technol. Lett.* **8** 921
[13] Sansoni L, Sciarrino F, Vallone G, Mataloni P, Crespi A, Ramponi R and Osellame R 2012 *Phys. Rev. Lett.* **108** 010502
[14] Roeloffzen C G, Zhuang L, Taddei C, Leinse A, Heideman R G, van Dijk P W, Oldenbeuving R M, Marpaung D A, Burla M and Boller K J 2013 *Opt.Express* **21** 22937
[15] Chang K W and Huang C C 2016 *Sci. Rep.* **6** 19609
[16] Jin T K and Park S 2014 *J. Opt.* **16** 218
[17] Guan X, Wu H, Shi Y, Wosinski L and Dai D 2013 *Opt. Lett.* **38** 3005
[18] Liu F, Li Y, Wan R, Huang Y, Feng X and Zhang W 2007 *Appl. Phys. Lett.* **90** 141101

Acknowledgments

This work was supported by Natural National Science Foundation of China (NSFC) (61307066 and 61450110442), Scientific Research Foundation of Graduate School of Southeast University (YBJJ1535, YBPY1603 and YBJJ1513), Fundamental Research Funds for the Central Universities and Graduate Innovation Program of Jiangsu Province (KYLX_0109) and Foundation of Key Laboratory of Micro-Inertial Instrument and Advanced Navigation Technology, Ministry of Education, China (201402).

IOP Conf. Series: Materials Science and Engineering **284** (2017) 012021 doi:10.1088/1757-899X/284/1/012021

Study on hydrothermal synthesis dynamics of nanoscale xonotlite fibers

F Liu[1,2,3], S Chen[1,3], Q Lin[1,2], X D Wang[2], and J X Cao[1,2,3,4]

[1] School of Chemistry and Chemical Engineering, Guizhou University, Guiyang, Guizhou 550025, PR China
[2] Key Laboratory of Green Chemical and Clean Energy Technology, Guiyang, Guizhou 550025, PR China
[3] Key Laboratory of Efficient Utilization of Mineral and Green Chemical Technology, Guiyang, Guizhou 550025, PR China
[4] jxcao@gzu.edu.cn

Abstract. The xonotlite crystals were synthesized via the hydrothermal synthesis manner from CaO and SiO_2 as the raw materials with their Si/Ca molar ratio of 1.0. Hydrothermal synthesis dynamics of nanoscale xonotlite fibers was explored by masterly measuring the electrical conductivities and the calcium concentrations of product slurries synthesized at various reaction temperature in this paper. The results indicated that the calculated values of the products' quality at various reaction temperatures were consistent with the measured values. Based on chemical reaction kinetic, using fourth-order Runge-Kutta method, spline interpolation and least-squares fitting method, the dynamic relationship of xonotlite fibers synthesized via hydrothermal synthesis process is of $-dc_A / dt = kc_A^{4/5}$.

1. Introduction

Xonotlite crystal belongs to monoclinic system (a=1.67nm, b=0.73nm, c=0.695nm), and it is a mineral of calcium silicate hydrate with the lowest crystalline water content, the best heat resistance and thermal stability (decomposition temperature of 1050-1100 ℃) [1]. In general, the xonotlite products used in industry were fabricated using dynamic process from CaO and SiO_2 as the raw materials with their Si/Ca molar ratio of 1.0[2-6]. Due to the dissolution characteristics of siliceous material, the xonotlite crystals from H_2SiO_3 and K_2SiO_3 as siliceous materials could grow to form the various morphologies of spherical particles and fibers respectively. Actually, it was difficult to synthesize xonotlite crystals (specially xonotlite fibers) because of the harsh conditions such as the high reaction temperature and pressure, resulting in the extremely low synthetic efficiency [7-11]. It can be confirmed by many scholars' research that the synthesis of xonotlite crystals was feasible based on the hydrothermal synthesis thermodynamics. But since the invisibility of the reaction and the harsh conditions of xonotlites' formation, the dynamics data of xonotlite crystals was hardly possible to obtain so that few people took the hydrothermal synthesis dynamics into account [12]. Accordingly, in this paper, with K_2SiO_3 solution and $Ca(OH)_2$ suspension as siliceous material and calcareous material respectively, hydrothermal synthesis dynamics of nanoscale xonotlite fibers was explored by masterly measuring the electrical conductivities and the calcium concentrations of product slurries, aiming to study on the dynamics of xonotlite fibers, to improve the synthetic efficiency of it and to make the morphologies of the xonotlite crystals be controllable.

2. Experiment

2.1 Materials and methods

The xonotlite fibers were synthesized via the dynamic hydrothermal synthesis method.

K_2SiO_3 solution and $Ca(OH)_2$ suspension were used as siliceous material and calcareous material respectively. K_2SiO_3 solution was prepared by mixing H_2SiO_3 solution with KOH 100g·L^{-1} at the molar ratio of 1:2. $Ca(OH)_2$ suspension was prepared by mixing $CaCl_2$ with KOH 100g·L^{-1} at the molar ratio of 1:2. The calcareous material, distilled water, and siliceous material were homogeneously mixed, and then the mixture was transferred into a reaction chamber with a stirrer at a heating rate of 1.5 ℃·min^{-1} and a stirring rate of 200 rpm. The hydrothermal treatment proceeded under the following conditions: the Ca/Si molar ratio of 1.0, the reaction temperature of 190 ℃, 210 ℃ and 230 ℃ respectively. After the hydrothermal treatment, the suspensions were filtrated and washed three times by distilled water. The resultant powders were dried at 75 ℃ for 12 h. Technological flow diagram of hydrothermally synthesizing xonotlite products was shown in Figure 1.

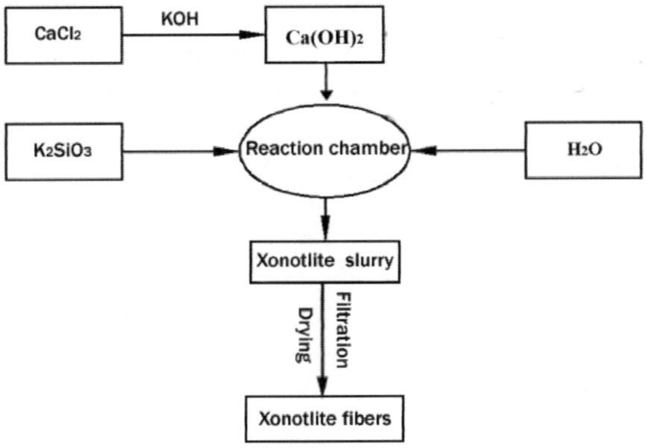

Figure 1. Technological flow diagram of hydrothermally synthesizing xonotlite fibers.

2.2 The Determination Method of Samples

The reaction equation for the formation of the xonotlite fibers could be simplified as:

$$6Ca(OH)_2 + 6K_2SiO_3 + H_2O \rightarrow 6CaO\cdot 6SiO_2\cdot H_2O + 12KOH \quad (1)$$

The above equation (1) was only a simplified final reaction equation for the preparation of xonotlite fibers, but the whole reaction was actually made up of a series of complex chemical reaction steps. Due to the invisibility of the hydrothermal synthesis reaction, it was impossible to observe the reaction process directly. When K_2SiO_3 solution was used as siliceous material, with increasing reaction time, the number of ions in the system would be reduced so that the reduction of these conductive ions would cause the changes in the conductivity of the reaction system. Therefore, the dynamic test of the hydrothermal synthesis process of the xonotlite can be achieved by measuring the change of the electrical conductivity during the reaction process.

The product slurries synthesized by hydrothermal synthesis process at various reaction temperatures and time were kept at 25 ℃ in thermostat water bath cauldron, and the electrical conductivities of various slurries were measured using a DDS-12A digital conductivity meter. The EDTA titration method was used to determine the calcium concentrations of various product slurries formed by hydrothermal synthesis process under various reaction temperature and time. The product slurries were accurately weighed by FA2004 electronic balance after drying. After the induction time (t_0), the qualities of the products obtained under various reaction times can be defined as m(Total)=m(KCl)+m($Ca(OH)_2$)+m(C-S-H(I))+m(xonotlite).

3. Results and discussion

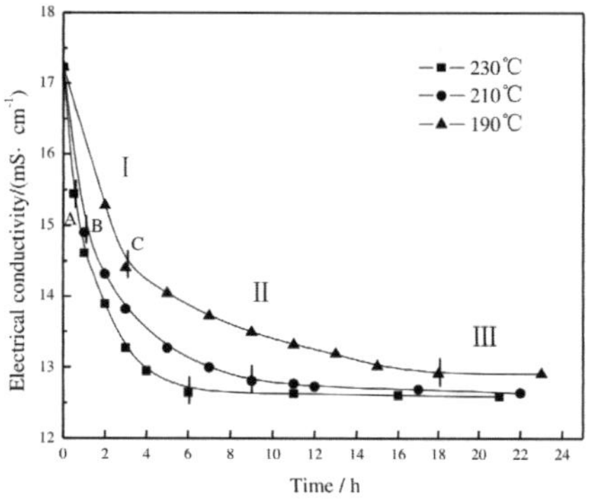

Figure 2. Time dependence of the electrical conductivities of hydrothermally synthesized products at various temperatures.

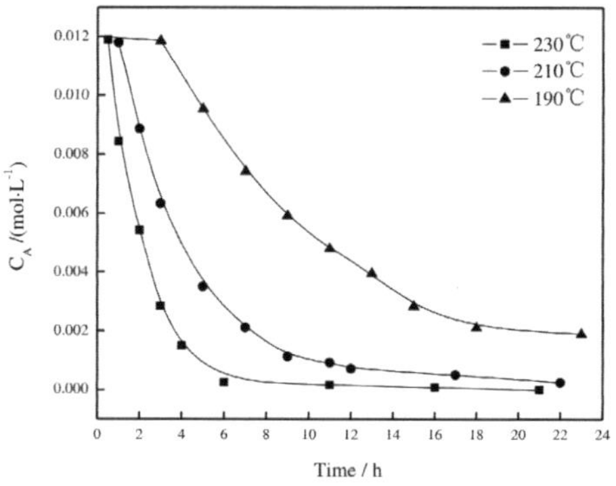

Figure 3. Time dependence of the calcium concentrations of the products hydrothermally synthesized at various temperatures.

The electrical conductivities and calcium concentrations of the system were respectively measured after various reaction times, shown in Figure 2 and Figure 3 respectively. It can be seen from Figure 2, with prolonging reaction time, the electrical conductivities of hydrothermally synthesized products at various temperatures decreased, and its descending trend can be divided into three stages. The descending trend of these three curves in the first stage was fastest, in the second stage was relatively slow, obviously, the slowest of the descending trend existed in the third stage. And in the third stage, the conductivities tended to a certain value. The descending trend of these three curves indicated that the reaction rate in the first stage was fastest, then it became slightly slow in the second stage, and the slowest reaction speed appeared in the third stage.

In the hydrothermal synthesis process of xonotlite fibers, the reaction process can be expressed by the following reaction steps:

$$Ca(OH)_2 + K_2SiO_3 + H_2O \rightarrow C - S - H(gel) \tag{2}$$

$$C - S - H(gel) + Ca(OH)_2 + K_2SiO_3 + H_2O \rightarrow C - S - H(\mathrm{II}) \tag{3}$$

$$C - S - H(\mathrm{II}) + Ca(OH)_2 + K_2SiO_3 + H_2O \rightarrow C - S - H(\mathrm{I})(5CaO \cdot 6SiO_2 \cdot 5H_2O) \tag{4}$$

$$Ca(OH)_2 + C - S - H(\mathrm{I})(5CaO \cdot 6SiO_2 \cdot 5H_2O) \rightarrow 6CaO \cdot 6SiO_2 \cdot H_2O + 5H_2O \tag{5}$$

Equation (2) represented the formation of amorphous C-S-H gels at the initial stage of the reaction. With the progress of the reaction, as shown in equation (3), C-S-H gels were transformed into calcium-rich hydrated calcium silicate C-S-H (II). Then due to the silicon-rich state in the reaction system, as shown in equation (4), C-S-H (II) would be transformed into the silicon-rich C-S-H(I) which owned a structure that similar to the tobermorite ($5CaO \cdot 6SiO_2 \cdot 5H_2O$). Finally, the silicon-rich C-S-H(I) directly transformed into the fibrous xonotlite crystals, as shown in equation (5). According to the reaction process of the xonotlite fibers and Figure 2, it can be assumed that the above-mentioned chemical equation (2, 3 and 4) occurred in the first stage of Figure 2 to form C-S-H. And there was no formation of the fibrous xonotlite crystals in the first stage. Moreover, the reaction time of the first stage was assumed to be the induction time (t_0) required for the formation of fibrous xonotlite crystals. Assuming that the chemical reaction of above equation (5) occurred in the second stage, at this moment, the nucleation and growth of xonotlite fibers took place on the surface of the C-S-H, and there was no C-S-H formation in this stage. As before, we also supposed that most of the xonotlite fibers had formed before the third stage. So in the third stage, the reaction rate decreased to the minimum, which led to the conductivities decreased slowly and tended to a certain value.

Based on the above assumptions and the data of calcium concentrations in Figure 3, the qualities of the products under various reaction temperatures and times during the second stage can be calculated, as shown in Tables 1, 2 and 3. According to the experimental method of 2.2, the quality-time values of the products at various reaction temperatures were measured, as shown in Tables 4, 5 and 6. At the same reaction temperature, the quality-time calculated values of the products were compared with the measured values, and the results were as shown in Figure 4.

Table 1. Calculated data of quality-time of products hydrothermally synthesized at 230 ℃.

t/[h]	m(KCl)/[g]	m(Ca(OH)₂)/[g]	m(C-S-H(I))/[g]	m(Xonotlite)/[g]	m(Total)/[g]
0.5	7.4500	0.8806	6.0833	0.0000	14.4139
1	7.4500	0.6241	4.3117	1.7328	14.1186
2	7.4500	0.4017	2.7749	3.2360	13.8625
3	7.4500	0.2101	1.4515	4.5304	13.6419
4	7.4500	0.1112	0.7684	5.1984	13.5281
6	7.4500	0.0185	0.1281	5.8247	13.4213
11	7.4500	0.0124	0.0854	5.8665	13.4142
16	7.4500	0.0062	0.0427	5.9082	13.4071
21	7.4500	0.0000	0.0000	5.9500	13.4000

Table 2. Calculated data of quality-time of products hydrothermally synthesized at 210 ℃.

t/[h]	m(KCl)/[g]	m(Ca(OH)₂)/[g]	m(C-S-H(I))/[g]	m(Xonotlite)/[g]	m(Total)/[g]
1	7.4500	0.8806	6.0833	0.0000	14.4139
2	7.4500	0.6557	4.5296	1.5197	14.1550
3	7.4500	0.4689	3.2392	2.7818	13.9399
5	7.4500	0.2592	1.7908	4.1985	13.6985
7	7.4500	0.1563	1.0797	4.8939	13.5800
9	7.4500	0.0839	0.5794	5.3833	13.4966
11	7.4500	0.0686	0.4740	5.4864	13.4790
12	7.4500	0.0534	0.3687	5.5894	13.4615
17	7.4500	0.0381	0.2633	5.6924	13.4439
22	7.4500	0.0191	0.1317	5.8212	13.4219

Table 3. Calculated data of quality-time of products hydrothermally synthesized at 190 ℃.

t/[h]	m(KCl)/[g]	m(Ca(OH)₂)/[g]	m(C-S-H(I))/[g]	m(Xonotlite)/[g]	m(Total)/[g]
3	7.4500	0.8806	6.0833	0.0000	14.4139
5	7.4500	0.7055	4.8734	1.1834	14.2123
7	7.4500	0.5498	3.7979	2.2354	14.0330
9	7.4500	0.4379	3.0249	2.9914	13.9042
11	7.4500	0.3552	2.4535	3.5503	13.8089
13	7.4500	0.2919	2.0166	3.9776	13.7361
15	7.4500	0.2092	1.4452	4.5365	13.6409
18	7.4500	0.1557	1.0755	4.8981	13.5793
23	7.4500	0.1556	1.0753	4.8983	13.5792

Table 4. Experimental data of quality-time of products hydrothermally synthesized at 230℃.

t/[h]	0	0.5	1	2	3	4	6	11	16	21
m(Total)/[g]	15.05	14.42	14.13	13.82	13.61	13.54	13.44	13.42	13.42	13.41

IOP Conf. Series: Materials Science and Engineering **284** (2017) 012021 doi:10.1088/1757-899X/284/1/012021

Table 5. Experimental data of quality-time of products hydrothermally synthesized at 210℃.

t/[h]	0	1	2	3	5	7	9	11	12	17	22
m(Total)/[g]	15.05	14.40	14.16	13.92	13.68	13.59	13.48	13.47	13.45	13.45	13.43

Table 6. Experimental data of quality-time of products hydrothermally synthesized at 190℃.

t/[h]	0	1	2	3	5	7	9	11	12	17	22
m(Total)/[g]	15.05	14.60	14.41	14.22	14.02	13.92	13.81	13.73	13.64	13.57	13.56

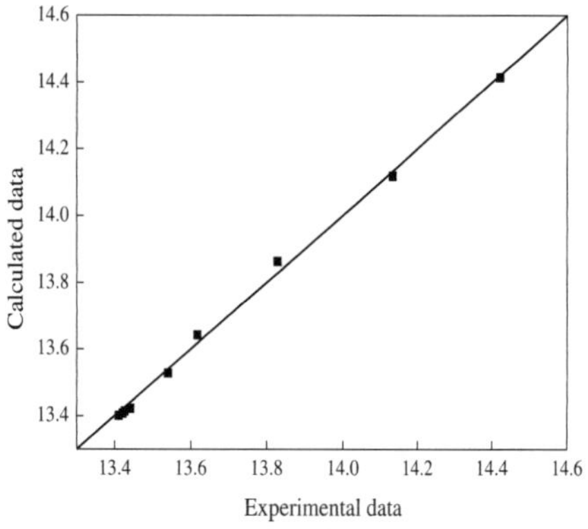

(a) 230 ℃

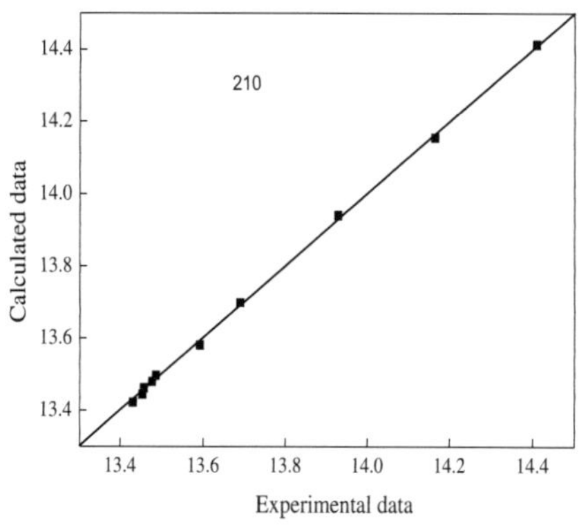

(b) 210 ℃

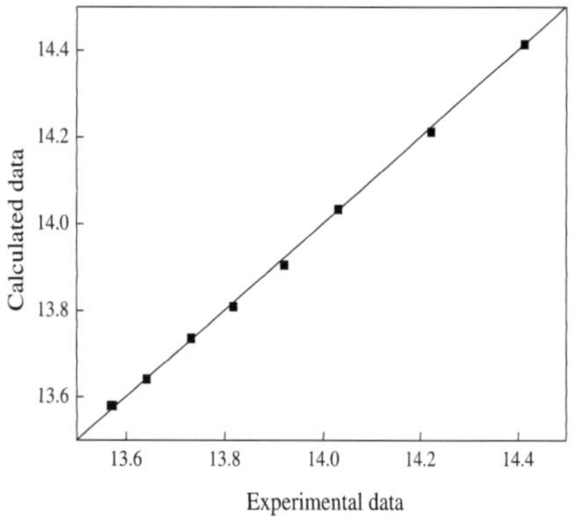

(c) 190 ℃

Figure 4. The experimental data comparison with calculated data of xonotlite fibers hydrothermally synthesized at different temperatures.

It can be seen from Figure 4 that when the reaction temperature was 230, 210 and 190 ℃ respectively, the quality-time calculated values were consistent with the measured values after the reaction time of 0.5 h, 1 h and 3 h respectively, so the reaction time of 0.5 h, 1 h and 3 h can be defined as the induction time required for the formation of fibrous xonotlite crystals at 230 ℃, 210 ℃ and 190 ℃ respectively, as indicated by points A, B and C in Figure 2. There was no formation of fibrous xonotlite crystals before the induction time. At the same time, the above analysis also showed that the above hypotheses were confirmed, that was, the second stage, beginning with the induction time, was the stage of the formation of xonotlite fibers, and there was no C-S-H produced in this stage.

According to the above analytical results, the equation (5) can be simplified as the following chemical reaction:

$$A + B \rightarrow C + 5D \tag{6}$$

In the equation： A— $Ca(OH)_2$

B— $C-S-H(\mathrm{I})(5CaO\cdot 6SiO_2\cdot 5H_2O)$

C— $6CaO\cdot 6SiO_2\cdot H_2O$

D— H_2O

According to the equation (6), the rate equation usually has the following form:

$$-dc_A / dt = k_A c_A^{n_A} c_B^{n_B} \tag{7}$$

Because A=B, that is, B=C. Thus, equation (7) can be simplified as:

$$-dc_A / dt = kc_A^n \tag{8}$$

In the equation, $k = k_A$ was the total rate constant of the reaction, $n = n_A + n_B$ was the total order of reaction.

In this equation, the kinetic parameters were only k and n, therefore, the rate equation can be established by determining the two parameters. Since there was only one constant in the equation, the form of the rate equation only depended on n, so the key to determine the rate equation was to determine the reaction order. In order to determine the reaction order, it was necessary to know the concentrations of the reactants (C_A) at different times (t) at a definite temperature, that was, to know the relationship of C_A-t in this chemical reaction.

In this paper, the internal performance function provided in Matlab were used, then the reaction order (n) and the rate constant (k) were determined and calculated respectively by dealing with the dynamic data (C_A-t) corresponding to the different temperatures in Figure 3 using the differential method. The values of n and k were used as the iterative initial values in the process of solving the fourth-order Runge-Kutta method, and the detailed solving process was shown in figure 5. It should be noted that a series of (t_i - $[c_A]_{ci}$) values obtained by the fourth-order Runge-Kutta method may not have a corresponding value ($[c_A]_i$), therefore, it was necessary to use spline interpolation to solve the values of $[c_A]_i$.

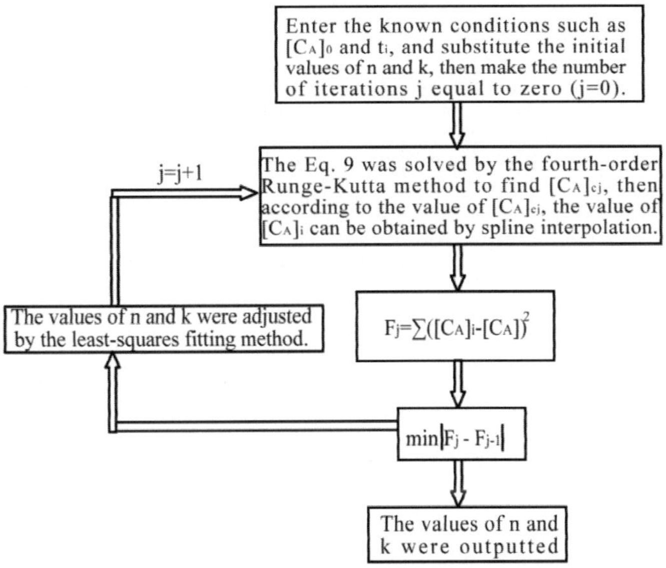

Figure 5. Flow chart for program of kinetic parameters.

IOP Conf. Series: Materials Science and Engineering **284** (2017) 012021 doi:10.1088/1757-899X/284/1/012021

Firstly, the values of $[c_A]_i$ were calculated by using Matlab which completed the loop iteration of the fourth-order Runge-Kutta method and the spline interpolation. Then according to $\min F = \sum([c_A]_i - [c_A])^2$, the optimum solution of k were obtained by the least-squares fitting method.

$$\begin{cases} y_{n+1} = y_n + \frac{h}{6}[k_1 + (2 - 2^{\frac{1}{2}})k_2 + (2 + 2^{\frac{1}{2}})k_3 + k_4 \\ k_1 = f_n \\ k_2 = f(t_n + \frac{1}{2}h, y_n + \frac{1}{2}hk_1) \\ k_3 = f[t_n + \frac{1}{2}h, y_n + \frac{2^{\frac{1}{2}} - 1}{2}hk_1 + (1 - \frac{2^{\frac{1}{2}}}{2})hk_2] \\ k_4 = f[t_n + h, y_n - \frac{2^{\frac{1}{2}}}{2}hk_2 + (1 + \frac{2^{\frac{1}{2}}}{2})hk_3] \end{cases} \tag{9}$$

According to the above-mentioned method, the initial values of kinetic parameters n and k at hydrothermal synthesis temperature of 230 ℃ were 0.7993 and 0.1976$(mol \cdot m^{-3})^{1/5} \cdot s^{-1}$ respectively. Since the errors of experimental measurement and the errors of calculation, n=0.8 and k=0.1976$(mol \cdot m^{-3})^{1/5} \cdot s^{-1}$ were taken as the iterative initial values after rounded the reaction order, then according to $\min F = \sum([c_A]_i - [c_A])^2$, the value of k that obtained by a least-squares fitting method was of 0.1976$(mol \cdot m^{-3})^{1/5} \cdot s^{-1}$.

Similarly, when the hydrothermal synthesis temperature was 210 ℃ or 190 ℃ respectively, according to $\min F = \sum([c_A]_i - [c_A])^2$, the values of k were 0.1109 $(mol \cdot m^{-3})^{1/5} \cdot s^{-1}$ and 0.0411$(mol \cdot m^{-3})^{1/5} \cdot s^{-1}$ respectively. When the hydrothermal synthesis temperature was of 230 ℃, 210 ℃ and 190 ℃ respectively, the value of n was of 0.8, and the values of k were of 0.1976$(mol \cdot m^{-3})^{1/5} \cdot s^{-1}$, 0.1109 $(mol \cdot m^{-3})^{1/5} \cdot s^{-1}$ and 0.0411$(mol \cdot m^{-3})^{1/5} \cdot s^{-1}$ respectively. The fitting results of the fourth-order Runge-Kutta method were shown in Figure 6.

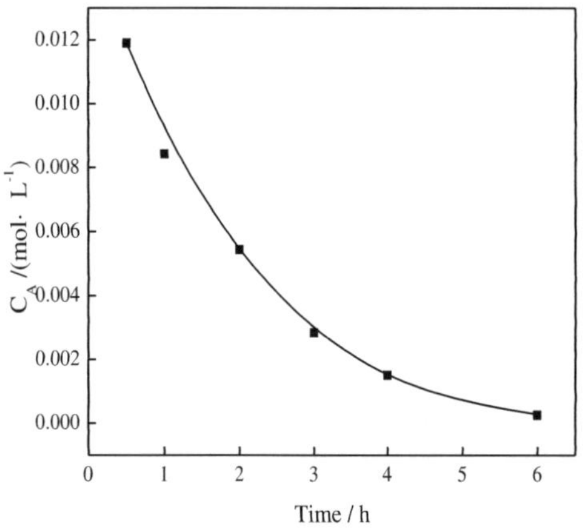

(a) 230 ℃

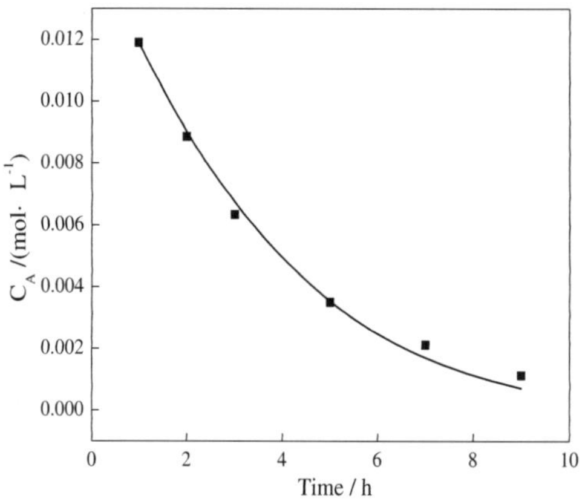

(b) 210 ℃

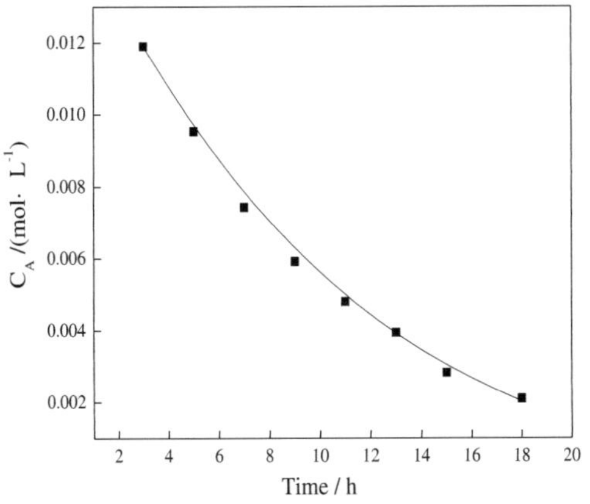

(c) 190 ℃

Figure 6. The curves of concentration-time fitted by fourth-order Runge-Kutta method at different temperatures.

It can be seen from the Figure 6 that the curves of reaction concentration-time fitted by fourth-order Runge-Kutta method at different temperatures were consistent with the experimental data, obviously, the results showed that the values of *n* and *k* obtained by aforementioned computational method had a high accuracy. Therefore, the rate equation of xonotlite fibers hydrothermal reaction could be confirmed as $-dc_A / dt = kc_A^{4/5}$.

4. Conclusions

(1) With K_2SiO_3 solution and $Ca(OH)_2$ suspension as siliceous material and calcareous material respectively, the dynamic test of the hydrothermal synthesis process of the xonotlite fibers can be achieved by measuring the change of the electrical conductivity during the reaction process.

(2) By measuring the electrical conductivities and the calcium concentrations of product slurries synthesized at various reaction temperature, it can be assumed that the hydrothermal synthesis process of xonotlite fibers can be divided into three stages, that was, the stage forming amorphous C-S-H gels, the stage forming the most of fibrous xonotlite crystals, the stage forming a little xonotlite fibers.

(3) According to the data of calcium concentrations, the calculated values of the products' quality at various reaction temperatures were consistent with the measured values, showing that the hypotheses dividing the formation process of fibrous xonotlite crystals into three stages were confirmed.

(4) Based on chemical reaction kinetic, using fourth-order Runge-Kutta method, spline interpolation and least-squares fitting method, the dynamic relationship of xonotlite fibers obtained by direct conversion of silica-rich C-S-H (I) is of $-dc_A / dt = kc_A^{4/5}$.

5. References

[1] Churakov S V and Mandalie P 2008 *Cement. Concrete* **38** 300-11
[2] Zou J, Guo C, Jiang Y, Wei C and Li F 2016 *Mater. Chem. Phys* **172** 121-8
[3] Liu F, Zhu B and Cao J X 2011 *Adv. Mater. Res* **148-149** 1755-8
[4] Yang J, Zhang X, Ma H W, Wang M W and Wu H 2014 *Key Eng. Mater* **633** 7-10
[5] Liu F, Wang X D and Cao J X 2013 *Mater. Int. J. Miner. Metall. Mater* **20** 88-93
[6] Hartmann A, Schulenberg D and Buhl J C 2015 *J. Mater. Sci. Chem. Eng* **3** 39-55
[7] Spudulis E, Šavareika V and Špokauskas A 2013 *Mater. Sci-medzg* **19** 190-6
[8] Zou J J, Guo C B, Wei C D and Jiang Y S 2015 *Res Chem Intermed* **18** 1-12
[9] Guo X Y, Ma S H, Lü S Q, Zheng S L and Zou X 2015 *Chin. J. Nonferrous. Metals* **25** 534-44
[10] Yue H Z, Wang X, Yang Z Z and Wei C C 2017 *Key. Eng. Mater* **726** 569-75
[11] Black L, Garbev K and Stumm A 2009 *Adv. Appl. Ceram* **108** 137-44
[12] Liang H X and Li M Q 2002 *J. Chin. Ceram. Soc* **30** 294-9

Acknowledgments
We gratefully acknowledge the supports of this work by the National Natural Science Foundation of China (Grant No. 21666007), Science Technology Foundation of Guizhou (Grant No. 2014-2007) and Program for "Hundred" High-level Talents in Guizhou (Grant No. 2016-5655).

IOP Conf. Series: Materials Science and Engineering **284** (2017) 012022 doi:10.1088/1757-899X/284/1/012022

Effect of calcium chloride concentration on output force in electrical actuator made of sodium alginate gel

Yuda Wu[1] Gang Zhao[1], Chengye Wei[1], Shuang Liu[1], Yu Fu[1] and Xvxiong Liu[1]
1 Harbin Engineering University, College of mechanical and electrical engineering, 15001 Harbin, China
E-mail: zhaogang@hrbeu.edu.cn

Abstract. As a kind of artificial muscle intelligent material, the biological gel electric driver has the advantages of low driving voltage, large strain, good biological compatibility, good flexibility, low price, etc. The application prospect is broad and it has high academic value. Alginate, as a common substance in sea, has characteristics of low cost, green and pollution-free. Therefore,this paper obtains biological gel electric actuator by sodium alginate and calcium chloride. Effects on output force of the electric actuator is researched by changing the crosslinking of calcium chloride concentration and the output force enhancement mechanism is analyzed in this paper.

1. Introduction

New materials have been constantly discovered and applied in the past decades. These can contract and elongate when the parameter condition changes, such as electric field intensity [1, 2], magnetic field intensity [3], illumination intensity [4, 5], temperature [6] and pH [7]. Due to the limitations of materials, artificial muscle research has developed slowly until the emergence of a new type of material-Electroactive polymers (EAP). The strain produced by EAP is two orders of magnitude larger than that of electroactive ceramics, moreover, it has faster response speed, smaller density and larger resilience than the shape memory alloy. In addition, it also has high tensile strength with similar biological muscle and inherent vibration damping performance [8]. The emergence of EAP has a new impact for artificial muscle. Artificial muscle actuators based on electroactive polymer materials have been developed rapidly from the beginning of the last century at the beginning of 90s. Under the influence of an electric field, the electric stimulus responsive smart gel can be produced reversible bending deformation or volume change, this is due to the transformation of electrical energy into mechanical energy. This feature can be used in artificial muscle actuators[9, 10]and drug controlled release system[11, 12, 13]. In this paper, sodium alginate and calcium chloride are used as natural polymer EPA materials due to their characteristics: green environmental protection, non-toxic, low cost, biodegradable etc, they have very important research value and play a significant role in the field of artificial muscle developing new materials.

2. The preparation of the electric actuator based on the crosslinking of sodium alginate and calcium chloride

Electric actuator preparation process is as follows：

Measured out 1.5g sodium alginate and 50ml distilled water into the beaker and put the beaker into water bath cup heating and stirring at 50℃, which sodium alginate is fully dissolved in distilled water.

b) More than the concentration of 1 mg/mL calcium chloride solution pour slowly in the concentration of 30mg/mL sodium alginate solution, which rapidly becomes gelatinous solid, calcium chloride and sodium alginate crosslinked excessively at this time. Less than the concentration of 0.75mg/ml calcium chloride solution pour slowly into the concentration of 30 mg/ml sodium alginate solution, the solution formed without gel phenomenon, which is not completely crosslinking. Therefore, we selected the concentration of 0.75 mg/ml, 0.8mg/ml, 0.85mg/ml, 0.9mg/ml, 0.95mg/ml and 1mg/ml calcium chloride and sodium alginate crosslinked. Put 0.075g, 0.08g, 0.085g, 0.09g, 0.085g and 0.1g anhydrous calcium chloride and 100ml distilled water into a beaker respectively. Put beakers at 50℃ water bath heating and stirring, until anhydrous calcium chloride was fully dissolved in distilled water.
c) The sodium alginate aqueous solution in the preparation of step a is slowly added to step b in different concentrations of calcium chloride aqueous solution making the alginate and calcium chloride crosslinked.
d) The fully crosslinking electric actuator film solution is put into the ultrasonic cleaning machine by shock and deaeration, then poured into a glass mold, put inside with the vacuum drying oven and set the drying temperature is 50℃ for dry. The actuator film can be formed with different concentrations of sodium alginate and calcium chloride after crosslinking.
e) Through the electrode and electrically actuated film paste assembly, electric actuator made by different concentrations of calcium chloride and sodium alginate gel were prepared which used in electric actuator output force test.
The preparation of the main experiment materials of electric actuator: sodium alginate powder (National medicine group chemical reagent limited company)、anhydrous calcium chloride (Analytical reagent, Heng Xing reagent)、glycerin (Analytical reagent, Tianjin Dong Li chemical reagent factory)、
distilled water.
Main experimental apparatus of the preparation of electric drive are used in table 1, in addition, the experimental instruments have also used glass mould, filter paper, measuring cylinder, a beaker and so on.

Table 1. The main instruments used in the experiment

Instruments	Model	Manufacturer
Magnetic Stirrer	MS300	Shanghai Special Instrument Manufacturing Co., Ltd.
Vacuum Drying Oven	DZF-6020	Shanghai Bo Xun Industrial Co., Ltd. Medical Equipment Factory
Analytical Balances	FA1004	Shanghai On Flat Instrument Co., Ltd.
Ultrasonic Cleaning Machine	LT-05C	Dragon Standard Electronic Co., Ltd.

3. Changes of calcium chloride concentration impact on electric actuator's output force

In this experiment, the output end of the electric actuator was putted on the analytical balance and applied voltage keeping for 120s. Due to the electric actuator had the tendency of macroscopic deformation and would produce an output force. Output force data can be measured by analytical balance, and reading data through the Matlab. The output force was selected as the stationary value in 120s period.
In this experiment, the assembled electric actuator film was cut into a rectangular structure with the width of 5mm and the length of 35mm, the DC voltage of 3 V was applied to the clamping end. The output data was read by Matlab and plotted by origin. The output force of the electric actuator produced by crosslinking the different concentrations of calcium chloride and sodium alginate was obtained. As can be shown in Figure 1.

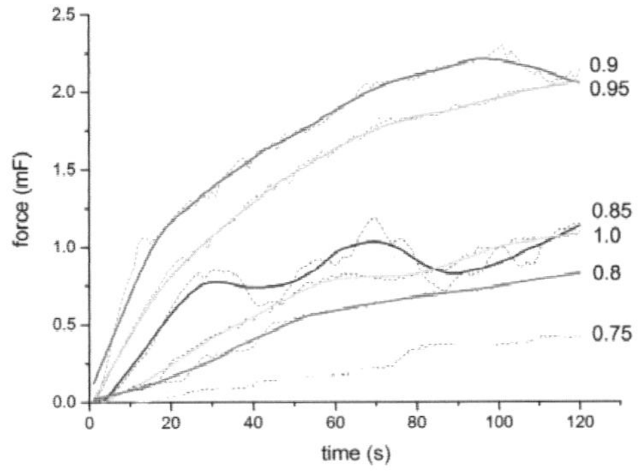

Figure.1 Output force and time change curve of electric actuator linear fitting chart

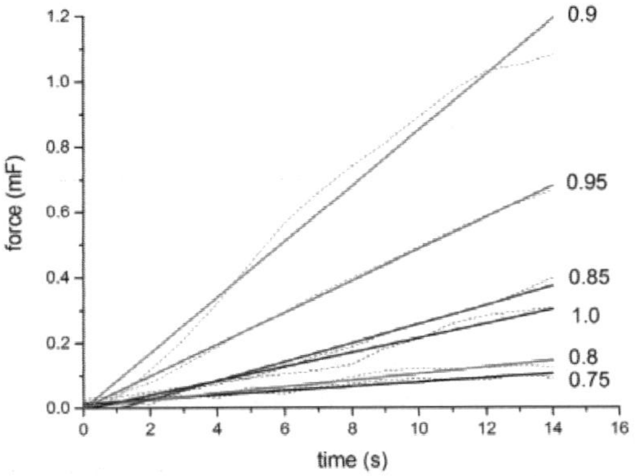

Figure.2 Electric actuator force output and time linear fitting chart

By analyzing the trend of curve Fig2. Crosslinking with different calcium chloride concentration had an effect on the output of electric actuator. When the concentration of calcium chloride was 0.85mg/ml, the output power performance curve of the electric actuator would have a significant shock phenomenon after 30s. The other calcium chloride concentration's electric actuator's output force would rise rapidly in 15s , gently rose range of 15s to 80s and kept the original level after 80s. Therefore, the output force of different calcium chloride concentrations at the front 15s are fitted in line, as shown in Figure 3. The slope was the response speed of the electrical actuator.

Through the data collection and processing, it could be obtained the change regulation of output force of different calcium chloride concentration crosslinked with sodium alginate. As shown in Table 2.

Table 2. Electric actuator output force changes with the different concentration of calcium chloride crosslinking

Calcium Chloride Concentration (mg/mL)	0.75	0.8	0.85	0.9	0.95	1
Output Force (mN)	0.422	0.827	1.148	2.304	2.075	1.104
Respond Speed (mN/s)	0.0066	0.0101	0.0290	0.0854	0.0486	0.0219
Quality (g)	0.118	0.102	0.103	0.105	0.117	0.110
Output Force Density (mN/g)	3.576	8.108	11.146	21.943	17.735	10.036

The change rule of the output force and response speed of the electric drive with different concentration of calcium chloride was drawn by Origin, as shown in Figure 3(a), the change rule of the output force density of the electric driver with the crosslinking of different calcium chloride concentrations is shown in Figure 3(b).

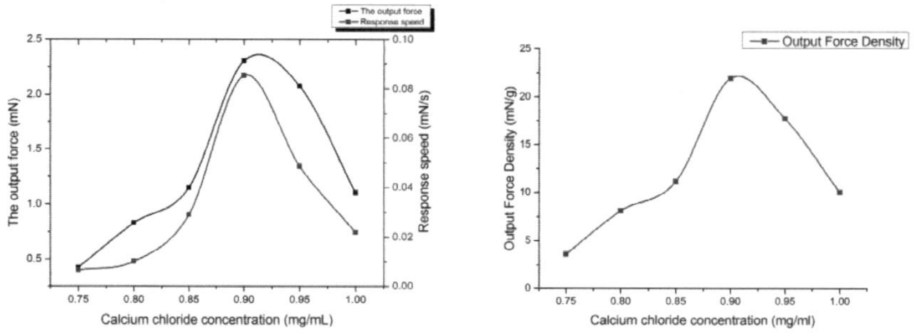

a) b)

Figure.3 The rule of different calcium chloride concentration crosslinked on the output force of the electric actuator: a)The change of the output force and its response speed with the crosslinking of calcium chloride, b)The change of the output force density with the crosslinking of calcium chloride

As can be seen from Figure 3 (a), when the calcium chloride concentration rose from 0.75mg/ml to 0.9mg/ml, the output force increases gradually, at 0.9mg/ml, the output force reached maximum, and the output force decreases when 0.95mg/ml concentration,at the concentration of 1mg/ml, the output force was almost equal to the output force at the 0.85mg/ml concentration. The response speed of the output force was almost the same as the variation trend of the output force. At the concentration of 0.9mg/ml, the response speed of the output force reached the sshowed a downward trend. At 1mg/ml concentration, the response speed of the output force was slightly slower than the response rate under 0.85mg/ml concentration, but faster than the 0.8mg/ml concentration. As can be seen from Fig. 3 (b), the variation trend of the output force density was almost the same as that of the output force. When the concentration changes from 0.75mg/ml to 0.9mg/ml, the output force density increases gradually, and the output force density reached the maximum at the 0.9mg/ml concentration. At 0.95mg/ml concentration, the output force density showed a downward trend. At 1mg/ml concentration, the output force density was slightly lower than the 0.85mg/ml concentration, and the output force density was slightly higher than that of 0.8mg/ml.

4. Conclusion

When the calcium chloride concentration increases gradually, the electric actuator is crosslinked fully and its membrane structure is more compact. Fully crosslinked at concentration of 0.9mg/ml, ion channels can be improved, exerting the same voltage, the ions which in the electric actuator film are more easily passed through the ion channels. It improves the response speed of the electric actuator output force. When the concentration of calcium chloride crosslinking is more than 0.9mg/ml, the electrically actuated film is crosslinked fully. Excess calcium ions block ion channels, resulting in free movement of ions's speed through the ion channel became slow when exerted the same voltage. It makes the output response speed slow down. It also reduces the number of free moving ions through the ion channels, resulting in reduced output forces. Therefore, it is concluded that the concentration of calcium chloride affects the output force in the process of cross-linking with sodium alginate, and the best proportion of calcium chloride is 0.9mg/ml concentration crosslinked with 30mg/ml concentration alginate.

Calcium chloride and sodium alginate as the material preparation method of crosslinked electric actuator is feasible. It has great research value to provide new materials for electric actuator made of EAP.

This paper was supported by C.N. National Science Foundation(51675112 and 51275102, and the Fundamental Research funds for the Central Universities (HEUCFP201730).

IOP Conf. Series: Materials Science and Engineering **284** (2017) 012022 doi:10.1088/1757-899X/284/1/012022

5. References

[1] Chen Q, Sun Y, Qin L, Wang Q M 2013 *IEEE Transactions on Ultrasonics Ferroelectrics & Frequency Control* **60** 2142.
[2] Zrinyi M, Nakano M, Tsujita T 2012 *Smart. Mater. Struct.* **21** 371-8.
[3] ZrõÂnyi M 2000 *Colloid Polym. Sci.* **278** 98-103.
[4] Lendlein A, Jiang H, Junger O and Langer R 2005 *Nature* **434** 879-82
[5] Jiang H Y, Kelch S and Lendlein A 2006 *Adv. Mater.* **18** 1471-5
[6] Ponomarev A F, Krasilnikov V A, Vasiliev M and Lachinov A N 2003 *Tech. Phys.* **48** 1493-5
[7] Ng S R, Pang H, Chen P, Li C M and O'Hare D 2016 *Electroanal.* **27** 2745-52
[8] SG Wax , RR Sands 1999 *Proceedings of SPIE* **3669** 2-10
[9] Schreyer H B, EGbhart N, K J M and Shahinpoor M 2000 *Biomacromolecules* **1** 642-7
[10] Osada Y, Okuzaki H and Hori H, 1992 *Nature* **355** 242
[11] Kagatani S, Shinoda T, Konno Y, UFkui M, Ohmura T and Y.Osada 1997 *J. Pharm. Sci.* **86** 1273-7
[12] Kwon I C, Bae Y H and Kim S W 1991 *Nature* **354** 291-3
[13] Kim S Y, Lee Y M 2015 *J. Appl. Polym. Sci.* **74** 1752-61

IOP Conf. Series: Materials Science and Engineering **284** (2017) 012023 doi:10.1088/1757-899X/284/1/012023

Effect of cutting fluids and cutting conditions on surface integrity and tool wear in turning of Inconel 713C

R HIKIJI[1]

1 Faculty of Education, Kumamoto University, Kumamoto, Japan
E-mail: hikiji@educ.kumamoto-u.ac.jp

Abstract. The trend toward downsizing of engines helps to increase the number of turbochargers around Europe. As for the turbocharger, the temperature of the exhaust gas is so high that the parts made of nickel base super alloy Inconel 713C are used as high temperature strength metals. External turning of Inconel 713C which is used as the actual automotive parts was carried out. The effect of the cutting fluids and cutting conditions on the surface integrity and tool wear was investigated, considering global environment and cost performance. As a result, in the range of the cutting conditions used this time, when the depth of cut was small, the good surface integrity and tool life were obtained. However, in the case of the large corner radius, it was found that the more the cutting length increased, the more the tool wear increased. When the cutting length is so large, the surface integrity and tool life got worse. As for the cutting fluids, it was found that the synthetic type showed better performance in the surface integrity and tool life than the conventional emulsion. However, it was clear that the large corner radius made the surface roughness and tool life good, but it affected the size error etc. in machining the workpiece held in a cantilever style.

1. Introduction
The automobile industry has been making better use of the engines equipped with turbochargers which control the amount of the air and make the exhaust gas clean in order to cope with the present environmental restrictions. The trend toward downsizing of engines helps to increase the number of turbochargers around Europe [1]. As for the turbocharger, it is necessary for measures of the lower pollution gas of the automotive engines. And the demand for the lower pollution gas may be increasing in the future [2], [3]. The temperature of the exhaust gas is so high that the parts made of nickel base super alloy are used as high temperature strength metals [4]. The alloy is one of difficult–to-cut materials and many studies on it have been carried out so far [5]–[12]. However, there are very few practical techniques for using the actual automotive parts. This study dealt with Inconel 713C as the nickel base super alloy. The workpiece was machined in the same size as the actual automotive parts. The effect of the cutting fluids and cutting conditions on the surface integrity and tool wear was investigated, considering global environment and cost performance.

2. Experimental

2.1. Workpiece Material
The valve stem bars of Inconel 713C were used as the workpiece material. They are part of the westgate type turbocharger. Inconel 713C which is the cast type nickel base super alloy is used so that it can endure the high exhaust gas temperature which amounts to 900 °C [4]. The suitable condition of the precise machining for Inconel 713C is required so that the valve stem bars may acquire high

quality of the surface integrity. Grinding has been carried out as finishing process. However, external turning was used in this experiment, given the running cost and machining time. Since the shape of workpiece disturbs its correct chucking, the cantilever holding has to be used. As a result, the dimension error may give a little damage to surface integrity. Tables 1, 2 and 3 show the nominal compositions, the physical properties and the mechanical properties of Inconel 713C, respectively [13].

Table 1. Nominal chemical composition of Inconel 713C.

C	Cr	Mo	Nb	Al	Ti	Zr	B	Ni
0.12	12.5	4.2	2.0	6.1	0.8	0.1	0.012	bal

2.2. Experimental Conditions

In the experiment, CNC lathe was used as shown in Fig. 1. Under the cutting conditions shown in Table 4, the PVD coating positive throwaway inserts were applied to the Inconel 713C test bar, 8.1 mm in the external diameter and 68 mm in length. This coating layer is super-multi layered coating with ultra-thin (nanometer) layers of TiAlN and AlCrN, alternately stacked up to 1,000 layers, and superior in heat and wear resistance [14]. The projected part in the cantilever state from lathe chuck was 35 mm. Generally, in the case of machining for the heat-resisting alloy, the positive rake angle throwaway insert with a sharp cutting edge equivalent to the smallest possible honing is used for machining so that the cutting edge may not be filled with cutting heat. Thus, the rake angle was held constant at 8 degrees. Table 4 shows the cutting conditions. In this experiment, three types of cutting fluids were used: synthetic soluble as the sample A, an emulsion including a sulfur type extreme pressure additive as the sample B, a synthetic emulsion including a sulfur type extreme pressure additive as the sample C. Each feature is shown in Table 5. As a cutting fluid, each of the types was diluted 10 times. The cutting fluid was added to the cutting point at the rate of 10 L/min with the external nozzle. After machining, arithmetic average roughness Ra in the direction of the tool feed motion were measured with the surface measuring instrument by stylus method, at the position at which the circumference was divided equally into three. Furthermore, size error and roundness were measured with the coordinate measuring machine. Tool wear was measured with the toolmakers microscope. The number of cutting times was one in the previous study [15], but by varying it in this study, surface integrity and tool wear were investigate.

Table 2. Physical properties of Inconel 713C.

Density	Melting range	Specific heat			Thermal conductivity		
		At 21°C	At 538°C	At 1093°C	At 93°C	At 538°C	At 1093°C
(g/cm^3)	(°C)	(J/kgK)	(J/kgK)	(J/kgK)	(W/mK)	(W/mK)	(W/mK)
7.91	1260-1290	420	565	710	10.9	17.0	26.4

Mean coefficient of thermal expansion		
At 93°C	At 538°C	At 1093°C
(×10-6/K)	(×10-6/K)	(×10-6/K)
10.6	13.5	17.1

Table 3. Mechanical properties of Inconel 713C.

Tensile strength		0.2 yield strength		Tensile elongation	
At 21°C	At 538°C	At 21°C	At 538°C	At 21°C	At 538°C
(MPa)	(MPa)	(MPa)	(MPa)	(%)	(%)
850	860	740	705	8	10

Dynamic modulus of elasticity	
At 21°C	At 538°C
(GPa)	(GPa)
206	179

IOP Conf. Series: Materials Science and Engineering **284** (2017) 012023 doi:10.1088/1757-899X/284/1/012023

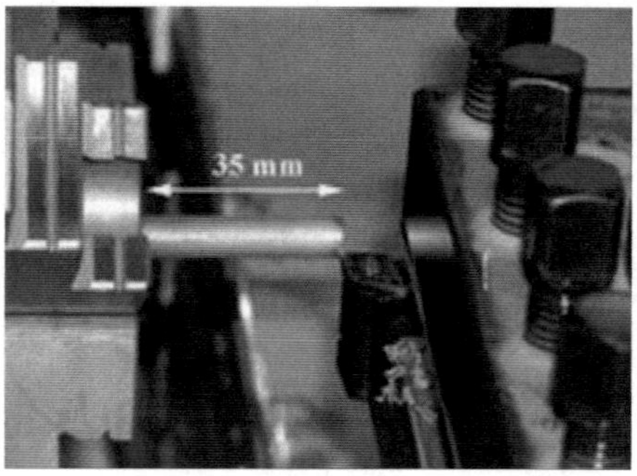

Figure 1. Cutting method.

Table 4. Cutting conditions

Diameter *D* (mm)	φ8.1
Rake angle γ (deg.)	8
Corner radius *rε* (mm)	0.2, 0.4
Depth of cut *a* (mm)	0.1, 0.5
Cutting speed *V* (m/min)	30
Feed rate *f* (mm/rev)	0.01
Cutting length *L* (m)	50, 100, 150, 200, 300

Table 5. Cutting fluids

Cutting fluids	Sample A	Sample B	Sample C
Type	Synthetic soluble	Emulsion	Synthetic emulsion
Density (g/cm^3(15°C))	0.98	0.938	1.012
Surface tension (mN/m(25°C))	32.9	34	25.2
S (Sulfur)	-	○	○
P (Phosphorus)	-	-	-
B (Boron)	-	-	-
pH	9.3	9.7	9.87

3. Results and Discussions

Figure 2 shows the relation between the cutting length and the arithmetic average roughness Ra. The figure (a) and the figure (b) show the depth of cut *a* is 0.1 mm and 0.5 mm, respectively. In this figure, the white mark and the black one show the corner radius *rε* is 0.2 mm and 0.4 mm, respectively. These display methods are the same in Figs. 2 to 6. In the cases of depth of cut *a* = 0.1 mm and 0.5 mm, it is found from this figure that the larger the cutting length *L* is, the worse Ra becomes. And the larger the corner radius *rε* is, the better Ra becomes. Under the same condition under which the feed rate *f* and depth of cut *a* are the same, it is found that the more the corner radius *rε* increases, the better the theoretical roughness becomes. As for the effect of the cutting fluids, the synthetic emulsion of the sample C was somewhat excellent. In addition, in the case of the sample C, the adhesion area cannot be confirmed in the cutting edge after machining, it seems to be superior in lubricity. It is considered that the wetting on the tip near the cutting edge and the surface of the work material performed efficiently, because of the smallest surface tension of the sample C in all the cutting fluids used this time. In the case of the large depth of cut, such as *a* = 0.5 mm, Ra became worse on the whole. However, the synthetic soluble showed as good results in the sample A as in the sample C. Therefore, in the case of these cutting fluids, it was found that Ra of *rε* = 0.4 mm settled within 1 μm even if the cutting length *L* was 300 m.

Figure 3 shows the effects of the cutting conditions on the size error. It means the difference in diameter between the workpiece measured by the end retainer setting with CNC lathe before and after machining. As for the effect of the cutting length, the more the cutting length *L* increases, the greater the size error becomes, like the previous result of the surface roughness, as seen from this figure. However, as for the effect of the corner radius *rε*, the more the corner radius *rε* decreases, e.g. at 0.2

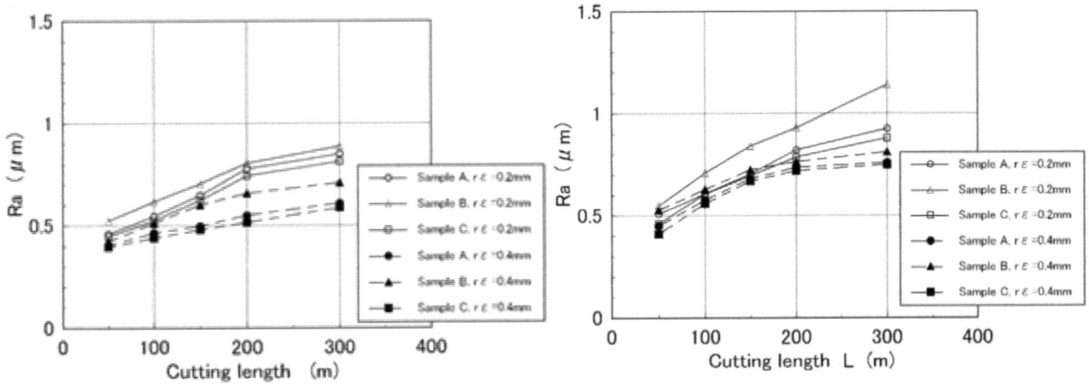

(a) a=0.1mm (b) a=0.5mm

Figure 2. Effects of cutting conditions on Ra.

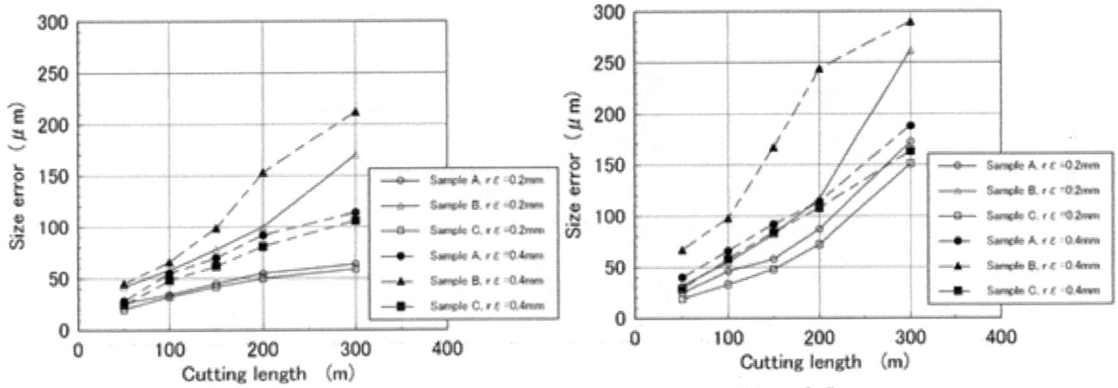

(a) a=0.1mm (b) a=0.5mm

Figure 3. Effects of cutting conditions on size error.

mm, the less the size error becomes, unlike the effect on the surface roughness. Furthermore, in the case of large depth of cut, a = 0.5 mm, it is found that the size error gets extremely worse. Since the workpiece was held and machined in the cantilever style, the deflection of the workpiece changed greatly. Thus, the more the corner radius increases, the more gradually the cutting force also increases. When the depth of cut a = 0.1 mm, the size error of $r\varepsilon$ = 0.4 mm by using emulsion is over 100 μm at the cutting length L = 150 m. However, the synthetic emulsion and the synthetic soluble enable the size error to keep below 100 μm up to the cutting length L = 200 m.

Figure 4 shows the effects of the cutting conditions on the roundness. It is found that the more the cutting length L increases, the more the roundness increases. In the case of the roundness, the more the corner radius $r\varepsilon$ decreases, the better roundness becomes, like the previously mentioned case of the size error. Furthermore, when the depth of cut a becomes larger, 0.5 mm, it is found that the roundness gets somewhat worse. As for the cutting fluids, the synthetic emulsion and the synthetic soluble are more effective to the roundness than the conventional emulsion. The synthetic emulsion and the synthetic soluble enable the roundness to keep below 20 μm up to the cutting length L = 300 m.

In this way, the effects of the cutting condition on the surface integrity after machining were showed. Figure 5 shows the photo of the tool wear in the case of the cutting length L = 200 m. From these figures, it is found that the tool wear by using the synthetic emulsion becomes smaller than by using the emulsion. Figure 6 shows the effects of the cutting condition on the tool wear. As for one of the causes that the large cutting length made the surface integrity get worse as shown in Figs.2 to 4, it is consider that the more cutting length increase, the more tool wear increases. At the cutting length L = 50 m, the tool wear of the depth of cut a = 0.5 mm tends to be somewhat larger than of a = 0.1 mm. In

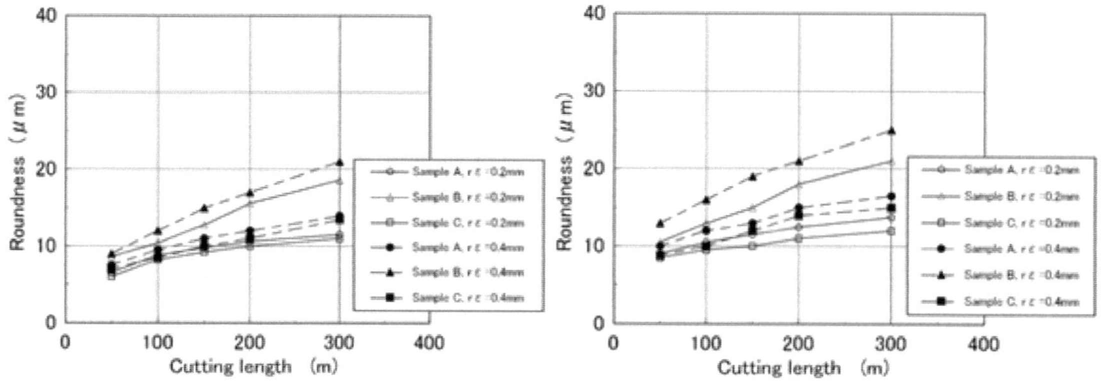

(a) a=0.1mm (b) a=0.5mm

Figure 4. Effects of cutting conditions on roundness.

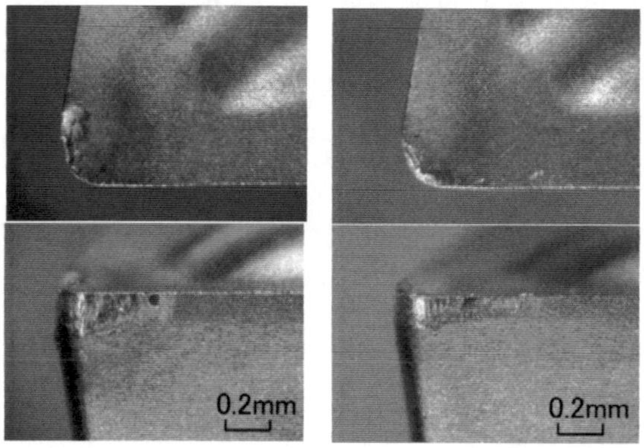

(a) Emulsion (b) Synthetic emulsion

Figure 5. Photo of tool wear (a = 0.5 mm, $r\varepsilon$ = 0.2 mm).

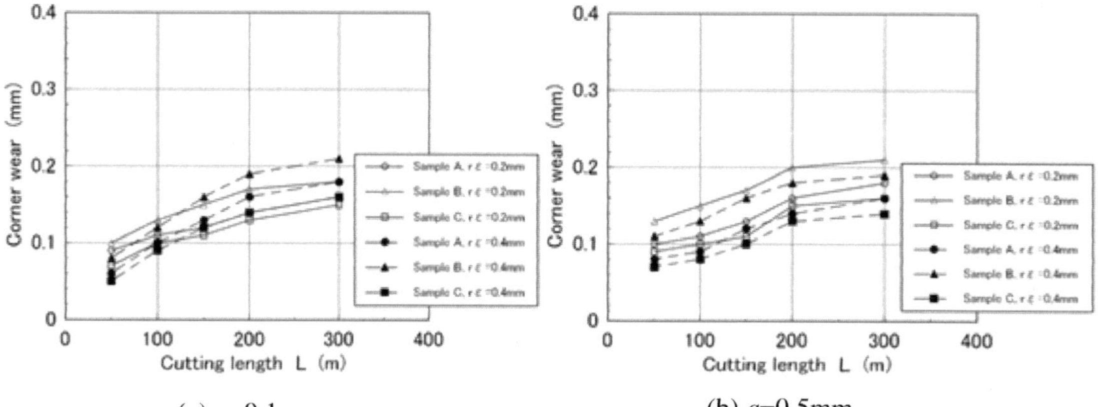

(a) a=0.1mm (b) a=0.5mm

Figure 6. Effects of cutting conditions on tool wear.

the case of the depth of cut a = 0.5 mm, on the whole, the more the corner radius increases, the more the tool wear decreases. In the case of a = 0.1 mm, when the cutting length increases, the tool wear of the large corner radius gets worse than of the small one. It is considered that when the cutting length increases, the cutting edge becomes dull, so the tool of the large corner radius tends to slip easily due to the small depth of cut. For all this, the synthetic emulsion and the synthetic soluble enable the tool wear to keep bellow 0.2 mm up to the cutting length L = 300 m.

4. Conclurion

External turning of Inconel 713C which is the cast type nickel base super alloy and used as the valve stem of the turbocharger parts was carried out. And the influence of the cutting condition on surface integrity was investigated. Furthermore, the surface integrity and tool wear were investigated experimentally by comparing the performance of the cutting fluids between the synthetic type and the conventional emulsion. As a result, in the range of the cutting conditions used this time, when the depth of cut was small, the good surface integrity and tool life were obtained. However, in the case of the large corner radius, it was found that the more the cutting length increased, the more the tool wear increased. When the cutting length is so large, the surface integrity and tool life got worse. As for the cutting fluids, it was found that the synthetic type showed better performance in the surface integrity and tool life than the conventional emulsion. However, it was clear that the large corner radius made the surface roughness good, but it affected the size error etc. in machining the workpiece held in a cantilever style.

5. References

[1] Mitsubishi Heavy Industries Ltd 2016 *Mitsubishi Heavy Industries Graph* 180 pp 14-15
[2] Ebisu M, Terakawa Y and Ibaraki S, 2004 Mitsubishi Turbocharger for Lower Pollution Cars *Mitsubishi Heavy Industries Ltd. Technical Review* **41** 1 pp 40-41
[3] Satake K, Yamada S, Yanagisawa M, Monaka T, Endo H and Abe T 2008 The Rapid Development of Diesel Engines Using an Optimization of the Fuel Injection Control *Mitsubishi Heavy Industries, Ltd . Technical Review* **45** 3 pp 6-9
[4] Nishino K, Kawaura H, Tanaka K, Horie T, Saito T and Uchida H 2000 High Performance Alloys Developed for Turbochargers *R&D review of Toyota* **35** 3 pp 27-34
[5] Choudhury I A and El-Baradie M A 1998 Machinability of nickel-base super alloys: a general review *J. Mater. Process. Technol.* **77** pp 278-284
[6] Ezugwua E O, Wanga Z M, and Machado A R 1999 The machinability of nickel-based alloys: a review *J. Mater. Process. Technol.***86** pp 1-16
[7] Hanasaki S, Fujiwara J, Sakagami K and Hasegawa Y 1989 Cutting of a High-Nickel Alloy (1st Report, Mechanism of Flank Groove Wear Formation) *J.Jpn.Soc.Mech.Eng. Ser. C* **55** 511 pp 780-786
[8] Ezugwu E O and Bonney J 2004 Effect of high-pressure coolant supply when machining nickel-base, Inconel 718, alloy with coated carbide tools," *J.Mat.Proc.Tech.* **153–154** pp 1045-1050
[9] Kamata Y and Obikawa T 2007 High speed MQL finish-turning of Inconel 718 with different coated tools *J.Mat.Proc.Tech.***192–193**, pp 281-286
[10] Shinde S N, Vankudre H V and Thakur D.G 2016 A comparative study of cutting force, feed force, surface roughness and tool wear in machining of Inconel 718 with uncoated and coated tungsten carbide inserts *Int. J. Appl. Sci. Eng. Res* **5** 1 pp 30-38
[11] Swain N, Venkatesh V, Kumar P, Srinivas G, Ravishankar S and Barshilia H C 2017 An experimental investigation on the machining characteristics of Nimonic 75 using uncoated and TiAlN coated tungsten carbide micro-end mills *CIRP J. Manuf. Sci. Technol.***16** pp 34-42
[12] Wang F, Li L, Liu J and Shu Q 2017 Research on tool wear of milling nickel-based superalloy in cryogenic *Int. J. Adv. Manuf. Technol.* pp 1-10
[13] AMS INTERNATIONAL 1990 *METALS HANDBOOK* **1** pp 982-985
[14] Shubata A, Fukui H and Tsuda K 2008 Development of ACE-COAT AC510U/AC520U for Machining of Exotic Materials *SEI TECHNICAL REVIEW* 67 pp 7-10
[15] Hikiji R 2014 Study on Improvement of Surface Integrity in Cantilever Bar Turning of Inconel 713C *Proc. 15th Int. Conf. Prec. Eng* pp 513-517

Acknowledgments
The author would like to thank Daido Precision Industries Ltd. for their support that enabled this work to be carried out.

IOP Conf. Series: Materials Science and Engineering **284** (2017) 012024 doi:10.1088/1757-899X/284/1/012024

Fabricating the spherical and flake silver powder used for the optoelectronic devices

Ju Wei[1], Ma Wangjing[2], Zhang Fangzhi[1], Chen Yixiang[2], Xie Jinpeng[2]

1 Advanced Material Institute of Shandong Academy of Sciences, Jinan, China
2 Technical Institute of Physics and Chemistry of Chinese Academy of Science, Beijing, China
E-mail: quwei8175@126.com

Abstract. The spherical and flake silver powder with different particle size for the optoelectronic devices was partly prepared by using chemical reduction and ball milling method, and charactered by scanning electron microscope (SEM), X-ray diffraction (XRD), laser particle size analyzer and thermo-gravimetric(TG) analyzer. The particle size of three series of spherical silver powder fabricated by chemical reduction is about 1.5μm, 1μm and 0.6μm, respectively; after being mechanical milling, the particle size of flake silver powder with high flaky rate is about 10μm, 6μm and 2μm respectively. Thermo gravimetric (TG) and XRD analyses showed that the silver powders have high purity and crystalline, and then the laser particle size and SEM analyses showed that the silver powders has good uniformity.

1. Introduction

As a key type of material of the optoelectronic devices, the silver powder (SP) is widely used in electronic industry, such as photovoltaic battery, membrane switch, wave filters and other electronic equipments [1-6] et al. The physical performance of SP has an important influence on the photoelectric devices. In recent years, researchers had done a lot of work in preparing the SP by various methods. Wang H. first prepared the spherical and mono-disperse silver powder with average particle size of about 1~2μm by chemical reduction [7].Liang H. Z. fabricated the hexagonal sheet silver with the thickness of less than 0.1μm, and the maximum projection length of 0.2~0.5μm in the medium of glycol [8]. Chen S H and his workers fabricated nanodisc with the thickness of 20~30nm and the diameter of 40~300nm in the CTAB solution containing silver crystal nucleus [9].

In the paper, the spherical and flake silver powder were prepared by chemical reduction and mechanical milling. And the influencing factors of fabricating the silver powder were investigated, such as surfactant concentration, reaction time etc.

2. Experiment

2.1 Reagent

Distilled water; Silver nitrate; Ammonia; Ascorbic acid; Ethanol; Polyvinylpyrrolidone (PVP, K30); Ethylene glycol; Oleic acid.

2.2 Fabricating the spherical silver powder

The PVP was dissolved in distilled water to form the reaction medium. Then the silver nitrate was dissolved in the PVP solution to form 0.1mol.L^{-1} silver nitrate solution. Finally the solution was confected into the silver ammonia solution. At the temperature of 80℃, the reducer solution was added into the silver ammonia solution at even rate and stirred at the same time. After the reaction finished, the silver powder was obtained by centrifugation, then washed by distilled water and ethanol successive for several times, finally dried in a vacuum oven.

2.3 The fabrication of the flake silver powder

During the ball milling, ethanol, oleic acid and the agate ball of a diameter of 1mm acted as the milling medium, the assistant, and the milling ball. The mass ratio of the milling ball to the spherical silver powder, and to the spherical silver powder was 12:1and1:1, respectively. The milling speed achieved 500rad/min.

When the ball milling finished, ethanol was used to clean the silver powder adhering to the surface of milling ball. Separated silver powder was dried in vacuum drying oven.

3. Result and discussion

3.1 The influence of the PVP concentration

PVP could chelate with silver ions, change the reduction potential of oxidation-reduction reaction, thus nucleation and growth rate of silver particles was reduced. In addition PVPs which adsorbed on the surface of silver particles reduce the surface energy of crystal nucleation and make it consistent in all directions. In this case the silver atoms could grow evenly on the surface of crystal nucleus, which results in that the SPs with uniform size and spherical morphology were easily prepared. Furthermore, there was steric hindrance effect which made the potential barrier among the particles increase rapidly due to the PVP adsorbed on the surface of the particles. Thus PVP could make the SPs be monodisperse during reaction, solid-liquid separation and vacuum drying. Reaction was shown as follows:

$$Ag^{+} + Vc- + H_2O \rightarrow Ag + Vc- + H^{+}$$

As shown in Figure 1 (a), (b) and (c), the SP was prepared when 2%, 5% and 10% of PVP was added to the solution, respectively. The SP's size decreases with the increase of PVP concentration. The reason was that with the increasing of PVP concentration, the more PVP covered on the surface of SPs to prevent the crystal particles from growth. It was estimated that the concentration of PVP was inversely proportional to the square of the SP's size.

The SP with a diameter of about 1.5μm was tested by the laser particle size analyzer, as was shown in Figure 2. Compared with SEM characterization, it also testified that the SP had the property of monodispersity.

3.2 The Influence of the milling time

There were SEM photographs of the three kinds of SP milled for 8h and 15h, respectively in Figure 3. The flake rate was higher and higher with the increase of the milling time. In addition, the forming process of the flake silver powder was analyzed from the photographs. Firstly, large bulk particles formed gradually under the external force. With the increase of the milling time, the number of spherical particle was getting smaller and smaller, and vanished eventually. Then, the bulk particles were squeezed by the milling ball, and began to be thin. Finally, the flake silver powder formed after enough milling time.

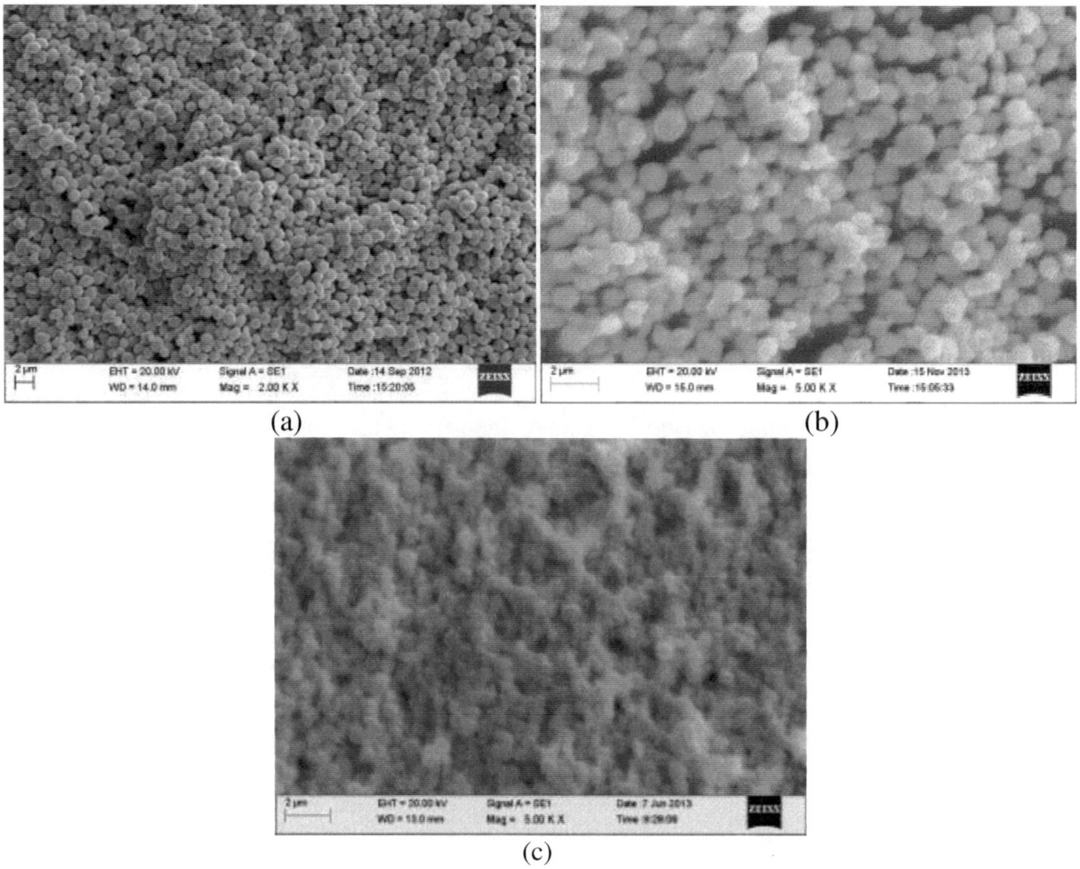

Figure 1. The SEM photographs of silver powder reduced in the case that the mass percentage concentration of PVP was 2% (a), 5% (b) and 10% (c)

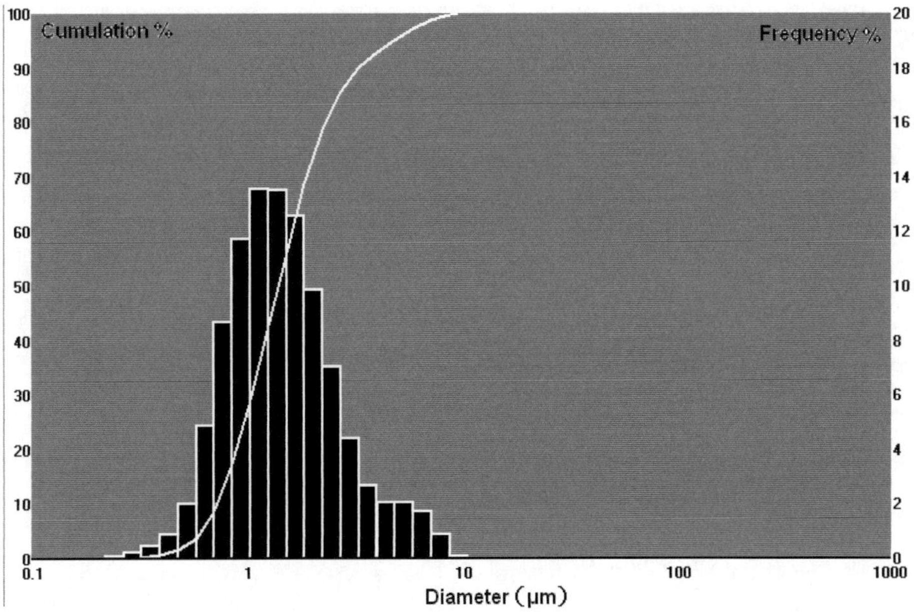

Figure 2. The result of silver powder by the laser particle size analyzer

(e) (f)

Figure 3. The SEM photographs of the silver powder with 1.5μm, 1μm and 0.6μm particle size milled for8h (a), (b) and (c); 15h(d), (e) and (f).

3.3 XRD characterization of silver powder

It indicates that characteristic spectrum of SP was completely accordance with the standard crystal card of silver in the graph (Ag65-2871), and there are no other obvious impurity diffraction peaks (Figure 4). It is illustrated that SP had the property of high purity and low impurity content.

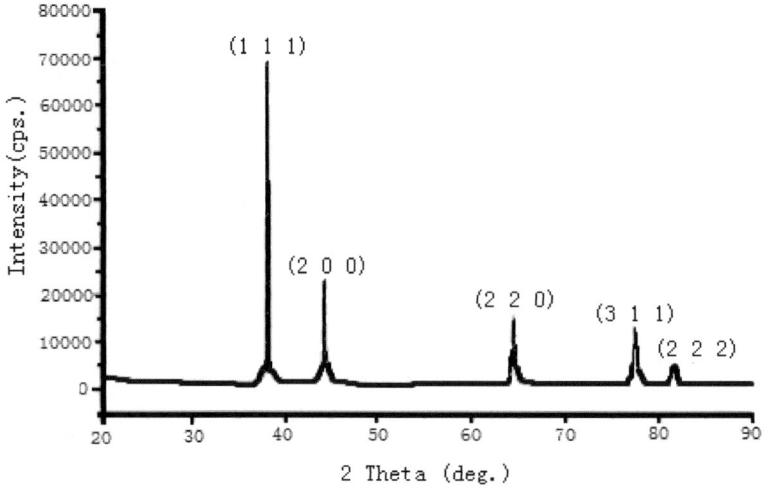

Figure 4. XRD pattern of silver particle

3.4 Thermo gravimetric analysis

The test result of thermogravimetric analyzer was shown in Figure 5. As seen, the quality loss of the flake silver powder was tiny during the process of heating-up, and the rate of slope of the differential of the heating-up curve was close to zero. It indicated that the amount of organic components adsorbed on the surface of SPs was very small. So, the flake silver powder was favorable to the preparation of conductive adhesive and the electrical performance of circuit.

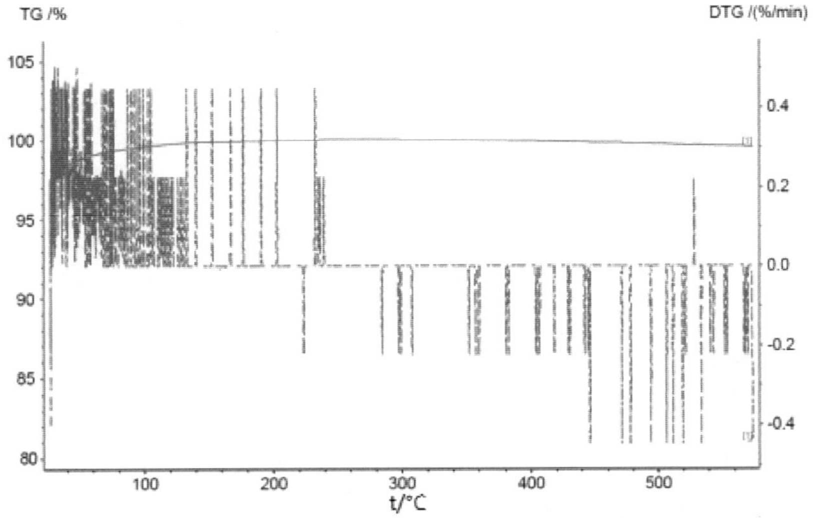

Figure 5. Thermogravimetric result of the flake silver powder

4. References

[1] Kushal D. Bhatte, Krishna M. Deshmukh, Yogesh P. Patil, Dinesh N. Sawant, Shin-Ichiro Fujita, Masahiko Arai and Bhalchandra M. Bhanage 2012 *Particuology* **10** 140

[2] P. Logeswari, S. Silambarasan, J. Abraham 2013 *Scientia Iranica* **20** 1049

[3] Chen D.M., Liang Z.C., Zhuang L., Lin Y.H. and Shen H 2012 *Appl. Energy* **92** 315

[4] Gholam Reza KHAYATI, Kamal JANGHORBAN 2013 *Transactions of Nonferrous Metals Society of China* **23** 1520

[5] Yaobin Lai, Hui Huang, Qian Huang, Hongyan Zhang and Zhongcheng Guo 2014 *Powder Technology* **263** 7

[6] Rajesh Warluji Raut, Vijay Damodhar Mendhulkar and Sahebrao Balaso Kashid 2014 *Journal of Photochemistry and Photobiology B: Biology* **132** 45

[7] Zhao Liu, XueLiang Qi and Hui Wang 2012 *Advanced Powder Technology* **23** 250

[8] Liang Huangzhen, Dong-jin Kim, Yu Kening, Hun S. Chung, Zhang Jie, Li Shaohua and Li Ruixing 2003 *Powder Metallurgy Technology* **21** 218

[9] Chen S, Carroll D L 2004 *The Journal of Physical Chemistry B* **108** 5500

Acknowledgments

The authors acknowledge the financial supports of Natural Fund of Shandong Provincial (ZR2015YL002). The authors are also grateful to Shandong Analysis and Test Center.

The efficiency analysis on assistant-grinding of lignosulfonate and its modified composites

Duan Hui, Mu Xiangfeng and Wang Ying
Zibo high tech Industrial Development Zone Refractory Engineering Research Institute, Zibo China
E-mail: hgdh2005@163.com

Abstract: The method of vibration ball milling is applied in this paper to make research on contents of lignosulfonate and modified composites as cementgrinding aids; and studied on the influence of regularity about grinding aids on products fluidity, particle size, specific surface area and powder particle size distribution. The study showed lignosulfonatehas had better effect on the cement grinding function; the cement's particle size and the distribution were improved, the average diameter was decreased, distribution narrowed and ultra-fine grain size increased, of the most importance, the contents of ultra-fine grain size were increased much. Among those, the effect of calcium lignosulfonate was relatively good while its modified composites performed better.

1.Introduction

The grinding efficiencycould be largely improved by cement grinding aids,reducing the cost of cement are the important research subjects in the field of Cement Science and Production Enterprise. During the production of cement, the grinding process is the most energy consuming procedure. (According to the report in America 50 billion kilowatt-hours was consumed in this field per year.) The energy usage was low. About 97% of its energy turned to be heat energy and is in vain. Only little proportion (0.6%-1%) is actually used in increasing surface energy of materials. As the development of dry production, the heat consuming is largely decreased, yet the electricity consuming has been increased. Therefore, how to raise the efficiency of grinding and reduce its electricity consuming are the focus attention for the cement workers and the important energy-saving subject.

Since 1930s, cement grinding aids had been used in cement industry[1,2]. After that, there are a large number of patents emerging. At present, the study of grinding aids is widespread in America, Germany, France, Japan and Russia. Rules and checking methods aboutgrinding aids have been also established in those countries. While in China, study and application of grinding aids are underdeveloped; the occupation coefficient of grinding aids is under 5%[3]. The application of lignin is in the form of lignosulfonate[4]. Lignosulfonate is a byproduct produced by the waste liquor of paper pulp after processing sulfonation and spray drying. It is widely used as a normal superplasticizer by most people. Lignosulfonate has better water-reducing property. If added in cement, it can improve its grinding effectand function as the superplasticizer.

2. Experiment

2.1 Raw material and regent

Done-material: rotary kiln clinker in a factory, mine components in Table1

Mineral waste residue: copper slag in a certain factory, the quality factor is 1.67.

Gypsum: SO_3 35.48%, crystallized water 17.86%.
Sodium lignosulfonate (SLSF), magnesium lignosulphonate (MLSF) and calcium lignosulfonate (CLSF) provided by Tian JinShengfu Jiang Chemical Industry Corp, modified composite lignosulfonate (self-made).

Table 1 The components of mine [%]

KH	SM	IM	C_3S	C_2S	C_3A	C_4AF	∑
0.894	2.65	1.58	53.60	24.35	8.27	7.06	93.28

2.2 Experiment devices, testers and methods

2.2.1 Experiment devices and testers

Mastersizer 2000 particle size analyzer made in Malvern Corp.; vibrating ball grinder; NRJ-411A cement glue-sand blender; SHBY-20B cement concrete maintaining standard container; the other devices in Table2

Table2 Devices used in experiment.

Name	Model	Further explanation
jaw crusher		broken to the size of clinker<7[mm]
roll crusher		broken to the size of clinker<3[mm]
shaking table	GZ-85	amplitude 0.85[mm], frequency 50[HZ], vibration time 2[min]
Electronicbalance		maximum range 600[g], minimum range 0.01[g]
slide caliper		accuracy 0.2[mm]

2.2.2 Analytical methods

0.08mmsquare hole sieveused in determination of sieve residue, dry sieveon the basis of GB45-77 Method.
The stop-angle was measured by methods of equal height injection.
Particle size analyzed by MS2000G laser particle size analyzer, results output by computers.
Normal consistency of cement water and setting timemeasured by GB1346-1999.
Cement glue-sand strength test by GBIT1 7671-1999.

2.2.3 Test sample preparation

First crush particle size below 2mm by jaw crusherandroll crusher, then the gypsum and slag are sieved to 2mm sizes. The weight of the ball must be calculated according to the ratio 7:1 to decide the weight of raw material, then refer to the ratio done-material 80%: slag 15%: gypsum 5% to get the components and dilute lignosulfonate by water according to the proportion 0.2% and 0.4%. Add 0.05% modifiedlignosulfonategrindingaids byburette and grind 1 h by ball grinder. The weight and ratio of the ballshould be remaining the same each time. LSF will be the shorten form of lignosulfonate in the following description. The quantity of mixture and test sample code will be found in Table 3.

Table 3 The quantity of mixture and test sample No.

grinding aids	Blank	SLSF		MLSF		CLSF		modified composites of lignosulfonate		
								SLSF	MLSF	CLSF
Dosage of grinding aids[%]	—	0.2	0.4	0.2	0.4	0.2	0.4	0.05	0.05	0.05
Sample No.	A	B1	B2	C1	C2	D1	D2	E1	E2	E3

3. Results and discussion

3.1 Fluidity of cement

The stop-angle is the largest angle formed by free surface layer of powder accumulation in balance with the water surface. It can also be realized in an easy way if there's no specialized device. Get the average value from tests more than once. See the results in Table4.

The results of test indicated that different types of grinding aids reduced cement's the stop-angle to 1-6°. After adding lignosulfonate grinding aids, there's no distinctive change in decreasing the stop-angle as it was added more. Big changes about MLSF can be seen in different lignosulfonate. 0.4 % reduced the cement's the stop-angle from 43° to 40.3°was reduced. The fluidity of cement became more powerful because of adding modified composites of lignosulfonate grinding aids. The stop-angle was diminished to 6°.

Table4 Results of cement powder stop-angle.

Sample No.	A	B1	B2	C1	C2	D1	D2	E1	E2	E3
thestop-angle[°]	43	42	40	41	40	41	39	37	37	37

3.2 Cement ' s dispersion

3.2.1 Sieve residue

See the results of sieve residue of 0.08mm square hole sieveinTable5. The cement fineness was raised after adding grinding aids. The sieve residue of 0.08mm square hole sieve was about 6.5% without any grinding aids, but after adding some grinding aids, the sieve residue reduced by 1~2.5%. As moresinglegrindingaids increased with dosage, the sieve residue was reduced afterwards. Among those three singlelignosulfonate, CLSF owned the smallest the sieve residue; thesieve residue of SLSF is as much as MLSF. Altogether, there's no big difference among these three. But modified composites of lignosulfonate grinding aids'sthesieve residue is much smaller, only 1% the sieve residue, reduced by 5%. It contributed to its good grinding effect and small diameter of powder particles. The results of experiment indicated that adding grinding aids can diminish the contents of coarse particles.

Table5 the sieve residue of 0.08[mm] square hole sieve[%].

Sample No.	A	B1	B2	C1	C2	D1	D2	E1	E2	E3
>0.08[mm]	6.5	5.6	4.7	5.1	4.4	4.9	4.1	1.0	0.9	0.8

3.2.2 Specific surface area

Figure1-Figure4 showed specific surface area of different sample after grinding 1h. From that, we can see the cement's specific surface area was increased after using some grinding aids. Grinding for 1 hour, the specific surface area of cement was 345m^2/kg without grinding aids. The differences of specific surface area among those three lignosulfonate single grinding aids were not clear. B1, C1, D1 are about 375m^2/kg, raised by 8%; B2, C2, D2 are about 400m^2/kg, raised by 16%. Compared with single grinding aids, the specific surface area of three modified composites of lignosulfonate grinding aids was increased obviously, reached 440m^2/kg, to 30%.

3.2.3 The average diameter of cement ' s powder particles and its distribution

The average diameter of cement's powder particles and particle distribution were changed various after adding grinding aids. Figure2 to Figure4 are comparison about distribution of different LSF with diverse dosage, modified composites of lignosulfonate and particle size of blank sample. After adding grinding aids, the average diameter of powder particles (D50) was decreased. Without grinding aids, D50 was 22.3 μm, but after adding single lignosulfonate for 0.2%, D50 decreased to about 19μm, and when the dosage increased to 0.4%, D50 reached to 18μm. However, because of using modified composites grinding aids, D50 decrease to 15μm. It can be showed from the distribution of powder particles, after using grinding aids, the percentage in each parts of cement was decreased. For lignosulfonate single grinding aids, cement's particle of 0.4% dosage got thinner one than that of

0.2%.

Compared with these three LSF, CLSF had betterassistant-grinding performance, while no significant difference between SLSF and MLSF. The contents of powder particles45μm for three LSF are almost the same. The main difference is about their particle size, that's 3-45μm, the cement's particle size was 3-45μm with adding 0.2% MLSF and SLSF. It occupied 67%. The percentage was 69% with 0.2% CLSF and particle size 3-45μm. With0.4% SLSF and CLSF's particle size was3-45μm. The percentage was 68%, and it was the same about dosage to be 70% CLSF. Figure 4 told that modified compositesgrindingaids had better grinding effect than single grinding aids. Powder particles with more than 60μm particle sizewas only less than 5%, while there was 9% of powder particles 0.4% CLSF and its particle size was 60μm. Modified composites grinding aids particle size was 3-45μm with about 74% powder particles which was 70% higher than CLSF. Modified composites grinding aids about every particle size'scumulative distribution was much higher than LSF. It illustrated that modified composites grinding aids's fine particles were more than LSF during every period of time.

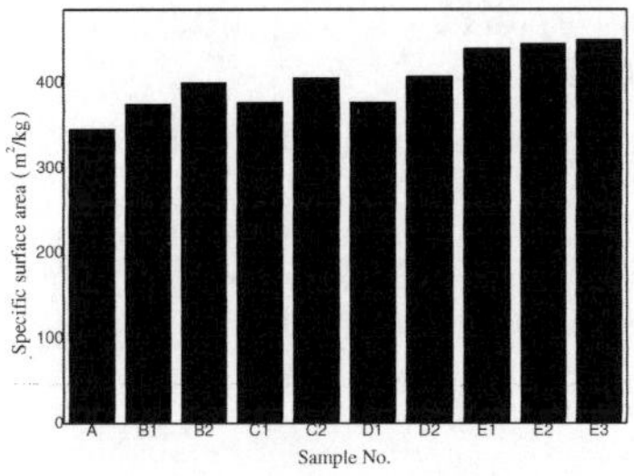

Figure 1 Comparison of specific surface area after grinding 1h.

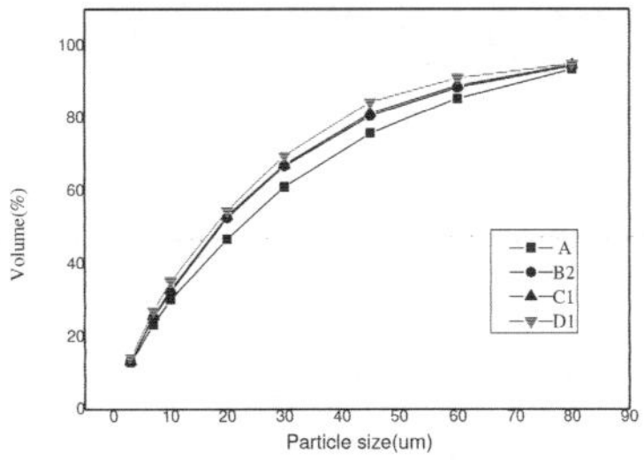

Figure 2 The comparison of cumulative distributionof 0.2% dosage LSF.

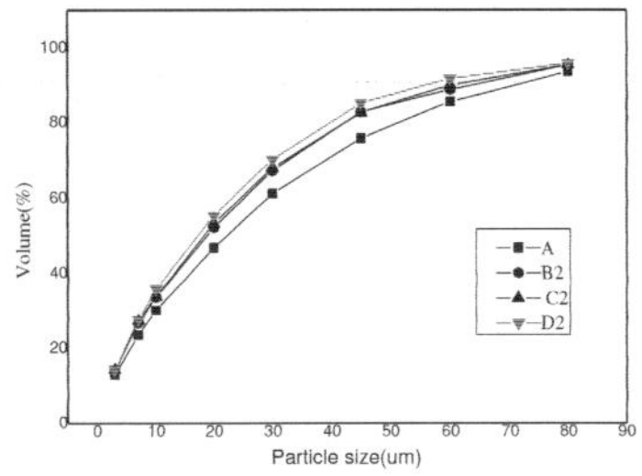

Figure 3 The comparison of cumulativedistributionof 0.4% dosage LSF.

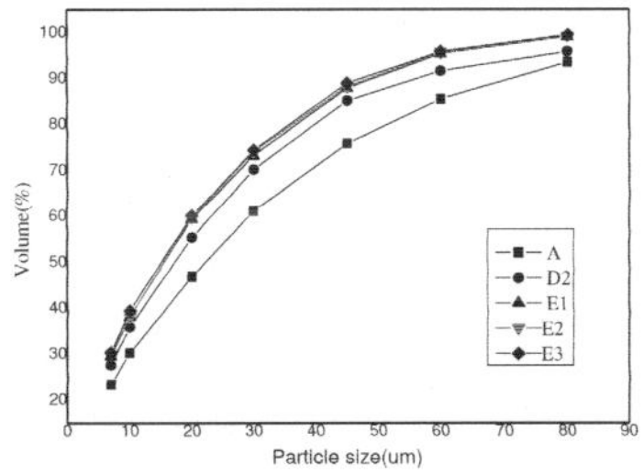

Figure 4 The comparison of cumulative distributionofmodified composites grinding aids.

4. Conclusion

Lignosulfonatesinglegrinding aids and its modified composites grinding aids can show an evident increasing fluidity of cement. The particle size of powder particles was raised to a new height, the cement's specific surface area was enlarged and especially the proportion of particle size between 3-30μm which played an important role in the strength of cement was enhanced. What's more, the cement particle size distribution of cement was improved a lot. As the increasing of dosage, lignosulfonate's grinding effect was more efficient accordingly. The best dosage was 0.3%-0.4%. Among those three lignosulfonate single grinding aids, CLSF got the best grinding effect; the grinding effect of modified composites grinding aids was better than single grinding aids . The modified compositesof CLSF got the best grinding effect.

References

[1] Leo Liberthson 1943 *Petroleum hydrocarbon emulsions, USA US2,307,744*

[2]Bruce A Lange, Ham pstead N H 1983 Additive combination for hydraulic cement compositions:*USA, US4,375,987*

[3] Zang Donglan, Xu Huayi, Zhang Yongtian2002Applied effectiveness analysis of several grinding aids.*Cement*, 4 p 23-25

[4] D. Touil, S. Belaadi, C 2006 Frances Energy efficiency of cement finish grinding in a dry batch ball mill, *Cement and Concrete Research*,36 p 416-421

IOP Conf. Series: Materials Science and Engineering **284** (2017) 012026 doi:10.1088/1757-899X/284/1/012026

Development of corrosion resistant surfaces via friction stir processing for bio implant applications

Gurvinder Pal Singh Sodhi[1,3] and Harpreet Singh[1]
[1]Department of Mechanical Engineering, Indian Institute of Technology Ropar
Rupnagar, Punjab, India
3Corresponding author email gurvinderpss@iitrpr.ac.in

Abstract. Current investigation presents the use of Friction Stir Processing (FSP) to improve the corrosion resistance of pure magnesium for biomedical applications. FSP has been used to incorporate hydroxyapetite (HAP) into Mg-surface so as to modify the chemical composition. FSP was done within a matrix of different parameters and conditions. Influence of various parameters on microstructure was also clearly observed. XRD analysis confirmed the presence of HAP, whereas SEM images revealed a uniform distribution of the imbedded phase. Micro-hardness and in-vitro corrosion studies were also performed. Influence of grain size on hardness was validated by Hall-Petch relationship. Corrosion behavior was explained on the basis of texture, which indicated better corrosion resistance in comparison to the pure Mg. Therefore, the study reveals that the proposed FSP methodology can be useful tool to improve mechanical and corrosion properties of pure Mg for biomedical applications.

1. Introduction

Biodegradable metals and alloys are being considered as the second generation of biomaterials. In this context, due to its close resemblance with the human bone in terms of mechanical properties, Mg has the potential to be amongst one of the best materials, which can be incorporated in human body. However, its rapid degradation is still a matter of serious concern. Many researchers [1-2] have been trying to address the problems pertaining to hydrogen evolution and high corrosion rates of Mg implants. According to literature [3], Mg implants are being widely used as cardiovascular stents, fixation plates etc. Due to the fact that, the properties of Mg can be engineered easily, it is highly beneficial to use it for customized biomedical applications. In this context, (FSP) approach has been chosen, which is a solid state process wherein the material is subjected to intense plastic deformation by a rotating tool. This in turn refines the grain structure of the metal/alloy by re-crystallization. In the recent past, agglomeration and uneven distribution of HAP powder in pure Mg during friction stir processing, have been identified as the key problems to be addressed [4-5]. It has been concluded that by playing with the FSP parameters and conditions, and/or by using various tool shapes, one can achieve the desired grain size and uniform distribution of secondary phase in the processed metal [6]. Mishra et al [7] in a review article, laid emphasis on mechanisms responsible for microstructural refinement, and effects of FSW/FSP parameters on resultant microstructure and final mechanical properties. Kang et al [8] reported, that grain refinement, homogeneity as well as porosity free microstructure can be achieved by FSP. Although literature reports various techniques other than FSP, such as laser cladding for improving wear and corrosion resistance properties of magnesium and its alloys, however cost and time are the important factors which are compromised [9]. There are some attempts to use cold spray technology also to modify the Mg-surfaces, for example depositing aluminum cold spray coatings on Mg-substrates, however premature corrosion and toxicity of aluminum poses a serious threat to use Al in human body [10]. Gray and Luan [11] in their review

IOP Conf. Series: Materials Science and Engineering **284** (2017) 012026 doi:10.1088/1757-899X/284/1/012026

articles, have concluded that various other surface coating techniques such as electrochemical plating (electroplating), conversion coatings, anodising, hybrid coatings, microarc oxidation and vapour-phase processes, have been considered for Mg alloys. All these surface modifications are either giving moderate corrosion resistance or slight changes in hardness, due to this fact they are not suitable for use in biomedical applications. Few researchers have tried to add TiC nanoparticles to magnesium alloy for strength and ductility enhancement and reported changes in microstructure which led to enhancement of tensile properties [12]. Although FSP alone can have significant effect on grain refinement, wear resistance, hardness, mechanical behavior of Mg alloys, however standalone Mg is not suitable for body implants, due to poor electrochemical performance. Thus comes an idea of incorporation of Hydroxyapatite (HA) particles ($Ca_5(PO_4)_3(OH)$) into the metal. An extensive review of the microstructural evolution during FSP and the effects of processing parameters is available [13]. It is largely accepted that dynamic recrystallization (DRX) governs grain refinement in FSP [14-17], although there is still a debate over the exact mechanisms that govern the resultant microstructure. Taking the advantage of available literature and research findings. FSP with a novel method of incorporating secondary phase has been used in this study.

2. Experimental details

Commercially pure magnesium ingot (Purity 99.9%) purchased from National Aerospace Laboratory Bangalore (India) was used as base material in this study. Hydroxyapatite powder (HA) was used as reinforcement to make a composite layer via FSP route. The HA was purchased from Sigma Aldrich with particle size in the range 25–60 µm . Fig. 1 & 2 show XRD analysis of pure magnesium and HA powder respectively.

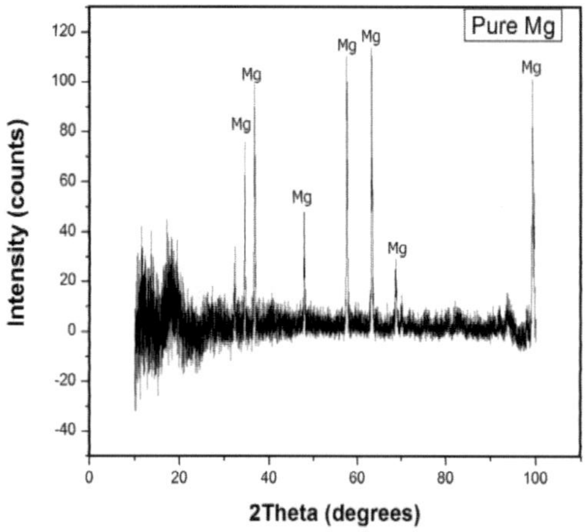

Figure 1. XRD analysis of pure magnesium.

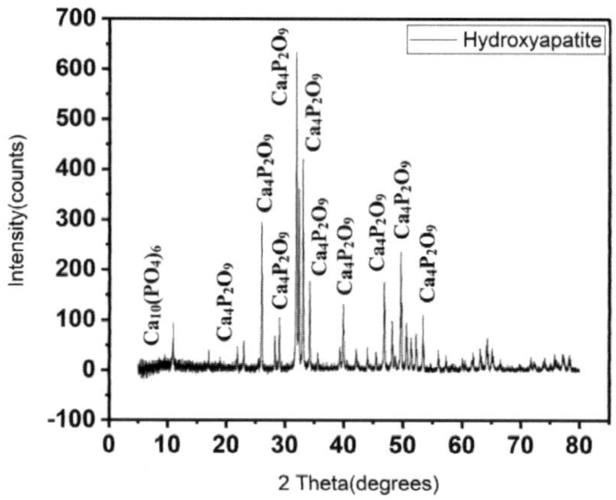

Figure 2. XRD analysis of HAP powder.

Three samples coupons (80 × 40 × 6 mm) of magnesium were cut by Abrasive Jet Cutter (Make: Chennai Metco; Model: Bainmount; India) for FSP. During FSP, the sample was fixed in FSP fixture, as shown in Fig.3. A conical shaped tool made from stainless steel of grade 202 was used for FSP. The tool had threads on its pin with shoulder diameter of 18mm, a tool-tip length 3 mm (Fig. 3). An improved material flow takes place in the samples processed with the tool consisting of a threaded pin due to downward movement of the material [18-19]. The workpieces were friction stir processed (FSPed) using 1 pass under constant rotational (w) and traverse (v) speeds of 2000 rpm and 60 mm/min respectively. The samples have been designated as per the nomenclature given in Table 1.

Table 1. Nomenclature of samples used in this study.

Designation	Description of processing
Mg	As-cast pure Magnesium without FSP
F-Mg	As-cast Magnesium subjected to FSP
F-Mg-H	As-cast Magnesium subjected to FSP with HAP as re-inforcements

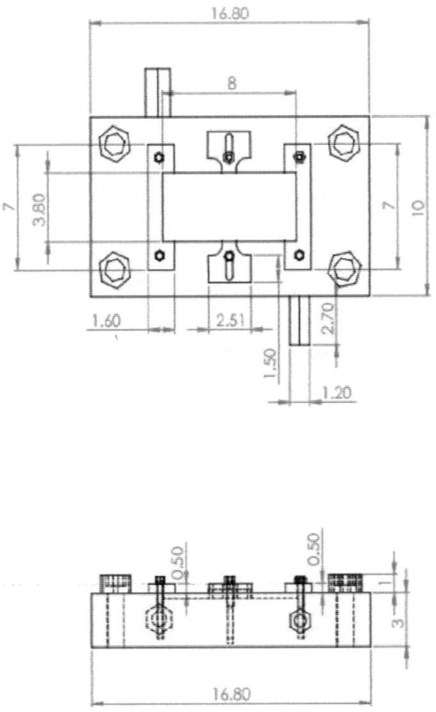

Figure 3. Schematic of fixture to hold Mg pieces for friction stir processing.

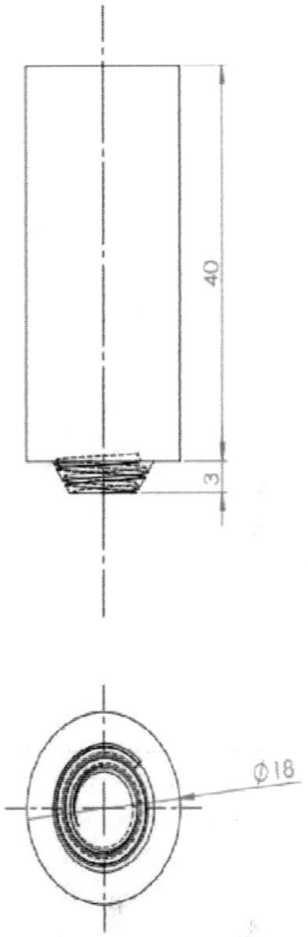

Figure 4. Sketch of conical tool used for friction stir processing of Mg.

To address the problems of agglomeration and non-uniform distribution of HAP powder in Mg substrate, a novel approach to fill HAP powder inside the sample has been proposed and utilized. A hole was drilled (vertically-2.5mm dia) throughout the length of the coupon, in which HAP powder (0.01% vol fraction was filled). Both sides of the hole were sealed by using a suitable fastener. This method of injection resulted in nearly zero wastage of powder, which is usually not achievable in any other methods of filling reinforcements during FSP. Fig. 5 shows the schematic illustration of the approach used for FSP.

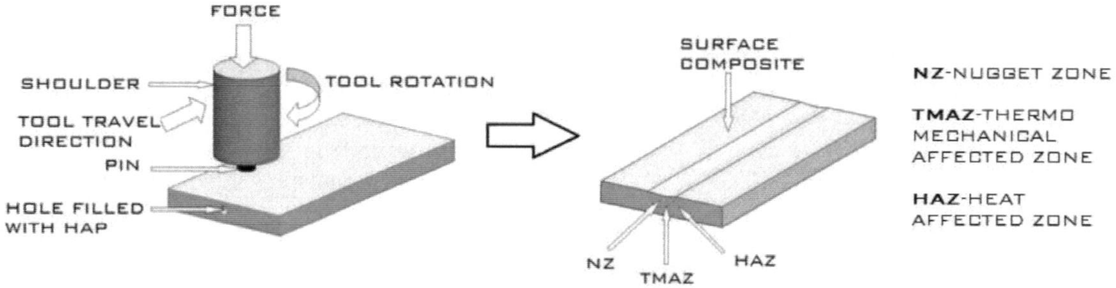

Figure 5. Schematic illustration of the complete FSP process, used to make Mg- composites reinforced with hydroxyapetite.

For the analysis of microstructure and mechanical properties, the FSPed samples were sectioned and mounted in transoptic powder at a hot mounting machine (Make: Chennai Metco; Model: Bainmount; India). Mirror polishing of the mounted samples was done by using standard metallographic procedure. For optical microscopy, etching was done with a solution consisting of 2.5 ml acetic acid, 1.5 ml picric acid, 2.5 ml acetic acid, 15 ml water and 50 ml ethanol. The cross-sections were examined using a Leica Microscope (Make: Leica Microsystems; Model: MEF4M; Germany) optical microscope. Subsequently, micro hardness measurements were made using a Hardness Tester (Make: Wilson Instruments; Model: 402 MVD; USA). Vickers indentor with a test force of 1 Kg and dwell time of 15 seconds was used. X-ray diffraction (XRD) analysis of FSPed samples was carried out using a PANalytical X-Pert Pro machine (Make: PANalytical; Model: X-Pert Pro; The Netherlands). All the XRD scans were carried out with a step size of 0.04° (2θ) and a sampling time of 3 s per step in 2θ range of 0° -180°. Scanning electron microscope (SEM), (Make: Jeol; Model: JSM6610LV; USA) equipped with energy dispersive spectroscopy, EDS (Make: Oxford, Model: INCAX-Act, USA) was used to characterize morphology and composition of the sectioned samples.
An Electrochemical Cell (Make: Gamry; Model: PTC1; USA) was used to perform in-vitro potentiodynamic studies, of three standard electrode configurations; Calomel electrode as reference electrode, Graphite electrode as counter electrode and test specimen as working electrodes. Mg, F-Mg and F-Mg-H samples were subjected to potentiodynamic polarization after 1 hr and 25 hr immersion in the Hank's solution, to stabilize the corrosion rate. The scan rate was 0.1mV/s and applied potential was varied ± 0.25 V about the open circuit potential. Initial delay of 1800 seconds was given for each sample before performing the potentiodynamic test. The obtained data was analyzed by Gamry Analyst software. All the tests were performed three times under ambient condition of room temperature (25±1°C) and relative humidity (55±5%).

3. Results and discussion

The initial micro structure of the Mg samples comprises large coarse grains with the average grain size of 820 μm (Fig. 6-a). Whereas the microstructure of F-Mg (Fig. 6-b) consists of homogenous and refined grains of about 150 μm average size, which could be attributed to dynamic re-crystallization [20-21]. The interface is shown in (Fig. 6-b) where the refinement is seen. With the rise in temperature and stirring effect while doing FSP finer grains are formed. In the third case, the microstructure of F-Mg–H composites got refined intensely, which is not possible to be revealed by OM (Fig. 6-c).

(a) (b) (c)

Figure 6. Cross-sectional optical micrographs of Mg-based surface composites fabricated by single pass of FSP at 2000 r.p.m with a traverse speed of 60 mm/min (a) Mg (b) F-Mg (c) F-Mg-H

The XRD analysis of all the samples is shown in Fig.7. which reveals the presence of Mg as the main phase in all the three cases. In F-Mg no oxidation took place which is a positive attribute. In case of F-Mg-H, elements such as magnesium, phosphorus, calcium and oxygen alongwith additional phases such as calcium phosphate and magnesium phosphate have also been found, which shows the presence of HAP in the sample.

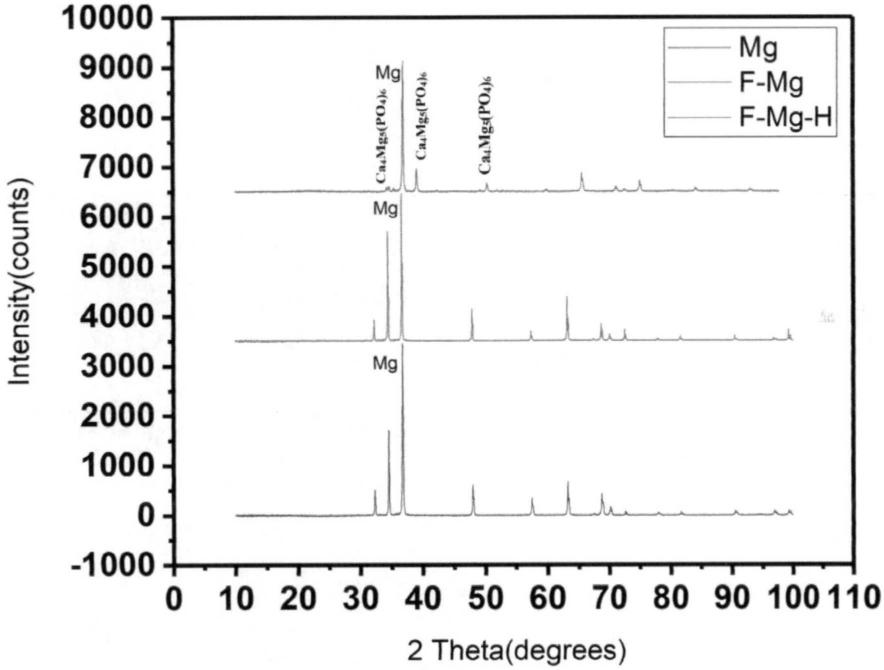

Figure 7. X-Ray diffraction of Mg-based surface composites fabricated by single pass of FSP at 2000 r.p.m with a traverse speed of 60 mm/min.

The cross-sectional SEM micrographs of F-Mg-H samples at different locations are shown in Fig.8, whereas Fig.9 shows the EDS maps of different elements. All the HAP particles are well distributed with the re-crystallized Mg after FSP. This can be attributed to the nature of FSP which forms the composite in solid state itself. Wherein it would have been very difficult to produce composites of Mg with HAP, due to the difference in the melting points of both materials. The element distribution of Mg and HAP particles are clearly seen in Fig. 9. The reinforced particles are dispersed and distributed evenly all over Mg matrix. There is no evidence, during the study, to report presence of clusters, agglomeration or element rich zones throughout the sample surface. Typical dispersion of HAP particles in the surface composites, ranging from 10-25μ has been recorded. The variation in element distribution in the figure is minimum, which leads to validation of homogenous distribution.

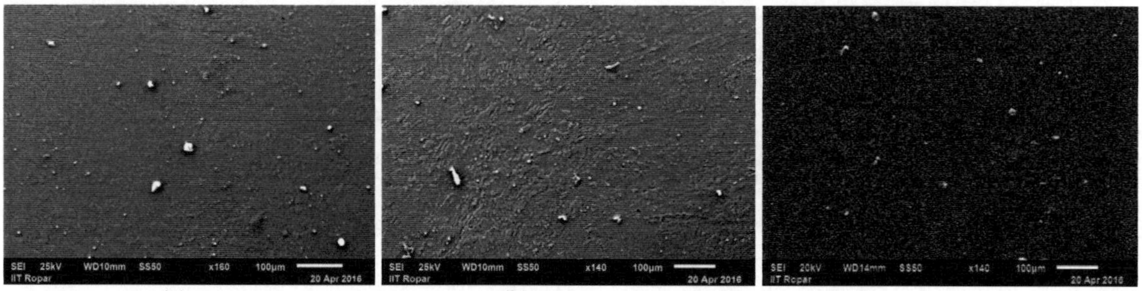

(a) (b) (c)

Figure 8. Cross-sectional scanning electron micrographs of Mg-alloy reinforced with HAP by single pass of friction- stir processing at 2000 r.p.m with a traverse speed of 60 mm/min, showing particle distribution at different locations.

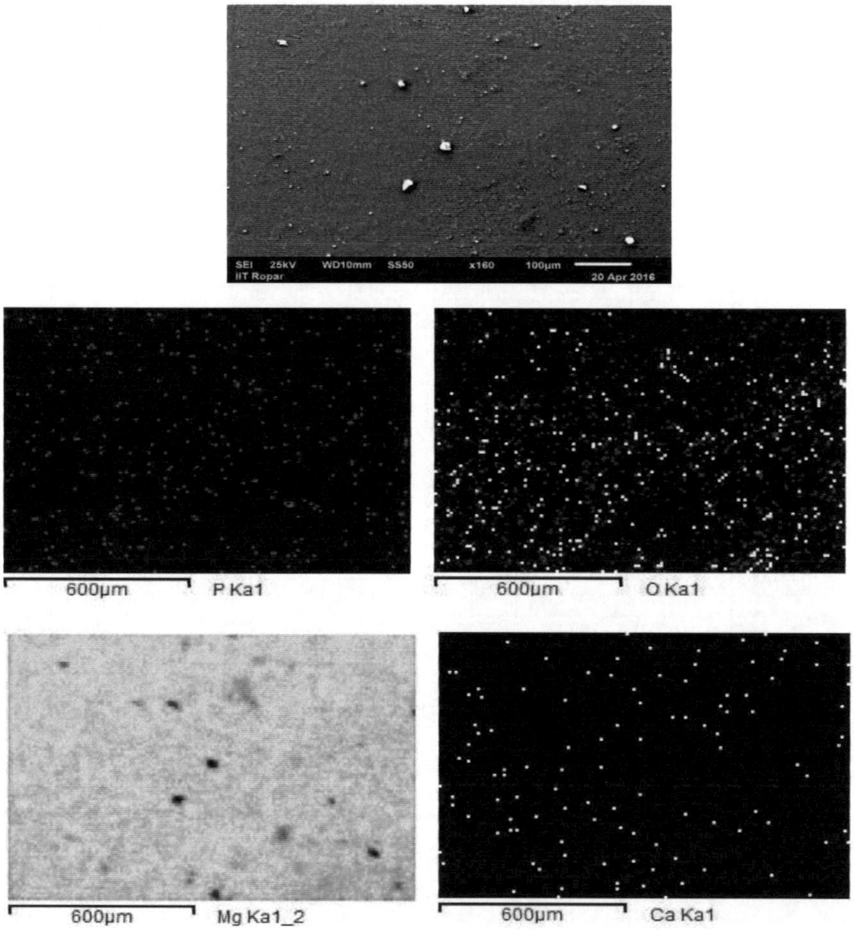

Figure 9. EDS maps of Mg-alloy reinforced with HAP by single pass of friction- stir processing at 2000 r.p.m with a traverse speed of 60 mm/min, showing presence of phosphorous, oxygen, magnesium and calcium after FSP.

The cross-sectional micro-hardness versus distance from the FSP zone- base material interface plots are shown in Fig. 10. The effect of tool rotational speed on the microstructure discussed by various researchers was verified in terms of getting very low surface hardness due to very high heat generations, due to high speeds and prolonged area of contact, which led to grain growth. But eventually when the tool traverse speed was increased it resulted in lower grain size and this higher hardness, which can be observed from the plots of three samples that the variation in micro-hardness is marginal in the base material regions for all the three samples. Higher values of micro-hardness in the heat affected zone (i.e. -25 to +25) were reported for all samples except non-FSPed pure Mg. Whereas as we approach towards the FSP region hardness starts increasing, becomes maximum at a point and then starts decreasing as we move towards the other side of FSP zone. The reason for high hardness in the centre is due to high temperatures in that region which results in dynamic re-crystallization (DRX) and lower hardness on the outer peripheries attributes to larger grain size in that region (due to lower temperature than that in the center) which is in accordance to the Hall–Petch relationship, which states that hardness is inversely proportional to (grain size)$^{1/2}$.The hardness values for all of the Mg/HAP surface composites have narrow distribution ranges, indicating a homogeneous distribution of the HAP phase in the Mg matrix. The average micro-hardness of all the three samples is tabulated below.

Table 1. Average cross-sectional micro hardness of Mg-alloy reinforced with HAP by single pass of friction- stir processing at 2000 r.p.m with a traverse speed of 60 mm/min.

Sr. No.	Sample	Hardness (HV)
1	Mg	36±5%
2	F-Mg	43±5%
3	F-Mg-HAP	47±5%

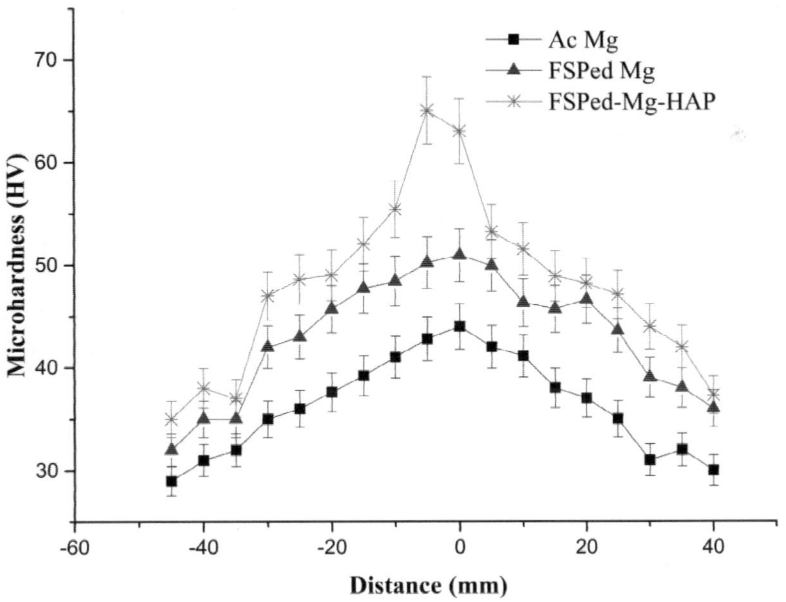

Figure 10. Variation in microhardness along the cross-section of Mg-alloy subjected to Friction Stir Processing (FSP) at 2000 r.p.m with a traverse speed of 60 mm/min.

During the in-vitro corrosion studies, the initial immersion results (Table 2) show equilibrium potential and corrosion rate for Mg, F-Mg and F-Mg-H samples for initial stage i.e 0 hours and after

25 hours. Initial corrosion rates for F-Mg-H (393.9 mpy) were greater than Mg (304.1 mpy), but after 25 hours F-Mg-H (122.9 mpy) shows best corrosion resistant properties, while the F-Mg (304.1 mpy at initial stage and 1529 mpy at 25 hr) shows highest corrosion rates for all time of immersion, which can be attributed to the grain refinement and increased number of grain boundaries. During initial hours of immersion the HAP particles act as cathodic sites on the F-Mg-H surface as a result micro-galvanic coupling occurs between the Mg matrix and HAP particles, similar behavior of micro galvanic coupling has been explained by Song et al [22] for the AZ91 Mg alloy where β particles (Mg17Al12) act as cathodic sites and Mg matrix act as anodic. As reported by Ralston et al [23] grain size of Mg and secondary phase has great significance on corrosion resistance therefore we have reported different corrosion resistances with the explanation of the fact that, due to "dislocation rearrangements" which results due to grain growth after dynamic re-crystallization [24].

Table 2. Equilibrium potential and corrosion rates of Mg-alloy subjected to Friction Stir Processing (FSP) at 2000 r.p.m with a traverse speed of 60 mm/min.

Time	0 hr (Initial)			25 hr (Final)		
Sample	Mg	F-Mg	F-Mg-H	Mg	F-Mg	F-Mg-H
Equilibrium Potential (Vs SCE)V	-1.541	-1.565	-1.564	-1.592	-1.530	-1.546
Corrosion rate (mpy)	304.1	367.7	393.9	165.5	1529	122.9
Icorr A/cm^2	349.8e-6	423.0e-6	453.1e-6	190.3e-6	1.759e-3	141.4e-3
Beta Cathode	364.9e-3	325.4e-3	349.5e-3	310.4e-4	377.7e-3	288.9e-3

4. Conclusion

A corrosion resistant, reinforced Mg composite has been developed via FSP route. With the novel approach in the method of introduction of secondary phase into the Mg has given us positive satisfactory results in terms of uniform distribution of HAP particle into Mg matrix. Improvements in perfect agreement with Hall-Petch relationship were observed which lead to enhancement in mechanical and electrochemical properties. Significant grain refinement has been done, the average grain size has been reduced to 150 μm from 820 μm. Even the hardness of the parent metal has been increased from 36 Hv to 46 Hv, which is attributed to the grain size reduction. Initial corrosion rates for F-Mg-H were greater than Mg but after 25 hours F-Mg-H shows best corrosion resistant properties while the F-Mg (304.1 mpy at initial stage and 1529 mpy at 25hr) shows highest corrosion rates for all time of immersion which can be attributed to the grain refinement and increased number of grain boundaries. FSP combined with innovative ways of incorporation secondary phase in Mg matrix is an effective way to enhance the electrochemical and mechanical properties of pure Mg. These results are of great technological importance since FSP processing can be chosen as a cost effective way to produce composites with adjustable properties.

5. References

[1] Chen, Z., Mao, X., Tan, L., Friis, T., Wu, C., Crawford, R., Xiao, Y., 'Osteoimmunomodulatory properties of magnesium scaffolds coated with β-tricalcium phosphate', *Biomaterials*, **Volume 35**, (2014), pp 8553–8565.

[2] Giridharan, V., Yun, Y., Hajdu, P., Conforti, L., Collins, B., Jang, Y., Sankar, J., 'Microfluidic Platforms for Evaluation of Nano biomaterials: A Review', *Journal of Nanomaterials,* **Volume 2012**, (2012), pp 17.

[3] Song, G., 'Control of biodegradation of biocompatable magnesium alloys', *Corrosion Science,* **Volume 49** (4), (2007), pp 1696–1701.

[4] Khalil, K.A., 'A New-Developed Nanostructured Mg/HAp Nanocomposite by High Frequency Induction Heat Sintering Process', *Int. J. Electrochem. Sci.*, **Volume 7** (2012), pp 10698 – 10710.

[5] Suchanek, W.L., Byrappa, K., Shuk, P., Riman, R.E., Janas, V.F., TenHuisen, K.S. 'Preparation of magnesium-substituted hydroxyapatite powders by the mechanochemical–hydrothermal method', *Biomaterials*, **Volume 25**, (2004), pp 4647–4657.

[6] Venkateswarlu, G., Davidson, M.J., Sammaiah, P., 'Effect of Friction Stir Processing Process Parameters on the Mechanical Properties of AZ31B Mg Alloy', *Journal of Manufacturing and Industrial Engineering*, **Volume 13**, (2014), pp 1-2.

[7] Mishra, R., Ma, Z.Y., 'Friction stir welding and processing', *Materials Science and Engineering: R: Reports*, **Volume 50**, (2005), pp 1–78.

[8] Kanga, J., Wilkinsonb, D.S., Mishra, R.K., Yuand, W., Mishra, R.S., 'Effect of inhomogeneous deformation on anisotropy of AZ31 Magnesium sheet', *Materials Science and Engineering: A*, **Volume 567**, (2013), pp 101–109.

[9] Carcel, B., Sampedro, J., Ruescas, A., & Toneu, X., 'Corrosion and wear resistance improvement of magnesium alloys by laser cladding with Al-Si', *Physics Procedia*, **Volume 12**, (2011), pp 353-363.

[10] Champagne, V., Gabriel, B., & Villafuerte, J., 'Cold spray coatings to improve the corrosion resistance of magnesium (Mg) alloys', *Corrosion Prevention of Magnesium Alloys*, (2011), pp 414.

[11] Gray, J. E., & Luan, B., *Journal of alloys and compounds*, 'Protective coatings on magnesium and its alloys—a critical review', **Volume 336** (1), (2002), pp. 88-113.

[12] Paramsothy, M., Chan, J., Kwok, R., & Gupta, M., 'Adding TiC nanoparticles to magnesium alloy ZK60A for strength/ductility enhancement', *Journal of Nanomaterials*, (2011), pp. 50.

[13] Mishra, R. S. and Ma, Z. Y., 'Friction stir welding and processing', *Mater. Sci. Eng., R*, **Volume 50**, (2005), pp. 1–78.

[14] Jata, K. V. and Semiatin, S.L., 'Continuous dynamic recrystallization during friction stir welding of high strength aluminum alloys' , *Scripta Mater.*, **Volume 43**, (2000), pp. 743.

[15] Su, J. Q., Nelson, T.W., Mishra, R. and Mahoney. M., 'Microstructural investigation of friction stir welded 7050-T651 aluminium', *Acta Mater.*, **Volume 51**, (2003), pp. 713.

[16] Jata, K. V., Sankaran, K.K. and Rushau. J. J., 'Friction-stir welding effects on microstructure and fatigue of aluminum alloy 7050-T7451', *Metall. Mater. Trans. A*, **Volume 31A**, (2000), pp. 218 .

[17] Sato, Y. S., Urata, M. and Kokawa, H., 'Parameters controlling microstructure and hardness during friction-stir welding of precipitation-hardenable aluminum alloy 6063', *Metall. Mater. Trans. A*, **Volume. 33A**, (2002), pp. 625.

[18] Padmanaban, G. and Balasubramanian, V., 'Selection of FSW tool pin profile, shoulder diameter and material for joining AZ31B magnesium alloy–an experimental approach', *Mater. Des*, **Volume 30**, (2009), pp. 2647–2656.

[19] Azizieh, M., Kokabi, A.H. and Abachi, P., 'Effect of rotational speed and probe profile on microstructure and hardness of AZ31/Al 2 O 3 nanocomposites fabricated by friction stir processing', *Mater. Des*, **Volume 32**, (2011), pp. 2034–2041.

[20] Forst, H. J., and Ashby, M. F., 'In: Deformation-mechanism maps', Oxford: Pergamon Press, (1982), pp. 21–44.

[21] Jata, K. V., and Semiatin, S. L., 'Continuous dynamic recrystallization during friction stir welding of high-strength aluminum alloys', *Scripta. Mater.*, **Volume 43**, (2000), pp. 743–749.

[22] Song, Y., Shan, D., Chen, R., Zhang, F. and Han, E.H., 'Biodegradable behaviors of AZ31 magnesium alloy in simulated body fluid', *Mater. Sci. Eng. C*, **Volume 29**, (2009), pp. 1039–1045.
[23] Ralston, K. and Birbilis, N., 'Effect of grain size on corrosion: a review', *Corrosion*, **Volume 66**, (2010), pp. 75005–75013.
[24] Ben-Haroush, M., Ben-Hamu, G., Eliezer, D. and Wagner, L., 'The relation between microstructure and corrosion behavior of AZ80 Mg alloy following different extrusion temperatures', *Corros. Sci.* **Volume 50**, (2008), pp. 1766–1778.

Acknowledgements
Authors are highly thankful to Indian Institute of Technology, Ropar, Punjab, India for its continuous financial support.

Failure analysis of single-bolted joint for lightweight composite laminates and metal plate

Linjie Li [1], Junli Qu[1] and Xiangdong Liu [1]

1 Department of Mechanics, Xi'an University of Science and Technology, 58 Yanta Road, Beilin District, Xi'an 710054, China
E-mail: linjie_happy@163.com

Abstract. A three-dimensional progressive damage model was developed in ANSYS to predict the damage accumulation of single bolted joint in composite laminates under in-plane tensile loading. First, we describe the formulation and algorithm of this model. Second, we calculate the failure loads of the joint in fibre reinforced epoxy laminated composite plates and compare it with the experiment results, which validates that our model can appropriately simulate the ultimate tensile strength of the joints and the whole process of failure of structure. Finally, this model is applied to study the failure process of the light-weight composite material (USN125). The study also has a great potential to provide a strong basis for bolted joints design in composite Laminates as well as a simple tool for comparing different laminate geometries and bolt arrangements.

1. Introduction

Composite is a state-of-art material in aircraft structure. To have a safer structural designs of aircraft with low weight, more and more advanced composites have been developed, which are not only light weight but also have higher strength to weight ratio [1]. One of the most useful applications of laminated composites as weight reducing elements is in aircraft wing where composite stringers has huge potential of being attached with the wing skin by means of a bolted joint [2]. Although composite materials are now being widely used as primary aircraft structures, the use is rather limited due to lack of research in composite mechanics and complicated failure mechanics of the composite joints, especially for the advanced composites [3].

2. Numerical simulations using ANSYS

2.1 The geometry of the component

The test specimen configuration is shown in Figure 1, and the specific values are shown in Table 1.

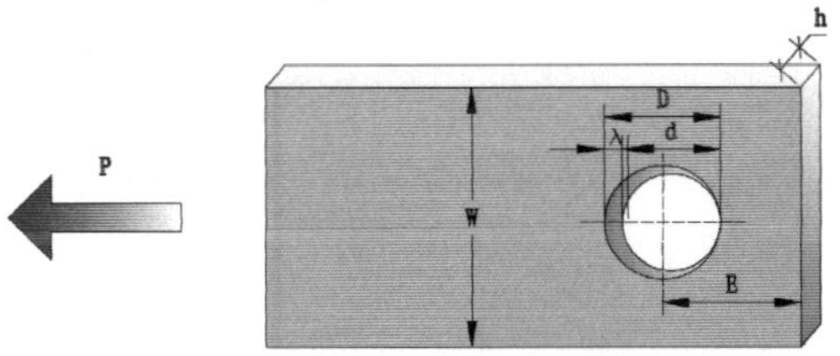

Figure 1. Geometries of the specimens

IOP Conf. Series: Materials Science and Engineering **284** (2017) 012027 doi:10.1088/1757-899X/284/1/012027

Table 1. Geometrical model of the composite joint.

Plate length (mm)	Plate thickness (mm)	Hole diameter (mm)	Plate width (mm)	Distance between the center of the hole and the right margin of plate(mm)
160.00	2.00	4.14	30.00	15.00

2.2 Finite element modeling

As shown in Figure 2, bolts, nuts and gaskets are simplified as a whole in this 3D model. The metal plate and bolt are modelled using Solid185 elements and the composite plate is modelled using Shell99 in ansys software. In order to make the simulation closer to the real situation, the impact of the contact model, frictional and preload were also considerate. The contact areas between bolt and plates are modelled using CONTA173 contact element. The magnitude of the frictional force between the contact surfaces is controlled by coulomb friction, the friction coefficient was 0.2. The preload applied on the bolt is 3 N□m .

The boundary conditions are shown in the Figure 3. The right edge of the model is fixed ($U_x = U_y = U_z = \phi_x = \phi_y = \phi_z = 0$) and a uniformly increasing tensile-load is applied on the left edge of the composite plate along the x-axial direction of the model ($U_y = U_z = \phi_x = \phi_y = \phi_z = 0$). The mesh convergence analysis is performed on a various number of elements to ensure the accuracy of calculations and freedom of the model to mesh size. The final mesh distributions through thickness and around the bolt hole are shown in Figure 3.

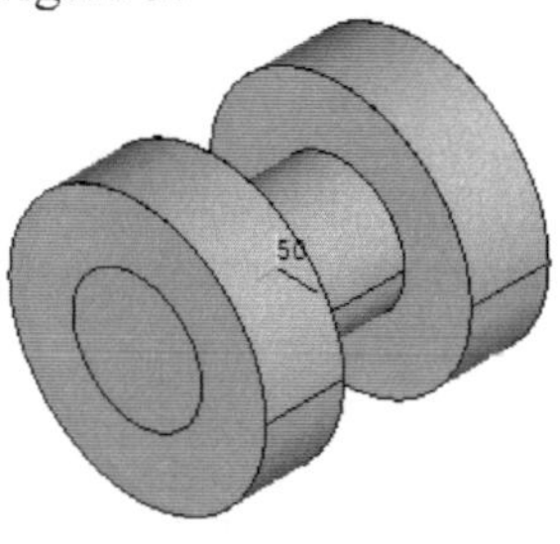

Figure 2. Simplified model of bolt

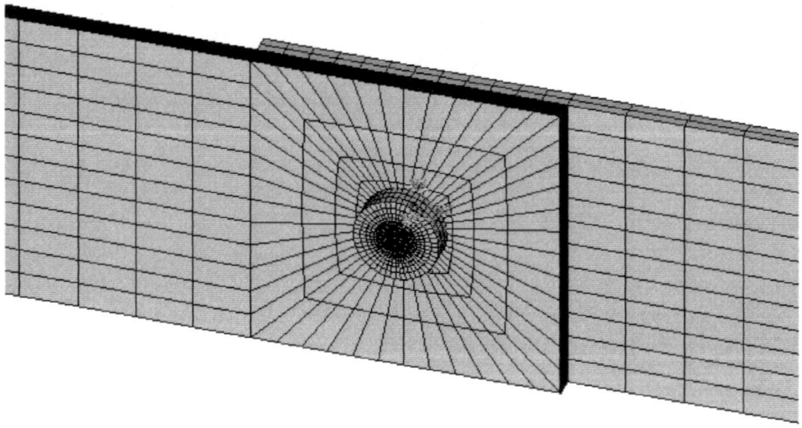

Figure 3. FEM mesh of composite joint

3. Analysis of failure

3.1 Laminated plate cumulative damage analysis processing

Dano et al. [4] developed a progressive damage model to investigate the effect of failure criteria and the material property degradation rules on the behaviour of joints in a graphite/epoxy composite laminate. A flow chart for the program is shown in Figure 4.

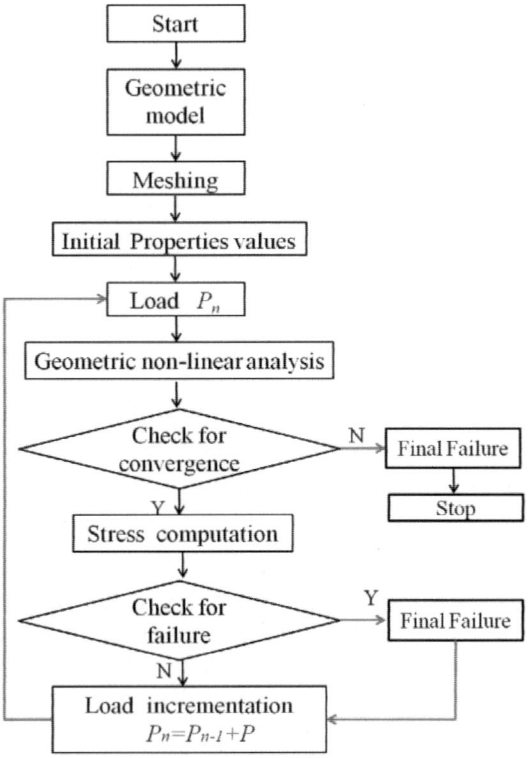

Figure 4. Progressive damage model algorithm

3.2 Stress analysis

Apply a constraint to one end of the composite laminates and apply load to another end. The load acting on the inner surface of the laminate increases gradually, when the load increases to a step *n*, i.e. $P_n=P_{n-1}+P$. Because there is no volume force and according to incremental form of virtual displacement principle and force boundary conditions, we have balance equation as follows.

$$\int_{n-1}\sigma_{ij}^{n}\delta(\Delta\varepsilon_{ij})dV-\int_{S_\sigma}\overline{T_i^n}\delta(\Delta u_i)dS=0 \qquad i,j=1,2,3, \tag{1}$$

where $\Delta\varepsilon_{ij}$ and Δu_i represent strain increment and displacement increment from the load step $n-1$ to the load step n ,respectively. σ_{ij}^n represent the laminates stress under load P_n, $\sigma_{ij}^n=\sigma_{ij}^{n-1}+\Delta\sigma_{ij}$, where $\Delta\sigma_{ij}$ is the incremental load from load step $n-1$ to the load step n . $\overline{T_i^n}$ represent the per unit area force acting on the boundary S_σ under load P_n . V_{n-1} is the volume deformation at step *n-1*. Due to the geometric large deformation, strain increment is shown below.

$$\Delta\varepsilon_{ij}=\Delta e_{ij}+\Delta\eta_{ij}, \tag{2}$$

where Δe_{ij} is the strain increment and $\Delta\eta_{ij}$ is the quadratic term. The stress-strain relationship of anisotropic elastomers is shown below [5].

$$\sigma_i=\sum_{j=1}^{6}C_{ij}\varepsilon_j \qquad (i=1,2,\cdots,6). \tag{3}$$

When the load from the step $n-1$ to step n, the increased load variable is $\Delta\sigma_i = C_{ijkl}^{n-1}\Delta\varepsilon_{kl}$, where C_{ijkl}^{n-1} is the composite material modulus under the load step $n-1$, which can be obtained by the following equation.

$$C_{ijkl}^{n-1} = C_{rs}^{n-1} = \sum_{k=1}^{m}\left[\frac{\overline{Q_{rs}^{(k)}}}{m}\right]^{n-1}, \quad (4)$$

where m is the number of composite layers. $\left[\overline{Q_{rs}^{(k)}}\right]^{n-1}$ is the composite material modulus of the k-th layer after converted by coordinates under the $n-1$ load step. $\Delta\varepsilon_{ij}$ and $\Delta\sigma_{ij}$ are liberalized: $\Delta\varepsilon_{ij} = \Delta e_{ij}$, $\Delta\sigma_i = C_{ijkl}^{n-1}\Delta e_{kl}$,then the equation (1) can be converted into equation (5):

$$\int_{n-1} C_{ijkl}^{n-1}\Delta e_{kl}\delta(\Delta e_{ij})dV + \int_{n-1} \sigma_{ij}^{n-1}\delta(\Delta\eta_{ij})dV = \int_{S_\sigma} \overline{T_i^n}\delta(\Delta u_i)dS - \int_{n-1}\sigma_{ij}^{n}\delta(\Delta e_{ij})dV. \quad (5)$$

The result of the variation equation on the above equation will give a linear system of equations for displacement increments, which can be further solved by finite element method. When the load increases to P_n, if the material damage occurs, it will lead to damage to the material properties of the damage zone, that is, the material modulus part of the damage area will change, the specific degradation will be described in the lower part. With the advent of damage, structural stress and strain will reassign, the equation (5) is calculated. The process is repeated until there is no new damage occurs.

3.3 Material degradation rule

With the increasing of load, the fibre and matrix of the laminated plate are gradually destroyed. The structural rigidity is decreasing, and the expansion of the damage can be characterized by the degradation of the elastic constants of the material. Tan [6] used parameters degradation to analysis the cumulative damage process as shown in Table2.

Table 2. Material degradation of Tan's rule

Failure Mode	Tan's rule
Matrix tensile cracking	$E_{22} \to 0.2E_{22}, G_{12} \to 0.2G_{12}, G_{23} \to 0.2G_{23}$
Matrix compressive cracking	$E_{22} \to 0.4E_{22}, G_{12} \to 0.4G_{12}, G_{23} \to 0.4G_{23}$
Fiber tensile cracking	$E_{11} \to 0.07E_{11}$
Fiber compressive cracking	$E_{11} \to 0.07E_{11}$
Fiber-matrix shear out	$G_{12} = \nu_{12} = 0$
Delamination in tension	$E_{33} = G_{23} = G_{13} = \nu_{23} = \nu_{13} = 0$
Delamination in compression	$E_{33} = G_{23} = G_{13} = \nu_{23} = \nu_{13} = 0$

3.4 Hashin's failure criteria [7]

Hashin's failure Criteria is popular in the composite industry due to its elegant form and accuracy matching the experimental results. The three-dimensional Hashin guideline expression is shown below:

(1) Fiber tensile cracking ($\sigma_{11} > 0$)

$$F_f^2 = \left(\frac{\sigma_{11}^2}{X_t}\right) + \left(\frac{\tau_{12}}{S_{12}}\right)^2 + \left(\frac{\tau_{13}}{S_{13}}\right)^2 \geq 1 \quad (6)$$

(2) Fiber compressive cracking ($\sigma_{11} > 0$)

$$F_f^{\ 2} = \left(\frac{\sigma_{11}^{\ 2}}{X_c}\right) \geq 1. \quad (7)$$

(3) Matrix tensile cracking ($\sigma_{22} + \sigma_{33} \geq 0$)

$$F_m^{\ 2} = \left(\frac{\sigma_{22} + \sigma_{33}}{Y_t}\right)^2 + \frac{(\tau_{23} - \sigma_{22}\sigma_{33})^2}{S_{23}^{\ 2}} + \left(\frac{\tau_{12}}{S_{12}}\right)^2 + \left(\frac{\tau_{13}}{S_{13}}\right)^2 \geq 1. \quad (8)$$

(4) Matrix compressive cracking ($\sigma_{22} + \sigma_{33} < 0$)

$$F_m^{\ 2} = \frac{1}{Y_c}\left(\left(\frac{Y_c}{2S_{12}}\right)^2 - 1\right)^2 (\sigma_{22} + \sigma_{33}) + \left(\frac{\sigma_{22} + \sigma_{33}}{2S_{12}^{\ 2}}\right)^2 + \frac{(\tau_{23}^{\ 2} - \sigma_{22}\sigma_{33})}{S_{23}^{\ 2}} + \left(\frac{\tau_{12}}{S_{12}}\right)^2 + \left(\frac{\tau_{13}}{S_{13}}\right)^2 \geq 1. \quad (9)$$

(5) Fiber-matrix shear out ($\sigma_{11} > 0$)

$$F_m^{\ 2} = \left(\frac{\sigma_{33}}{Y_C}\right)^2 + \left(\frac{\tau_{12}}{S_{12}}\right)^2 + \left(\frac{\tau_{13}}{S_{13}}\right)^2 \geq 1, \quad (10)$$

where $\sigma_{ii}(i = 1,2,3)$ represent the normal stress in the i direction and $\tau_{ij}(i, j = 1,2,3)$ is the shear stress in the i, j plane. X_t , X_c , Y_t , Y_c represent the tensile strength and compressive strength of the fibre direction and in the plane perpendicular to the fibre direction, respectively. $S_{ij}(i, j = 1,2,3)$ represent the shear strength in-plane. $F_j(j = f, m)$ is the failure factor which can be used to characterize the degree of the material damage. The greater the failure factor is, the more serious the damage is. When the failure factor is greater than 1, the material will be failure.

4. Validation of the numerical simulations

In order to verify the accuracy of the model, specimens with the same size and material (CCF300/QY8911) as that used in Ref 8 are calculated and analysed. The mechanical properties of the composite are shown in Table3. The stacking sequences for the composite ply is [45/0/-45/90/45/0/-45/0]$_s$ and the thickness of each ply is 0.125mm.

Table 3. Material properties of composite material (CCF300/QY8911).

Material properties	E_{11}(GPa)	E_{22}(GPa)	E_{33}(GPa)	G_{12}(GPa)	G_{23}(GPa)	X_t (MPa)	G_{13}(GPa)
CCF300/QY8911	133	9.9	9.9	6.67	6.67	2134.2	3.9

X_c (MPa)	Y_t (MPa)	Y_c (MPa)	S_{12} (MPa)	S_{13} (MPa)	S_{23} (MPa)	v_{12}	v_{13}	v_{23}
1121.2	33.0	191.3	87	87	110	0.27	0.27	0.27

In Table 3, E_{11} , E_{22} and E_{33} mean the elastic modulus in longitudinal direction and transverse direction, respectively. G_{12} , G_{13} and G_{23} are the in plane and the out-of plane shear modulus, respectively. X and Y are the strength in direction of warp and fil, respectively. S is shear strength. Subscript t and c mean the tensile and compressive properties. v is Poisson's ratio.

The material properties of metal plate and bolt are $E_1 = 115$ GPa, $v = 0.31$ and $S = 675$ MPa, where S represents the shear strength. The configuration of single-lap bolt joint was shown in Figure 5.

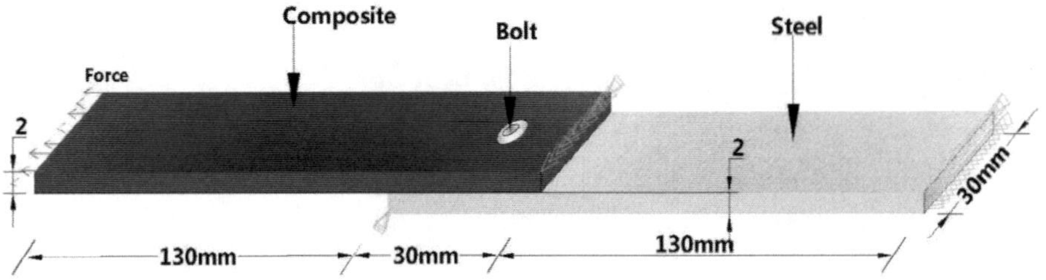

Figure 5. Configuration of single-lap bolted joint

Figure 6 and Table 4 show the comparison between the results of the simulation and experimental results under static tensile loading, respectively.

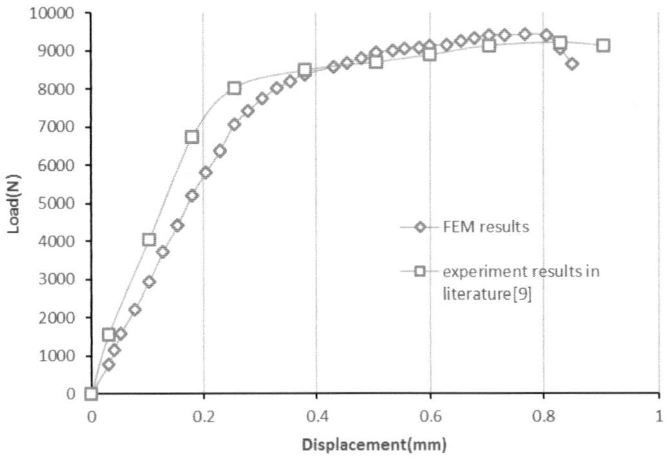

Figure 6. Comparison of FEM simulation result and experiment result

Table 4. Comparison of testing and numerical results

Items	Numerical result	Test result [8]	Relative Errors/%
Failure load(kN)	9.414	9.2	2.32
Stiffness($N\cdot mm^{-1}$)	142.195	137.8	3.19

According to the comparison presented above, the error between simulated prediction value and experimental results is less than 5%, indicating that this model can successfully be applied to predict the strength of advance composites, as well as to comparison of different joint configurations and effect of design parameters.

5. Failure analysis for lightweight composite laminates

Use our model to simulate the damage process and failure of light-weight composite (USN125) laminates. The material properties are listed in Table.5 and the ply is $[45/0/-45/90/45/0/-45/0]_s$.

Table 5. Material properties of composite material (USN125). [9]

Material Properties	E_{11}(GPa)	E_{22}(GPa)	E_{33}(GPa)	G_{12}(GPa)	G_{23}(GPa)	G_{13}(GPa)	X_t (MPa)
USN125	162	9.6	9.6	6.1	3.5	6.1	2552

X_c (MPa)	Y_t (MPa)	Y_c (MPa)	S_{12} (MPa)	S_{13} (MPa)	S_{23} (MPa)	v_{12}	v_{13}	v_{23}
2552	43	43	94	94	40	0.298	0.298	0.470

IOP Conf. Series: Materials Science and Engineering **284** (2017) 012027 doi:10.1088/1757-899X/284/1/012027

Figure 7 illustrates the initial stress states of the composite laminate. It can be seen from the stress component contour diagram that the lattice around the bolt hole of the composite material changes more obviously, especially in the direction of -45°,which indicates that the fiber orientation of the -45° layer would firstly reach the maximum bearing strength. In high compression zone, due to high values of S33, i.e. of principle stress along laminate thickness, there exists a strong potential of delamination due to compression, thus causing a typical bearing failure. It has also been observed that shearing stress in yz-plane (S23) and xz-plane (Sl3) are very low as compared to shearing stress in xy-plane (S12) and the transverse compressive stress of the fiber is lower than the shearing stress.

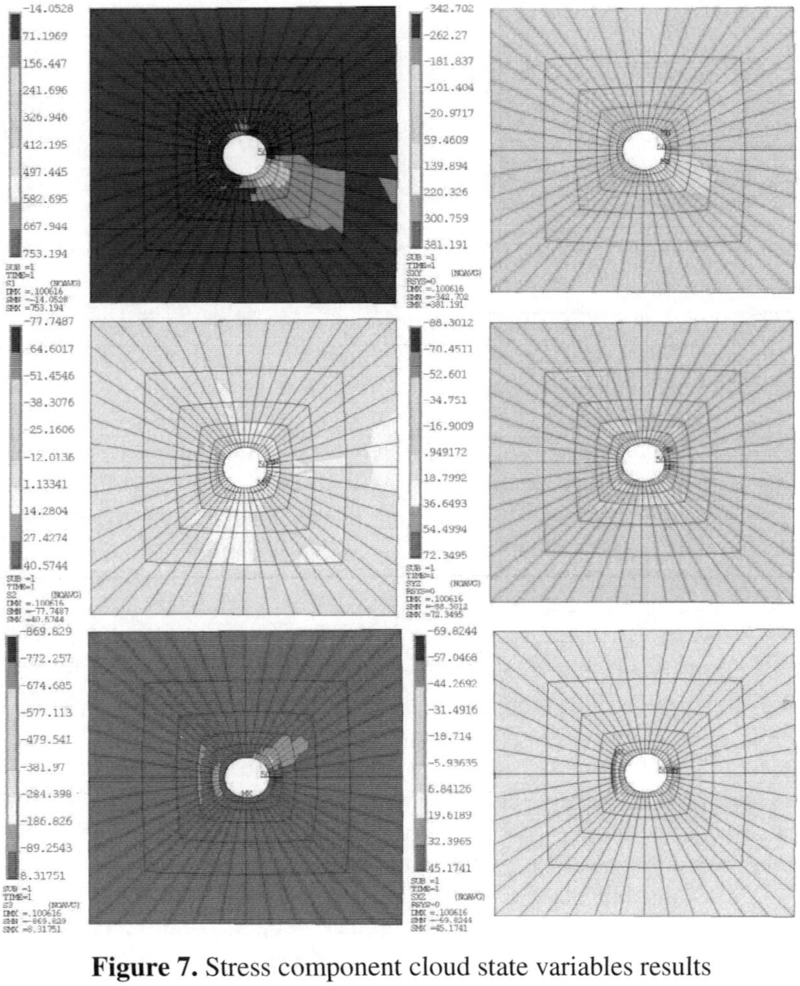

Figure 7. Stress component cloud state variables results

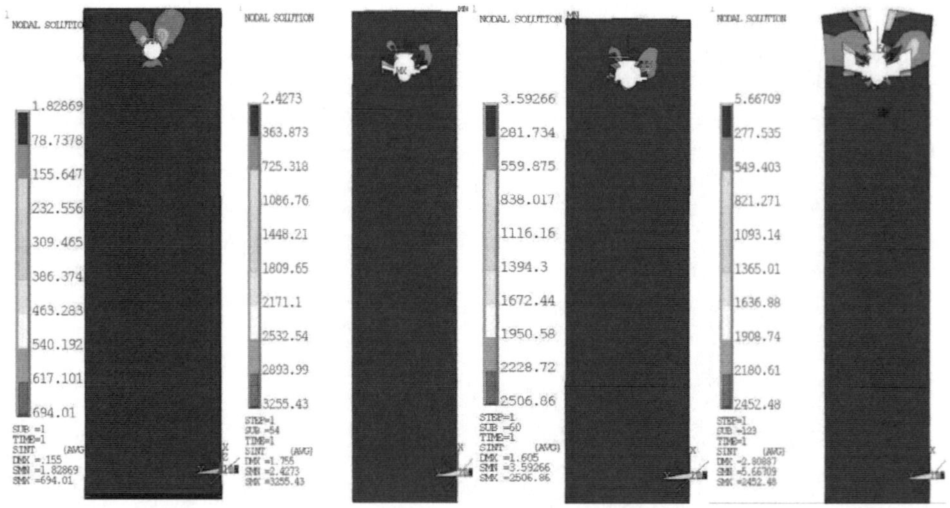

Figure 8. Stress distribution of bolt-hole and progressive damage of laminate

From Figure 8, we visualize the stress and deformation progressively. It is observed that most of the applied load is contributed towards compression along fiber directions. Remarkable tensile stress transverse to the direction of laminate is near the hole which may cause the effect of matrix cracking. However, because the region under compressive load is reasonably more as compared to region in tensile stress, there are more chances of bearing failure of the joint. The inter lamina stresses of both 90° plies with their neighboring plies may cause delamination, thus to effect the ultimate failure load. There is a warp between the composite laminates and the metal plates during the stretching process, and there is a tendency to be separated, and the bolt is inclined in the stretching direction.

6. Conclusions

In this study, a three-dimensional Progressive Damage Model was developed to compute the tensile strength and the maximum load of the composite metal joint damage accumulation failure analysis model, considering the preload, friction and complex contact area of connection and contact deformation. According to the study, conclusions are as follows.

(1) The tensile strength and the maximum load of the composite (CCF300/QY8911)-metal joint are 142.195MPa and 9.414kN, respectively. The results of the experimentation are 124.5MPa and 9.2kN. The error between simulation and experiment result is less than 5%, indicating that this model can be applied to predict the strength of advance composites.

(2)The failure process and failure load of the lightweight composite (USN125)-metal joint were obtained by the established model. The initial failure crack occurs in the direction of -45° and 45° of the fiber layer and the matrix separation of the stress generated in the direction of -45 °, so that the direction of -45° fiber bear the main load.

(3)Since the error is very stable between simulated strength prediction value and experimental value of strength so the model in the paper can successfully be applied for comparison of different joint configurations and effect of design parameters for single bolted joint but it's necessary to validate the three-dimensional finite model and calculation method in this paper in multiple-bolted joints in further study.

7. References

[1] Starikov R, Schon J. Quasi-static Behaviour of Composite Joints with Protruding Head Bolts. *Compos Struct* 2001; 51:411-25.

[2] Saqib Mehmood, Zhao Libin and Huang Hai. Failure Analysis of Single and Double Bolted Joints for Composite Laminates. *12th International Bhurban Conference on Applied Sciences & Technology* 2011;15: 49-54.

[3] Thoppul S D, Finegan J, Gibson. Mechanics of Mechanically Fastened Joints in Polymer-matrix Composite Structures -a review. *Compos Sci Techno* 2009; 69: 301-29.
[4] Dano ML, Kamal E, Gendron G. Analysis of Bolted Joints in Composite Laminates: strains and bearing stiffness predictions. *Compos Struct* 2007; 79(4):562-70.
[5] Shen GL, Hu GK. *Mechanics of Composite Materials*[M]. Tsing Hua University Press 2006.
[6] Tan SC, Perez J. Progressive Failure of Laminated Composites with a Hole under Compressive Loading. *J Reinf Plas Compos* 1993; 12:1043-1057.
[7] Hashin Z. Failure Criteria for Unidirectional Fibre Composites. *Applied Mechanics* 1980; 4**7**(2): 329-334.
[8] Liu X K, Zhang WW, Li Y Z. Investigation on the Deformation Failure of Single-bolted Metal to Composite Joints. *Science Technology and Engineering* 2014; 22(14):307-311.
[9] Seong MS, Kim TW, Nguyen KH, Kweon JH and Choi JH. A Parametric Study on the Failure of Bonded Single Lap Joints of Carbon Composite and Aluminium. *Compos Struct* 2008; 86:135-145.

Frequency response improvement of a two-port surface acoustic wave device based on epitaxial AlN thin film

Junning Gao[1, 2], Zhibiao Hao[3], Yi Luo[3] and Guoqiang Li[1, 2]

1 China State Key Laboratory of Luminescent Materials and Devices, School of Materials Science and Engineering, South China University of Technology, Guangzhou, 510641, China
2 Engineering Research Center on Solid-State Lighting and its Informationisation of Guangdong Province, South China University of Technology, Guangzhou, 510641, China
3 Tsinghua National Laboratory for Information Science and Technology, Department of Electronic Engineering, Tsinghua University, Beijing, 100084, China
Email: msjngao@scut.edu.cn, zbhao@tsinghua.edu.cn, luoy@tsinghua.edu.cn, msgli@scut.edu.cn

Abstract. This paper presents an exploration on improving the frequency response of the symmetrical two-port AlN surface acoustic wave (SAW) device, using epitaxial AlN thin film on (0001) sapphire as the piezoelectric substrate. The devices were fabricated by lift-off processes with Ti/Al composite electrodes as interleaved digital transducers (IDT). The impact of DL and the number of the IDT finger pairs on the frequency response was carefully investigated. The overall properties of the device are found to be greatly improved with DL elongation, indicated by the reduced pass band ripple and increased stop band rejection ratio. The rejection increases by 8.3 dB when DL elongates from 15.5λ to 55.5λ and 4.4 dB further accompanying another 50λ elongation. This is because larger DL repels the stray acoustic energy out of the propagation path and provides a cleaner traveling channel for functional SAW, and at the same time restrains electromagnetic feedthrough. It is also found that proper addition of the IDT finger pairs is beneficial for the device response, indicated by the ripple reduction and the insertion loss drop.

1. Introduction

Surface acoustic wave (SAW) with two-port structure consisting of two interleaved digital transducers (IDT) is of paramount importance in the SAW device family because it can be the basic configuration for the SAW delay line and bandpass device [1, 2]. The applications are wide spread in wireless communication and signal processing system such as radar, Wifi, Bluetooth, and navigation. When applying a sensing medium between the two IDTs, the two-port SAW device can work as sensor, which has been under thrust interest due to its great potential in medical, bio, and materials analysis [3-6].

The center frequency of a SAW device is the outcome of the sound velocity divide by sound wave length. AlN has the highest sound velocity among piezoelectric materials, which makes it a great candidate for high frequency application and less challenging for lithography process. It is favored among other piezo materials also for its superior physical, chemical and mechanical properties [7-12]. In addition for wireless communication, high operation frequency is favored for sensors as well because the penetration depth of SAW decreases with increasing frequency allowing more adequate interaction between the target and the SAW, thus improve the sensitivity [3].

For a highly functional two-port SAW device, great transmission properties mainly featured as large stop band rejection and small pass band ripple are crucial. Materials imperfections and improper device structure are generally the source of inferior frequency response. Single crystalline AlN film that is solely (0001) oriented along the surface normal and has small roughness is highly desirable for device fabrication. This is because the piezoelectric axis of AlN is along (0001) orientation and surface fluctuations and grain boundaries behave as acoustic wave reflection centers [13-15].

Delay length (DL) between the two IDTs is designed normally out of two considerations. The first is that it needs to be kept short to minimize the material-induced insertion loss and impedance. The second is that for delay line device it determines the delay time because the length between the two IDTs determines the travelling distance of the induced acoustic wave [12, 16]. To the best of our knowledge, the impact of DL on the overall frequency response of the device has not been reported, though we find it has significant influence. Besides, the number of finger pairs is also very important since it affects the resonance amplitude and determines the band width of the device. However, the addition of fingers enlarges the device size which increases the cost. Therefore, these parameters must be thoroughly explored and carefully leveled to reach optimized device performance and keep high cost-effectiveness.

In the presented research, two-port SAW devices have been fabricated on (0001) AlN single crystalline thin films grown by molecular beam epitaxy (MBE), and the device frequency response with an emphasis on the influence of DL and the number of finger pairs have been investigated in detail.

2. Experiments

The epitaxial AlN thin film was grown on (0001) sapphire substrate by using MBE with the similar procedure described in [17]. The surface morphology of the epi-wafer was assessed by a PSIA XE-100 atomic force microscopy (AFM). The films are (0001) oriented having HWHM of 131 arcsec for their (0001) X-ray diffraction rocking curves. All of the films in use have the same thicknesses of 1 μm. The two-port SAW devices were fabricated by lift-off processes. The AlN/sapphire wafer was first cleaned to remove organic and inorganic contaminants, then spin coated with AZ5214 photoresist. After being baked on a hot plate, the wafer was subjected to lithography using a SUSS MJB3 lithography machine. After exposure, the wafer was immersed into the developer solution to develop the pattern. Then it was placed onto the stage of a DC and RF bi-functional electrode sputtering equipment, where Ti/Al composite electrode was sputtered. The Ti bottom layer was RF sputtered and the thickness was controlled to be 10 nm, and the Al top layer was DC sputtered and the thicknesses were controlled to be around 110 nm. The process was completed on forming the IDT electrodes through removing the left photoresist in acetone. The optical microscopic image was taken by OLYMPUS microscope and the frequency response was measured by Agilent 8722ET network analyzer.

3. Results and discussion

3.1. Morphology of the device

Figure 1 (a) shows the AFM image of the AlN film surface. The root mean square roughness of the film is 1.8 nm. Figure 1 (b) gives an overlook of the device structure showing by microphotograph image. The SAW device is symmetric having one input and one output uniform IDT with each consisting of interleaved fingers, bus bar, and GSG pad for electric loading and detecting. The pads are spaced with standard 150 μm center-to-enter distance. Figure 1 (c) exhibits a magnified part of the IDT with details of the fingers and the bus bar. For all devices discussed in this paper, the aperture is 560 μm, and the finger width is uniformly 2 μm, the metallization ratio is 0.5, the IDT period is therefore 8 μm which defines the acoustic wave length as 8 μm.

3.2. Delay length influence on SAW device performance

Figure 2 shows the frequency response of the two-port SAW devices having 30 pairs of fingers with different DL. Figure 2 (a) exhibits the global transmission loss S21 spectra of the devices in the range from 50 MHz to 5 GHz, where all three exhibiting two pass bands with one centered at 714 MHz and the other at 1392 MHz. Figure 2 (b), (c) and (d) give detail profile of the three fundamental mode pass bands. The stop band rejection ratio of the device with 15.5λ DL is 12.4 dB, it grows 8.3 dB with the first 50λ of DL extension and another 4.4 dB with 50λ more. The insertion loss of the device with 15.5λ DL is 34.4 dB, it rises slightly by 1.1 dB when DL increases 50λ, and 2.6 dB further with another 50λ distance hopping. It is therefore can be concluded that the stop band rejection ratio increase with DL with slight sacrifice of the pass band insertion loss. It is also worth noting that the pass band ripple is reduced significantly when DL is increased. Following the ripple decrease, the peak shape in Figure 2 (d) is able to match with the theoretically expected Bell shaped function for symmetrical two-port IDT SAW device [19].

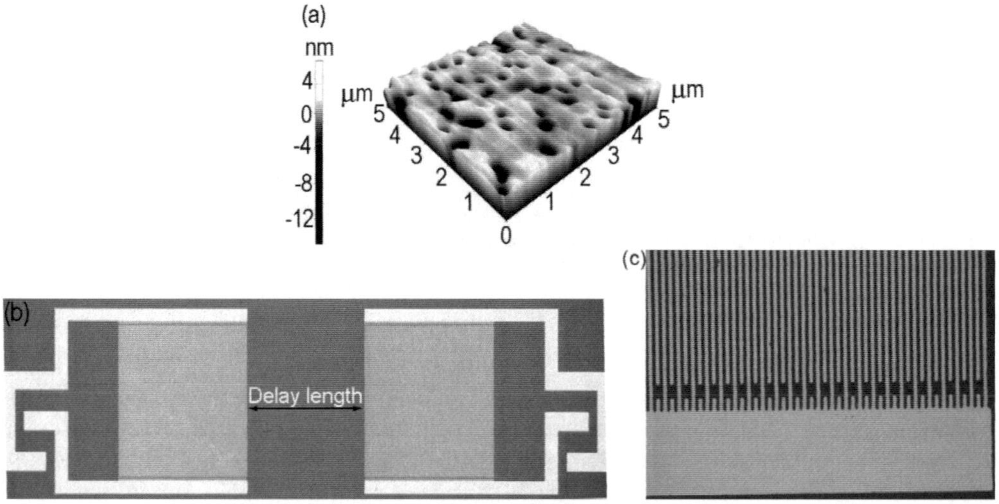

Figure 1. (a) AFM image of the AlN film surface,
the microscopic images of the device (b) an overall view and (c) a local detailed view.

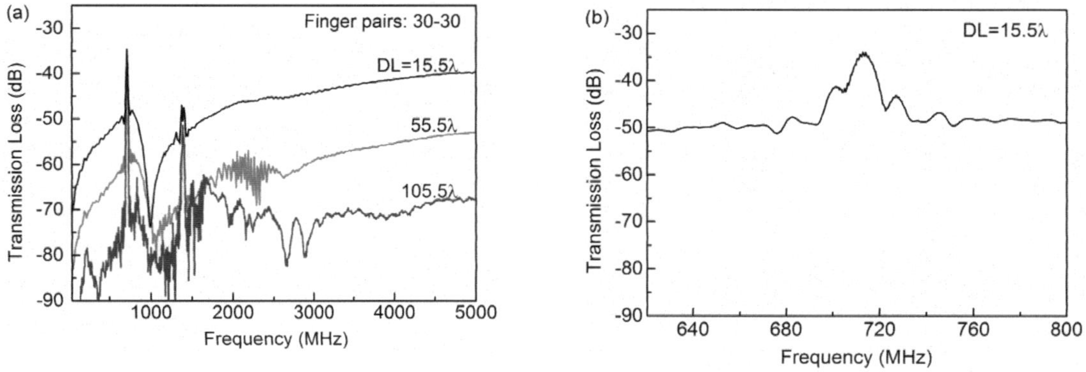

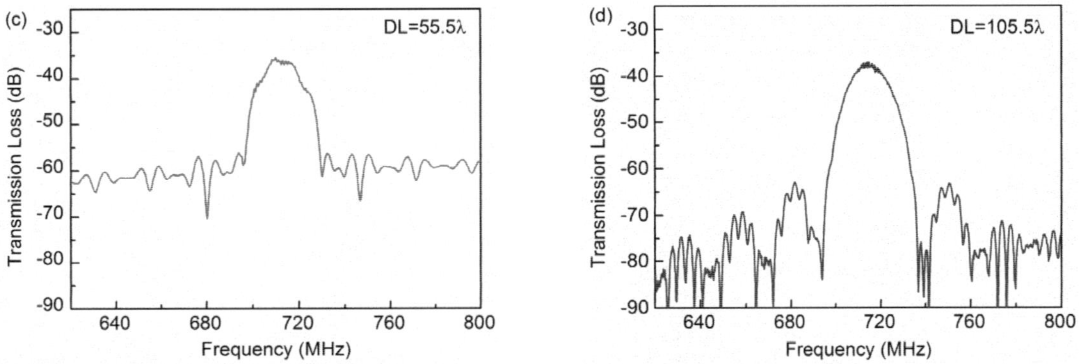

Figure 2. Transmission loss of the two-port SAW devices with 30 pairs of fingers: (a) the global transmission loss with different DLs; detailed frequency response of the fundamental mode resonances with DL of (b) 15.5λ, c) 55.5λ and (d) 105.5λ.

Figure 3 presents the pass band profile comparison between 1/8λ DL shifting, namely, π/4 phase shift. Illustrated by the almost overlapping spectrum in both Figure 3 (a) and Figure 3 (b), the frequency response shows highly resemblance before and after 1/8λ shifting of DL, indicating that the relative phase variation of the two IDTs has trivial impact on the frequency domain of the resonance.

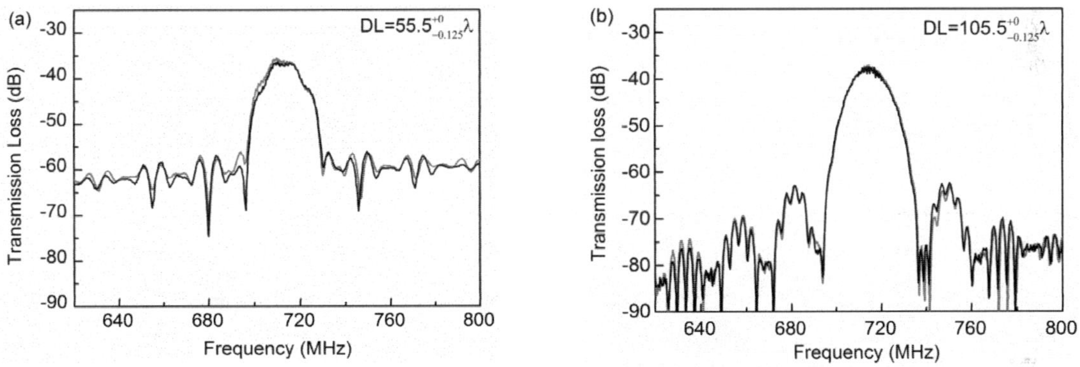

Figure 3. Frequency responses comparison of the fundamental mode resonance with 1/8λ of DL shifting: (a) 55.5λ and (b) 105.5λ.

The output of a two-port SAW device is the convolution of the frequency responses of the input IDT and the output IDT. The input IDT converts the incoming electric signal to acoustic wave through inverse piezoelectric effect, the wave then travel to the output IDT where it is converted back to electric signal by piezoelectric effect. The input signal with frequencies that match the eigen frequencies of the device will be picked up and passed, while the rest will be attenuated. When loaded, IDT generates both surface and bulk waves, those that are able to induce resonance are enhanced while the rest dissipate. When the travelling wave meet the edge of the IDT, a part of its energy will be reflected due to the acoustic impedance discontinuity, which can cause damaging second order effect. Bulk wave energy density dissipation is inversely proportional to the distance from the source in the order of -0.5, hence it decreases rapidly with distance [19, 20]. Larger DL can provide more adequate space for the dissipation of the detrimental acoustic energy such as bulk wave and reflected surface wave and thus repel them out of the travelling path of functional surface wave, providing a cleaner cavity for surface wave resonating. That is why SAW device with larger DL shows improved pass band frequency response. Another important factor that could influence on the properties of SAW delay line device is the electromagnetic feedthrough between the two IDTs, which should be restrained by longer DL between the two IDTs as well. It has found proof in the fact that when having

the same device configuration, device with smaller center frequency always reaches its optimum response at smaller DL distance as reported in another research by the authors [21].
However, larger DL elongates the propagation distance for SAW, which results in larger insertion loss, as shown by the insertion loss increase tendency with DL, and longer delay time. Besides, size augment brought by DL elongation may harm the flexibility and increases the cost of the device. It is therefore suggested that DL should be set at optimized length weighing both the performance and cost of the device.

3.3. Influence of the finger pairs number on the device performance

A very important factor impacting on the performance of SAW device is the number of finger pairs. The curves in Figure 4 present the rejection ratio and insertion loss (Figure 4 (a)) as functions of the finger pairs number. It is learned that the out of band rejection increases as the fingers number increases, and it rises to a plateau eventually. The insertion loss, on the other hand, exhibits a slower overall tendency of drop as the number of fingers is added, and reaches a saturation stage in the end. Both two curves state that the device performance grows better with increasing finger pairs and there is a saturation state for this growth. As indicated in the last section, the device performance can be affected by both stray acoustic wave energy and electromagnetic feedthrough. When keeping the number of finger pairs as the only variation parameter, the influence of feedthrough is ruled out. This behavior shown in Figure 4 illustrate that the proper increase of the fingers number is suitable for improving the performance of SAW delay line devices, especially when considering that DL decides the delay time. Further, the bandwidth of the SAW delay line varies inversely with the number of the pairs, as shown by curves in Figure 4 (b). The theoretically expected value was calculated by equation (1) [19], where N is the pair number, and f_0 is the center frequency of the device. The bandwidth of the device with 10 pairs is left out because the passband ripple is too large to provide the necessary precision for data extraction.

$$BW_{-3dB} = \frac{0.866}{N} f_0 \tag{1}$$

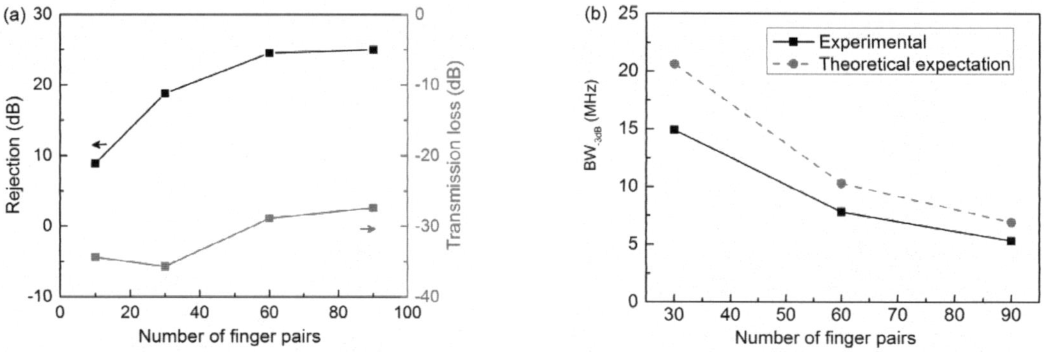

Figure 4. The influence of the number of finger pairs on the frequency response: (a) transmission loss and rejection, and (b) 3-dB bandwidth. The devices are all with center frequencies of 714 MHz and DLs of 55.5λ.

4. Conclusion

Two-port symmetrical AlN SAW device using epitaxial AlN thin film on (0001)sapphire has been fabricated by lift-off process with Ti/Al composite electrodes as IDT. The overall properties of the device can be greatly improved by DL elongation, indicated by the reduced pass band ripple and increased stop band rejection ratio. This is because larger DL provides larger room for nonfunctional acoustic energy dissipation and thus providing cleaner path for functional SAW transmission and better cavity for resonance, and restrains electromagnetic feedthrough as well. It is also found that proper addition of the IDT finger pairs improves the device response, and there is a plateau for this improvement. The reason for this behavior also lies in the fact that more fingers provide better resonance for acoustic waves and rule out stray resonance out of the device.

Since longer distance increase delay time and the size of the device, while more fingers burden fabrication process and increase size as well, there should be a careful weighting in the design of SAW delay line devices.

5. References

1. W. C. Wilson and G. M. Atkinson, in 1st Order Modeling of a SAW Delay Line using MathCAD, [Online]. http://ntrs.nasa.gov/archive/nasa/ casi.ntrs.nasa.gov/20070016024.pdf: Nasa_Ntrs, 2007, p.6.
2. E. Blampain, O. Elmazria, T. Aubert, et al., IEEE Sensors J. **13**, 4607-4612 (2013).
3. T. M. Gronewold, Anal Chim Acta **603**, 119-128 (2007).
4. J. Luo, P. Luo, M. Xie, et al., Biosens Bioelectron **49**, 512-518 (2013).
5. T. Nomura and A. Saitoh, in *Proc. IEEE Sensors 2002*, Tokyo, Japan, 507-510 (2002).
6. M. Jo, K. J. Lee, and S. S. Yang, Sensor Actuat. A-Phys. **210**, 59-66 (2014).
7. S. Fujii, S. Kawano, T. Umeda, et. al, in *Proc. IEEE International Ultrasonics Symposium*, Beijing, 1916-1919, (2008).
8. O. Elmazria, V. Mortet, M. E. Hakiki, et. al, _IEEE T Ultrason. Ferr._**50**, 710-715 (2003).
9. C.-M. Lin, Y.-Y. Chen, V. V. Felmetsger, D. G. Senesky, and A. P. Pisano, Adv. Mater. **24**, 2722–2727 (2012).
10. T. Manzaneque, J. Hernando, L. Rodríguez-Aragón, et al., Microsyst. Technol. **16**, 837-845 (2010).
11. E. Blampain, O. Elmazria, T. Aubert, et. al, IEEE Sensors J. **13**, 4607-4612 (2013).
12. J. Zhou, M. DeMiguel-Ramos, L. Garcia-Gancedo, et al., Sensor Actuat. A- Chem. **202**, 984-992 (2014).
13. V. Mortet, M. Nesladek, J. D'Haen, et al., Phys. Stat. Sol. A **193**, 482–488 (2002).
14. A. Ababneh, M. Alsumady, H. Seidel, et al., Appl. Surf. Sci. **259**, 59-65 (2012).
15. W. C. Shih and Z. X. Zoh, Ferroelectrics **459**, 52-62 (2014).
16. S. Fujii, Phys. Stat. Sol. A **208**, 1072-1077 (2011).
17. R. Fan, H. Zhi-Biao, Z. Chen, et. al, Chin. Phys. Lett. **27**, 068101 (2010).
18. M. B. Assouar, O. Elmazria, M. Elhakiki, et. al, J. Vac. Sci. Technol. B **22**, 1717 (2004).
19. A. A. Oliner, "Acoustic Surface Waves" *in Topics in Applied Physics,* (Springer-Verlag Berlin Heidelberg, New York, 1978).
20. J. W. Grate and G. C. Frye, Sensors Update **2**, 37-83 (1996).
21. J. Gao, Z. Hao, Y. Luo. Func. Mater. Lett. **9**, 1650034 (2016).

Acknowledgments

The author appreciates the financial support by the National Basic Research Program of China (Grant No.2013CB632804), the National Natural Science Foundation of China (Grant Nos. 51602105, 61176015, 61176059, 61210014, 61321004 and 61307024), Natural Science Foundation of Guangdong, China (Grant No. 2017A030313331), the Fundamental Research Funds for the Central Universities (Grant No. F040402), the High Technology Research and Development Program of China (Grant No. 2012AA050601), and the China Postdoctoral Science Foundation.

The thermally stimulated discharge of ion-irradiated oxide films

Qiuru Wang[1], Huizhong Zeng[1] and Wanli Zhang[2]

1 University of Electronic Science and Technology of China, Chengdu, 610054, P. R. China
2 Corresponding author, State Key Laboratory of Electronic Thin Films and Integrated Devices, University of Electronic Science and Technology of China, Chengdu, 611731, P. R. China
E-mail: 201311030114@std.uestc.edu.cn

Abstract. The ion irradiation technique is utilized to modify the surface structure of amorphous insulating oxide films. While introducing defects, a number of surface charges are injected into the films and captured in the traps during ion irradiation. The variation of surface morphology and the enhancement of emission spectrum corresponding to vacancy defects are respectively verified by atomic force microscopy and photoluminescence measurements. The surface charges trapped in the shallow traps are easy to release caused by thermal excitation, and discharge is observed during heating. Based on the thermally stimulated discharge measurements, the trap parameters of oxide films, such as activation energy and relaxation time, are calculated from experimental data.

1. Introduction

Thermally stimulated discharge (TSD) is a powerful technique to study the motion of molecular and the structures of trap. It also provides useful information on the charge-storage and transport process controlled by the traps [1]. This technique was first proposed by Frei to study the motion of frozen charge during raising temperature [2], and Bucci et al. gave its theoretical basis in 1966 [3]. From then on, TSD is widely used to investigate the basic characteristics of materials, including crystal and amorphous insulators [4], especially polymer community [5, 6]. Through analyzing the space charge distribution in solid, the trap level and charge storage mechanism can be acquired from the measurement of TSD. For space charges, they usually originate from ionization, impurity, electrode emission, physical and chemical adsorption. In this paper, we utilize ion irradiation technology to introduce surface charges into the insulating oxide films, and analyze the variation of surface topography and component caused by ion irradiation. In the process of thermal excitation, the discharge behavior is clearly detected as a result of the release of surface charges during heating, and the trap parameters are obtained from the discharge curves.

2. Experimental

In this experiment, the amorphous Ta_2O_5 and TiO_2 films, deposited on the silicon substrates with dimensions of 10 × 10 mm^2, were provided by the 55th Research Institute of China Electronics Technology Group Corporation (CETC55). These insulating samples were irradiated with an Ar^+ plasma with high RF powers in a plasma cleaner (PDC-32G-2). The surface morphology and roughness of samples were acquired using atomic force microscopy (AFM, SPI3800N) with the non-contact mode. The photoluminescence (PL) spectra of oxide films were measured on a photospectrometer (Hitachi F-7000) at room temperature. The exciting wavelength of an argon ion laser was 486 nm for Ta_2O_5 and 325 nm for TiO_2, respectively.

In order to monitor the process of TSD, the patterns of interdigital electrodes were prepared on the surface of samples by lithography. As schematically shown in figure 1, the length, width and distance of each electrode are 150 μm, 10 μm and 10 μm, respectively. The Pt electrodes were sputtered with the thickness of 50 nm. During the measurement, the charges releasing from materials as a result of heating in the vacuum chamber was recorded by an electrometer (Keithley model 6517B), and the temperature was controlled by cryogenic temperature controller (Lake Shore 350) from room temperature to 423 K at a rate of 10 K/min.

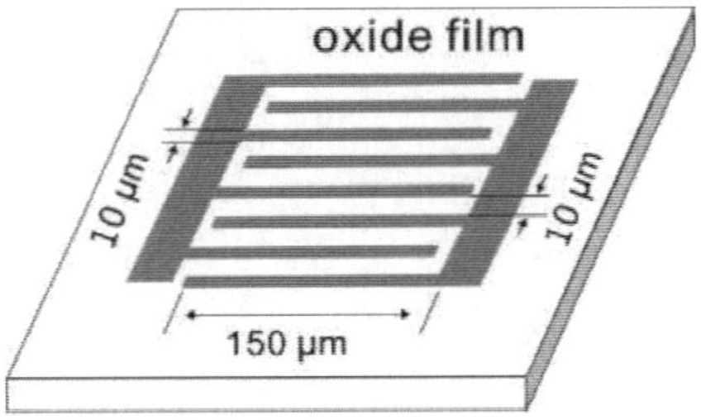

Figure 1. Schematic illustration of the interdigital electrodes structure on the oxide film.

3. Results and discussion

Ion irradiation is a common tool for cleaning the surface of materials, by which the atoms on the whole surface may get removed. Simultaneously, various types of defects can be generated at the surface because of the different diffusivity for atoms and some new physical performances are introduced into the materials during the process of ion irradiation [7-9]. Nevertheless, the obvious effect of ion irradiation on materials is the variation of surface morphology. By using the AFM, we investigate the topography on the oxide films irradiated by Ar^+ plasma with different time. Before irradiation, the AFM images of virgin films show the relatively smooth surface with the low root mean square (RMS) surface roughness < 0.7 nm (figure 2(a) and (d)). As displayed in figure 2(b) and (e), the irradiation under 8 min originally leads the surface to become smooth. With increasing irradiation time, the surface roughness rises gradually because of the damage induced by Ar^+ irradiation, and even plenty of cavities spread over the film surface are observed as illustrated in figure 2(c) and (f). This consequence is also exhibited obviously in the RMS surface roughness for oxide films with a size of 5 × 5 μm^2. As can be seen in figure 3, the RMS for both amorphous Ta_2O_5 and TiO_2 films decrease in the initial stage and then increase with the increment of irradiation time. The dynamic process of ion irradiation, like ion-milling, can account for this phenomenon. In the initial stages of irradiation, the intrinsic defects on the film surface, which are demonstrated in the PL measurements, get removed by Ar^+ plasma bombardment, while a number of defects is introduced into the surface in this process. That is the reason for the decrease of the RMS at a short period of irradiation time. However, the defects generated by ion irradiation continually accumulate with increasing treated time, which results in the roughening of film surface.

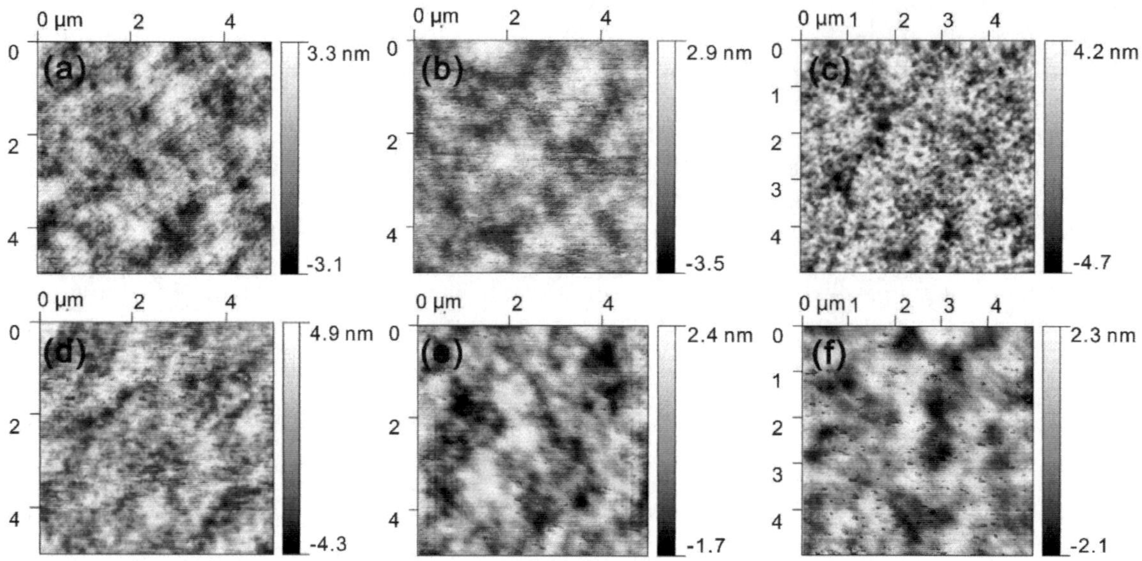

Figure 2. AFM images of (a), (b), (c) Ta_2O_5 and (d), (e), (f) TiO_2 films irradiated with 0 min, 8 min and 20 min, respectively.

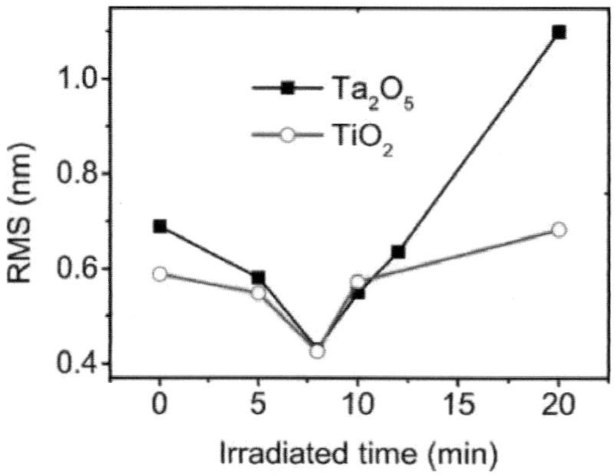

Figure 3. The irradiated time dependence of the root mean square surface roughness for oxide films.

PL spectrum is a well-developed optical method for the detection of impurities and defects in materials. As shown in figure 4, both the virgin and ion-irradiated samples were evaluated by PL spectroscopy using an excitation of Ar^+ laser line at room temperature. Figure 4(a) presents the PL spectra for Ta_2O_5 films, and a broad luminescence peak is revealed at around 600 nm for both virgin and ion-irradiated samples, which is attributed to the existence of defects such as oxygen vacancies [10]. The emitted visible light with limited intensity observed in virgin samples indicates the oxide films used in this work have vacancy defects virginally. It is seen that the PL intensity distinctly increases after ion irradiation, which means that more oxygen-vacancy defects are induced at the surface of oxide films through ion irradiation as confirmed in the irradiated $SrTiO_3$ single crystal in our previous work [11]. Similarly, this phenomenon also appears in the PL behavior for TiO_2 films. As can be seen from the figure 4(b), the intensity of emission peaks, correlated with the self-trapped excitons and oxygen vacancies [12, 13], gets enhanced greatly for the samples irradiated in vacuum. In general, the results of PL spectra indicate the increase of the defects via ion irradiation at the surface of amorphous oxide films.

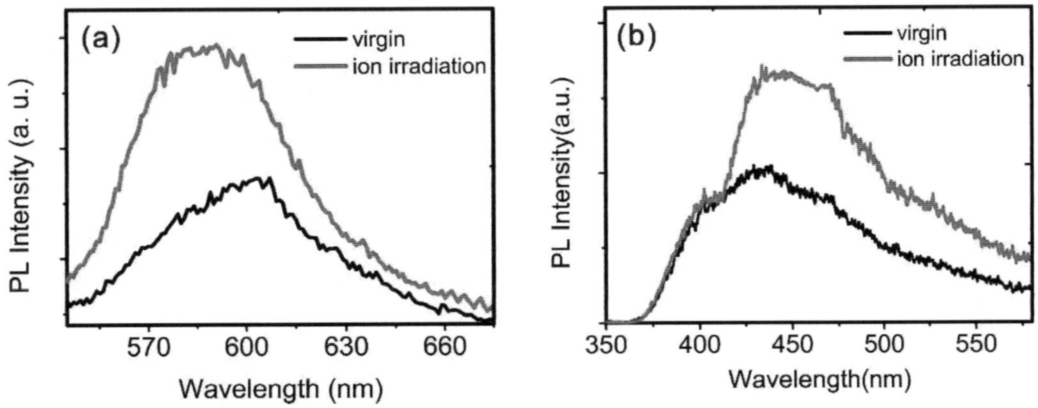

Figure 4. Room temperature photoluminescence spectra of (a) Ta_2O_5 and (b) TiO_2 before and after ion irradiation

During ion irradiation, various types of charge traps may be formed on account of structural damage induced by the generation of defects, and surface charges injected into the film surface through ion bombardment are partly captured by these physical traps. When heating the samples at a linear rate, this part of charges gaining enough energy releases from shallow traps (<1.5 eV), and a current can be detected in the external circuit, which is known as the TSD. Figure 5 shows the variation curves of charge and temperature in the whole process of thermally stimulated measurement. Each sample was heated from room temperature to 423 K at a uniform rate of 10 K/min, and cooled down naturally after kept at 423 K for a few minutes. As plotted in figure 5(a), the discharge caused by thermal excitation does not appear in the virgin films during heating, which implies there is no effective surface charges captured by shallow traps at the film surface before ion irradiation. Nevertheless, the discharge peak is obviously observed in irradiated oxide films at high temperature (figure 5(b)), and the maximum quantity of charge detected in the external circuit is about 0.28 nC. It indicates that a number of surface charges are injected into the surface via irradiated by Ar^+ plasma and the formation of defects during this bombarding process may contribute to the storage of charges.

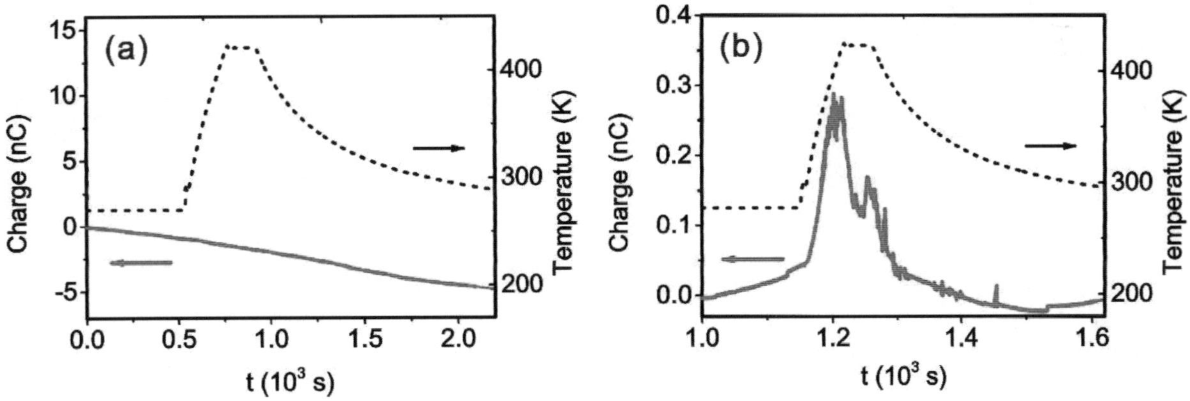

Figure 5. The results of thermally stimulated discharge measurement for (a) virgin and (b) ion-irradiated oxide film.

The results of TSD measurement are very sensitive to the structure of trap, and the trap parameters can be obtained from the current density curve as a function of temperature [3, 14]

$$j(T) = Aexp\left(\frac{-E_a}{kT}\right)exp\left[-\frac{B}{\beta}\int_{T_0}^{T}\exp\left(\frac{-E_a}{kT'}\right)dT'\right], \quad (1)$$

where E_a, k and β denote the activation energy of the process, the Boltzmann's constant and the heating rate, respectively. A and B in this equation are the constants related to the initial polarization and the characteristic relaxation time τ_0 at infinite temperature. As shown in figure 6, the surface charges release from shallow traps as a result of thermal excitation. However, the activation energy E_a is linear with *ln j(T)* measured by the following relation for the low temperature part [15, 16]

$$\ln j(T) = \text{const.} - \frac{E_a}{kT}. \quad (2)$$

Based on the fitting of experimental data (the insets of figure 6), the activation energy for Ta_2O_5 and TiO_2 are calculated as 0.121 eV and 0.228 eV, respectively, which implies the surface charges are mainly trapped in these traps with low energy level and are easy to be removed from the film surface. Furthermore, the relationship between the temperature T_m corresponding to the maximum current and activation energy can be written as [17]

$$T_m = \left[\beta\tau_0 \frac{E_a}{k} \exp\left(\frac{E_a}{kT_m}\right)\right]^{1/2}. \quad (3)$$

It can be seen that the value of T_m is fixed for a given heating rate. As displayed in Figure 6(a) and (b), the peaks of current curve are 0.242 pA at 370 K and 0. 343 pA at 356 K, respectively. Thus, their characteristic relaxation time τ_0 obtained from equation (3) are about 13 s for Ta_2O_5 and 0.17 s for TiO_2. Based on the Arrhenius-type relationship between the relaxation time τ and τ_0 [14], the values of τ for irradiated Ta_2O_5 and TiO_2 films at the maximum temperature T_m can be calculated as 585 s and 287 s, which implies the speedy attenuation of charges due to the low-level traps in our case. In a word, the measurements and analyses of TSD demonstrate that these two kinds of oxide films have certain significance in the research on energy storage, and ion irradiation is an effective method to the injection of surface charges.

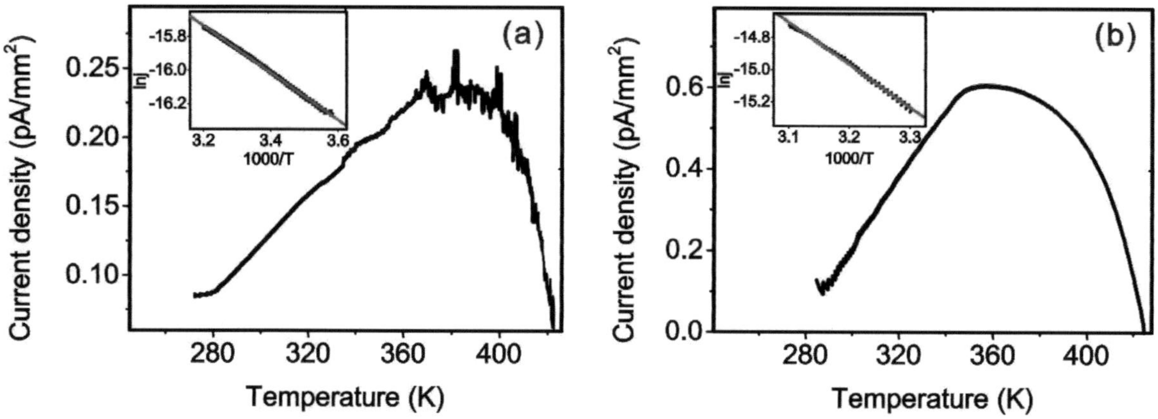

Figure 6. The temperature dependence of current density for (a) Ta_2O_5 and (b) TiO_2. The insets show respectively the fit to equation (2) in the initial rise of curves.

4. Conclusions

In summary, defects were introduced into the surface of insulating Ta_2O_5 and TiO_2 films by Ar^+ ion irradiation, which were characterized by the increase of roughness and the enhancement of PL spectra corresponding to vacancy defects. Simultaneously, space charges were injected into the film surface and captured by the traps during ion irradiated, which were easily removed by heating at a linear rate. According to TSD measurements, the activation energy and relaxation time of the trap in two oxide films were estimated respectively, which provides evidence for the evaluation of trap states. This work demonstrates the combination of ion irradiation and TSD technique is the simple and effective method to help people understand the charge storage characteristics and attenuation process in oxide films.

5. References

[1] R. Chen and Y. Kirsh, *Analysis of Thermally Stimulated Process* (Pergamon Press, Oxford,

1981).
[2] H. Frei and G. Groetzing 1936 *Phy. Z* **37** 720-724
[3] C. Bucci, R. Fieschi and G. Guidi 1966 *Phys. Rev.* **148** 816-823
[4] H. Amjadi 1999 *IEEE. Trans. Electr. Insul.* **6** 236-241
[5] N. V. Malm, J. Steiger, T. Finnberg, R. Schmechel and H. V. Seggern 2002 *Proc. SPEI-The Int. Soc. Opt. Eng.* **4800** 164-171
[6] E. J. Kim, T. Takeda and Y. Ohki 1996 *IEEE Trans. Electr. Insul.* **3** 386-391
[7] M. Chassé and G. G. Ross 2002 *J. Appl. Phys.* **92** 5872-5877
[8] D. Kan, R. Kanda, Y. Kanemitsu, Y. Shimakawa and M. Takano 2006 *Appl. Phys. Lett.* **88**, 191916
[9] M. Rogala, Z. Klusek, C. Rodenbucher, R. Waser and K. Szot 2013 *Appl. Phys. Lett.* **102** 131604
[10] M. Zhu, Z. Zhang, W. Miao, and W. Miao 2006 *Appl. Phys. Lett.* **89** 021915
[11] Q. Wang, Wanli Zhang, W. Zhang and H. Zeng 2016 *Appl. Surf. Sci.* **365** 84-87
[12] L. V. Saraf, S. I. Patil, S. B. Ogale, S. R. Sainkar and S. T. Kshirsager 1998 *Int. J. Mod. Phys. B* **12** 2635
[13] Y. Lei, L. D. Zhang, G. W. Meng, G. H. Li, X. Y. Zhang, C. H. Liang, W. Chen and S. X. Wang 2001 *Appl. Phys. Lett.* **78** 1125
[14] E. R. Neagu, J. N. Marat-Mendes, D. K. Das-Gupta, R. M. Neagu and R. Igreja 1997 *J. Appl. Phys.* **82** 2488- 2496
[15] T. Hino 1980 *IEEE Trans. Electr. Insul.* **15** 301-311
[16] W. Liu and C. A. Randall 2008 *J. Am. Ceram. Soc.* **91** 3245-3250
[17] J. Vanderschueren and J. Gasiot, "Field-induced thermally stimulated currents," in *Thermally Stimulated Relaxation in Solids*, edited by P. Bräunlich (Springer-Verlag Berlin Heidelberg, New York, 1979), pp. 135-223

Acknowledgments
This work was supported by the NSFC projects (No. 61471095 and No. V01435208) and "863"-projects (No. 2015AA03130102).

IOP Conf. Series: Materials Science and Engineering **284** (2017) 012030 doi:10.1088/1757-899X/284/1/012030

High precision slotted cavity measurement of a novel ceramic state polymer electrolyte

Wei Quan [1] **and Mohammed NurulAfsar** [1, 2]

1 Electrical and computer Engineering Department, Tufts University, USA
2 Fellow, IEEE
E-mail: wei.quan@tufts.edu

Abstract. Thin film materials are already used in a variety of microwave and higher frequency applications such as electrically tunable microwave devices, integrated circuits like MMICs, radomes, and radar absorbing coating. The determination of the dielectric properties of these films is thus of significant importance. The measurement of complex dielectric permittivity of thin films is very difficult at microwave, millimeter, and THz frequencies because both the amplitude change and phase shift are not large enough to evaluate the real part of the dielectric permittivity. A specially designed transverse slotted cavity for X-band microwave measurement has been designed and constructed to employ with a vector network analyzer to evaluate the real part of dielectric permittivity of thin films accurately and conveniently. Commercially available polymer thin films are measured to validate the methods.

1. Introduction

Although there are various well documented methods for determining the dielectric properties of low-loss ultra-thin films, finding methods that are applicable and sensitive enough for use on ultra-thin materials is a challenge. Close-ended coaxial probe technique is developed based on measuring the reflection coefficient from a coaxial transmission line. References [1, 2] take advantage of microstrip transmission lines and resonators. Dielectric resonator methods are also implemented on measuring properties of thin films [3-5].

The newly developed slotted cavity technique described in this paper can measure very thin materials around 1 mil (25.4 μm) thickness over microwave frequencies. A slotted cavity measurement scheme is shown here that employs a transverse slot in the narrow wall through which the sample can be inserted and removed. This method is much simpler and the sample occupies the entire cross section of the cavity. The transverse slot is more sensitive than longitudinal slot because the sample is oriented parallel to the wavefront, which can provide a reduced transmission and higher reflection.

The measurement techniques described in this paper on thin films in microwave frequency range are based on waveguide theory. The in-waveguide measurement, cavity measurement, transverse slotted cavity measurement are all employing the rectangular waveguides with Agilent 8510C vector network analyzer (VNA) to record the scattering parameters, frequency shift and quality factor. From the scattering parameters, frequency shift and quality factor, the complex permittivity of the thin films are determined.

2. Measurement technique

Two ports Vector Network Analyzer is used to characterize two port networks. In this study, it was employed to characterize the thin film material under test, which was loaded in waveguides or transverse slotted cavity to form the two-port network. An accurate measurement involves standardization and calibration of the system at the required frequencies and ambient conditions. Here, measurement was carried out from 2 to 40 GHz. The network analyzer includes a signal source, a receiver and a display unit to process the signals. The block diagram of network analyzer is shown in Figure 1.

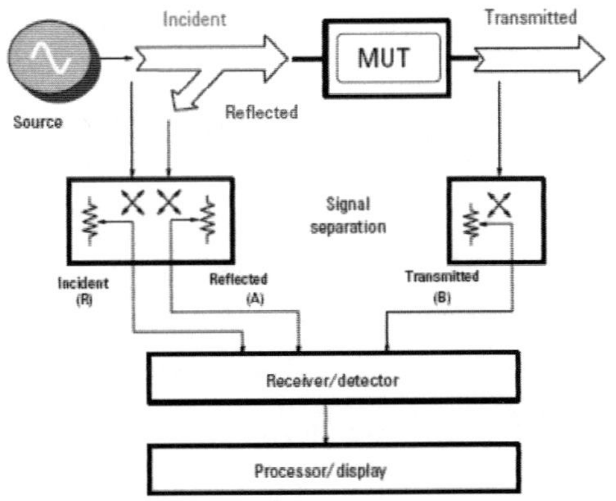

Figure 1. Signal flow inside the vector network analyzer showing how the data is measured and recorded.

The source launches a RF signal at a single frequency towards the sample under test. The test set separates the signal produced by the source into an incident signal, sent to the sample under test, and a reference signal against which the transmitted and reflected signals are later compared. The test set routes the transmitted and reflected signals from the material under test to the receiver. The receiver is tuned to the frequency of the RF signal provided by the source. The source is then stepped to the next frequency and the measurement is repeated to display the reflection and transmission response as a function of frequency. The measured response produces the magnitude and phase for each scattering parameter at that frequency.
The following techniques can be used with the network analyzer for electromagnetic measurements: Free space measurement, Resonant Cavity Measurement, In-Waveguide Measurement, Open-ended waveguide measurement. In this dissertation, in-waveguide measurement method and transverse slotted cavity measurement are showed.

2.1. In-waveguide measurement

The Here, a transmission reflection (T/R) based waveguide technique was used to carry out the measurements. The T/R method is a category of non-resonant methods that are widely used for the measurements of electromagnetic properties of materials. In this method, the sample under test is inserted into a segment of transmission line, such as waveguide or coaxial line, which forms the two port network shown in Figure2. The cables from the network analyzer are connected to across this network. The Vector Network Analyzer records the s-parameter values. Scattering equations are used to analyze the fields at the sample interfaces. These equations relate the s-parameters of the segment of transmission line filled with the sample under study to the permittivity and permeability of that sample. In T/R method, all the four s-parameters can be measured, so we have a record of more data than what we have in reflection measurements.The modified sparameters used in the measurements in this work are given below,

$$\widetilde{S}_{11} = S_{11} e^{j\left(0 \times \sqrt{k_0^2 - k_c^2}\right)}; \widetilde{S}_{21} = S_{21} e^{j\left((l-d) \times \sqrt{k_0^2 - k_c^2}\right)} \quad (1)$$

where, l is the quarter wavelength difference between thru and line (in air), d is the thickness of the sample inside the waveguide, k0 is the wavenumber of the sample and kc is the cutoff wavenumber. These equations take into account the effect of using samples with thickness (d) values that are smaller than the waveguide shim used in the experimental setup. [6] Return losses of less than -50 dB from the air inside the waveguide are easily achieved using these calibration techniques. This enables us to neglect any unwanted reflections from the inner walls of the waveguide when analyzing the S-parameters. The reflection and transmission by the scattering parameters inside the waveguide, in which the transmission and reflection resemble the free space formulation, can now be presented as follows:

$$\Gamma = K \pm \sqrt{K^2 - 1}; K = \frac{\tilde{S}_{11}^2 - \tilde{S}_{21}^2 + 1}{2\tilde{S}_{11}}; T = \frac{\tilde{S}_{11} + \tilde{S}_{21} - \Gamma}{1 - (\tilde{S}_{11} + \tilde{S}_{21})\Gamma} \quad (2)$$

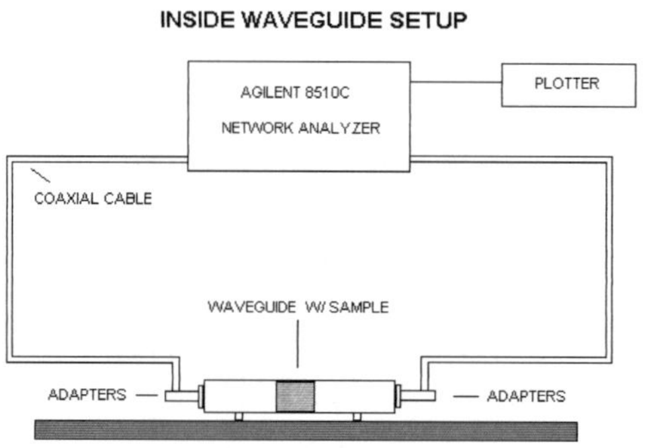

Figure 2. Setup of in-waveguide measurement method.

The transmission coefficient through the material may also be written as $T = e^{-\gamma d} = e^{-(a+j\beta)d}$.
The propagation constant through the material inside the waveguides has been derived to be:

$$\gamma_{TE_{10}} = \frac{\ln\left(\frac{1}{|T|}\right)}{d} + j\left(\frac{2\pi n - \varphi_T}{d}\right) \quad (3)$$

The complex permeability and permittivity associated with the propagation constant are then:

$$\mu = \frac{\eta \gamma_{TE_{10}}}{j\gamma_{TE_{10}}^0} = -j\left(\frac{1+\Gamma}{1-\Gamma}\right)\left(\frac{1}{2\pi d}\right)\left(\frac{\ln\left(\frac{1}{|T|}\right) + j(2\pi n - \varphi_T)}{\sqrt{\left(\frac{1}{\lambda_0}\right)^2 - \left(\frac{1}{2a}\right)^2}}\right)$$

$$\varepsilon = \frac{\mu}{Z_{TE}^2} = \frac{\mu}{\eta^2}\left(\lambda_0^2\left(\left(\frac{1}{\lambda_0}\right)^2 - \left(\frac{1}{2a}\right)^2\right)\right) \quad (4)$$

In our waveguide measurement technique the propagating wave was assumed to be the TE10 mode. Permittivity is then calculated as follows:

$$\varepsilon = -j\left(\frac{c}{f}\right)^2\left(\frac{1+\Gamma}{1-\Gamma}\right)\left(\frac{1}{2\pi d}\right)\left(\ln\left(\frac{1}{|T|}\right) + j(2\pi n - \varphi_T)\right)\left(\sqrt{\left(\frac{1}{\lambda_0}\right)^2 - \left(\frac{1}{2a}\right)^2}\right) \quad (5)$$

It was also noticed that the permeability and permittivity of the loaded sample affect the cut-off frequency for the waveguide band. This was accounted for in the calculations by including the cut-off frequency for each band in the derivation of permeability and permittivity from the data for s-parameters. The divergence in data was eliminated by using the electrical delay function of the network analyzer. The experiments carried out in the lab suggest that these modifications are necessary and known materials were measured to confirm the accuracy of the measurement technique. The derived permeability and permittivity data is very reliable and not affected by the scattering voltage ratios of the vector network analyzer.

3. Slotted cavity measurement

The In this section, a special designed transverse slotted cavity is built in X-band. It was applied to determine the complex permittivityof thin film materials by employing perturbation theory.

The slotted cavity technique is a combination of in-waveguide method and cavity method. The cavity method employs waveguide with a hole where the cylinder sample can be inserted in the waveguide. Each cavity has resonant frequenciesdepending on the dimension of the cavity. The inserted sample will slightly change the distribution of electromagnetic fields inside the cavity thus change the resonant frequency and quality factor.

However, a key difference in this transverse slotted cavity method compared to the traditional cavity method is that slots can be made in the waveguide wall transverse to wave propagation direction through which the sample can be fed into the waveguide [7]. This allows a very accurate measurement of the difference in scattering parameters with and without the sample in place which, in turn, allows for much greater measurement sensitivity. It is similar to a perturbation technique in that differences in scattering parameters are used in the calculations rather than the scattering parameters themselves.

Figure3 shows the WR-90 slotted X-band waveguide used for the measurement. The waveguide is 1” long with the sample placed at the center. The sample is fed through narrow slots in both side walls (rather than the longitudinal slot[7]) prior to the first measurement. It is held taut by pulling on both ends of the sample. This novel approach enabled the sample to stay in place without foam or Teflon supports. After the firstmeasurement is taken, the sample is removed through one of the slots, and a second measurement is taken of the system without the sample.

Figure 3. Transverse slotted X-band Waveguide. A transverse slot was made at the center transverse section of the waveguide.

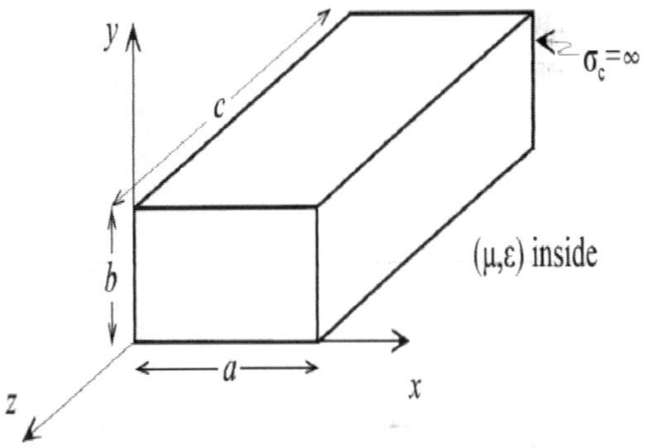

Figure 4. Dimension of the resonant cavity

The transverse slotted cavity comes from the rectangular waveguide resonator. This resonator is basically a section of rectangular waveguide which is enclosed on both ends by conducting irises to form an enclosed conducting box. We assume the same cross-sectional dimensions as the rectangular waveguide (a, b) and define the longitudinal length of the resonator as c shown in Figure4. Given the conducting walls on the ends of the waveguide, the resonator modes may be described by waveguide modes which are reflected back and forth within the resonator (+z and -z directions) to form standing waves.

Given a resonator made from a conductor such as copper or aluminum, the ohmic losses are very small and the quality factor is large (high Q, small bandwidth). Thus, resonators are used in applications such as oscillators, filters, and tuned amplifiers. Comparing the modes of the rectangular resonator with the propagating modes in the rectangular waveguide, we see that the waveguide modes exist over a wide band (the rectangular waveguide acts like a high-pass filter) while the rectangular resonator modes exist over a very narrow band (the rectangular resonator acts like a band-pass filter).

In the past, many researchers have reported the theoretical [8-9] and experimental [10-12] results of the cavity perturbation techniques. The measurements of permittivity and permeability of the dielectric materials are performed by inserting a small and appropriately shaped sample into a cavity from a pinhole or longitudinal slot and determining the properties of the sample from the resultant change in the resonantfrequency and loaded quality factor of the cavity. The basic idea of the cavity perturbation is the change in the overall geometric configuration of the electromagnetic fields with the insertion of a small sample must be small. Based on this assumption, a detailed derivation of the perturbation equation for the frequency shift upon the insertion of a sample into a cavity was given by Harrington [13].

In practice, most cavity measurements are also subject to certain pitfalls. The thin material must be self-supporting or supported in the proper place within the cavity. Typically, this is done with some low permittivity material such as foam or Teflon, whose effects must usually be calibrated out. This introduces sources of error to the measurement if the cavity is not assembled in exactly the same way for both the sample-loaded measurement and the unloaded reference measurement (i.e., with and without the thin material of interest). Problems can arise if the sample holder shifts position or a fastening bolt has even slightly different tension on it. Even the process of handling the cavity may cause temperature variations which could affect the geometry enough to be significant to the measurement, which requires measuring frequency shifts well under 1%.

To solve this problem, a novel slotted cavity is employed that allows the sample to be inserted and removed without perturbing the measurement setup. The same waveguide section used in the waveguide method discussed above is used again here with the addition of two coupling irises. Each iris is 0.03" thick and 0.08" radius and serves tocouple energy from WR-90 input waveguide into the dominant TE101 mode of the cavity. Two-port measurements were selected over one port because very good signal-to-noise ratio can be obtained in S21 by increasing the input source power. Figure5 shows the cavity with irises.

Figure 5. Slotted Cavity with Coupling Irises. Two irises are added to the slotted X-band waveguide.

From Equation of the cavity resonant frequencyeq 18, we can observe that the change of dielectric permittivity and magnetic permeability will lead the resonant frequency changing.

$$f_{resonant} = \frac{1}{2\sqrt{\mu\varepsilon}}\sqrt{\left(\frac{m\pi}{a}\right)^2 + \left(\frac{n\pi}{b}\right)^2 + \left(\frac{p\pi}{c}\right)^2} \quad (6)$$

In the unperturbed state, the empty cavity,By some mathematic manipulations, we obtain

$$\varepsilon' = 1 + \left(1 - \frac{f}{f_0}\right)\frac{c}{t}; \varepsilon'' = \frac{1}{2}\left(\frac{1}{Q} - \frac{1}{Q_0}\right)\frac{c}{t} \quad (7)$$

where the quality factor (Q) of a waveguide resonator isdefined as

$$Q = \frac{f_{resonant}}{\Delta f_{3dB\_bandwidth}} \quad (8)$$

The Agilent 8510C vector network analyzer is used to make the slotted cavity measurement. Figure6 shows the special slotted cavity setup. The thin films can be placed inside the waveguide. TRL calibration was used in order to minimize the systematic errors in the measurement process. The advantage of a transverse slotted cavity for X band connected at the other port to measure the thin films was also utilized here. The transverse slotted waveguide method is a novel technique and a number of known samples such as Teflon, Mylar, and black polyester are measured for validation of this new technique.

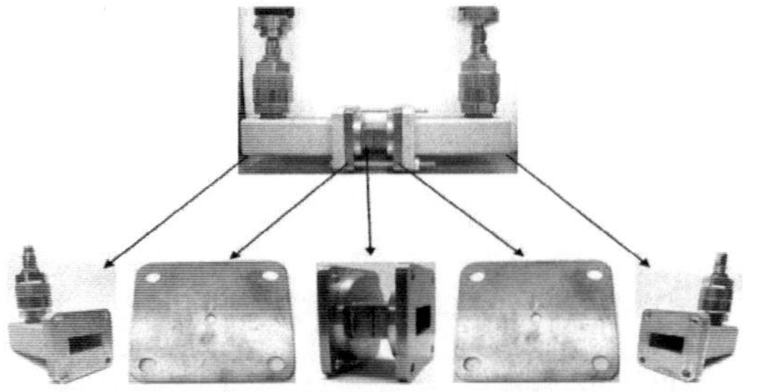

Figure 6. Setup for transverse slotted cavity measurement for the measurement at X–band. The cavity has a slot in the transverse direction.

The advantage of a transverse slotted cavity for X band connected at the other port to measure the films was also utilized here. The uncertainty in imaginary part was minimized by repeating measurements for more than 10 times. TRL calibration was used in order to minimize the systematic errors in the measurement process. The sample thickness was increased by adding multiple layers. The systematic error was significantly reduced with increasing number of layers. The final uncertainty of measured data is in ±5%.

4. Measurement Results

Real and imaginarypermittivity from 8.2 GHz to 40 GHz are shown in Figure7 and Figure8.

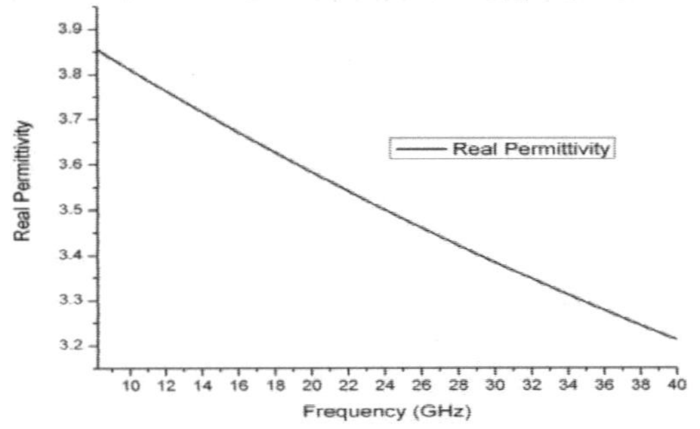

Figure 7. Real part of relative permittivity of solid polymer electrolyte from 4 GHz to 40 GHz. The dielectric constant keeps reducing through the frequency range as the ionic relaxation weakens.

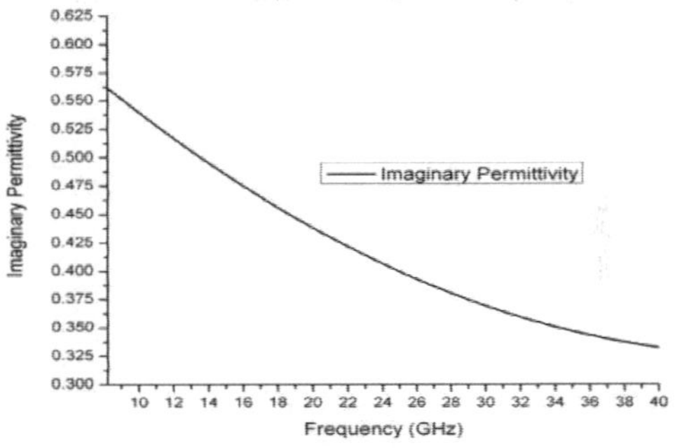

Figure 8. Imaginary part of relative permittivity of solid polymer electrolyte from 4 GHz to 40 GHz. The field direction changes faster than the ions fully diffuse through the electrolyte. Thus the imaginary part of dielectric permittivity decreases.

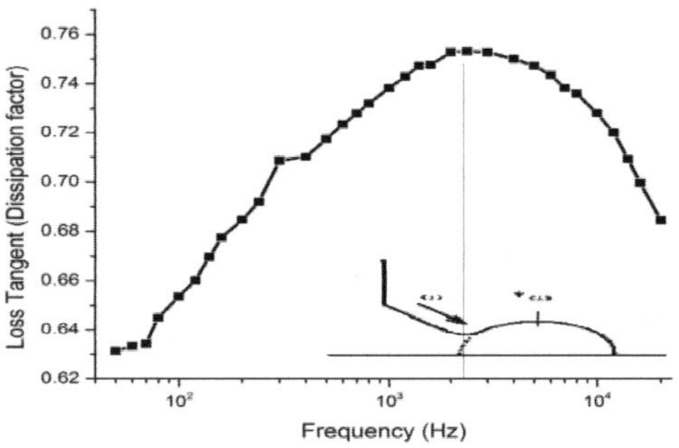

Figure 9. Real part of relative permittivity of solid polymer electrolyte from 4 GHz to 40 GHz. The dielectric constant keeps reducing through the frequency range as the ionic relaxation weakens.

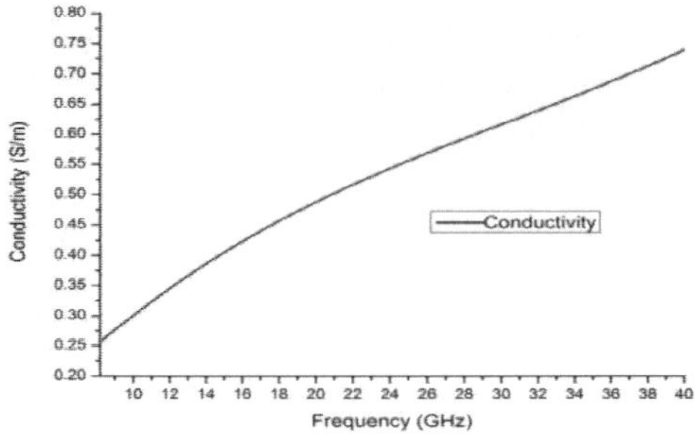

Figure10. Imaginary part of relative permittivity of solid polymer electrolyte from 4 GHz to 40 GHz. The field direction changes faster than the ions fully diffuse through the electrolyte. Thus the imaginary part of dielectric permittivity decreases.

Loss tangent spectra in low frequency from 50 Hz to 20 kHz are shown in Figure9. Real conductivity spectra measured by in-waveguide method in 4 GHz to 40 GHz are shown in Figure 10. The Cole-Cole plot is shown in Figure11.

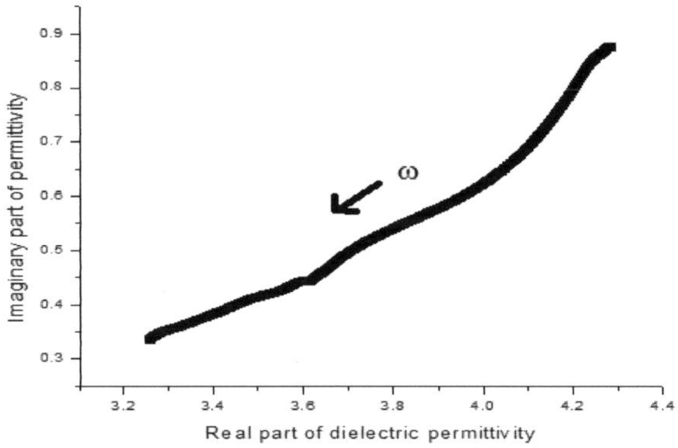

Figure 11. Cole-Cole plot of the electrolyte in microwavefrequencies. The x-axis is the real ε',the y-axis ε''.Essentially it is arepresentation of the Nyquist plot at microwavefrequencies. It is the higher frequency tail of a traditionalsemicircle Cole- Cole plot.

The comprehensive result is showed in TABLE1.

TABLE 1. Dielectric Measurement Result

Frequency (GHz)	ε''	ε'	Loss Tangent	Conductivity (S/m)
9.019	0.55178	3.835256	0.1438708	0.276856998
10.006	0.53999	3.811258	0.1416830	0.300591531
11.014	0.52827	3.78703	0.1394958	0.323694431
12.001	0.51712	3.76358	0.1374018	0.345256059
13.016	0.50598	3.739747	0.1352989	0.3663907
14.024	0.49525	3.716363	0.1332627	0.386392639
15.004	0.48514	3.693899	0.1313342	0.404948834
16.012	0.47505	3.671072	0.1294047	0.42317381
17.02	0.4653	3.648528	0.1275315	0.4405802
18	0.45614	3.626881	0.1257658	0.45677094

19.02	0.44693	3.604634	0.1239875	0.472911275
20.04	0.43806	3.582676	0.1222715	0.488382842
21.0175	0.42987	3.561905	0.1206868	0.502635377
22.0375	0.42166	3.540515	0.1190971	0.516963689
23.015	0.41411	3.520287	0.1176365	0.530226167
24.035	0.40657	3.499463	0.1161794	0.543631569
25.0125	0.39965	3.479778	0.1148487	0.556115153
26.0325	0.39276	3.459521	0.1135302	0.568817968
27.04	0.38629	3.439796	0.1122996	0.581096097
28.0525	0.38012	3.420258	0.1111365	0.59322192
29.065	0.37428	3.401005	0.1100485	0.605190077
30.01	0.36912	3.383292	0.1091022	0.616267686
31.0225	0.36393	3.364591	0.1081640	0.62809038
32.035	0.35906	3.346174	0.1073056	0.639920244
33.0475	0.35453	3.328043	0.1065284	0.651813454
34.06	0.35033	3.310197	0.1058341	0.663826259
35.005	0.34671	3.293797	0.1052622	0.675195622
36.0175	0.34316	3.276503	0.1047324	0.687598916
37.03	0.33993	3.259493	0.1042898	0.700286606
38.0425	0.33704	3.242768	0.1039360	0.713314864
39.055	0.33448	3.226329	0.1036725	0.726739927
40	0.33239	3.211243	0.1035092	0.739677551

5. Conclusion

Real part and imaginary part of complex permittivity of this thin film samples were accurately determined in frequencies. The novel ceramic state polymer electrolyte is measured and analyzed and exhibits high conductivity as desired.The newly developed slotted cavity method, which allows the sample to be inserted and removed without perturbing the measurement setup,is thus established as useful methods to characterize complex permittivity of thin film materials. The slotted cavity method is very convenient for it is easy to insert and the thin film material does not need any external support. And the measurement precision under shift position in dielectric measurement is improved. The work on thin films composed of known materials validates the ability of precise determination of dielectric permittivity by such technique. In the future, new samples can be studied using these techniques to determine their material properties and hence suitable applications at this frequency range. An extensive repeated measurements and signal averaging will increase the signal to noise ratio and the Q-factor significantly to allow us the determination of the imaginary part of complex dielectric permittivity and the loss tangent of thin film materials at microwave frequencies.

6. Reference

[1] J. Baker-Jarvis, R. G. Geyer, J. H. Grosvenor Jr, M. D. Janezic, C. A. Jones, B. Riddle, C. M. Weil, and J. Krupka, "Dielectric characterization of low-loss materials a comparison of techniques," Dielectrics and Electrical Insulation, IEEE Transactions on, vol. 5, pp. 571-577, 1998.

[2] B. Chung, "A convenient method for complex permittivity measurement of thin materials at microwave frequencies," Journal of Physics D: Applied Physics, vol. 39, p. 1926, 2006.

[3] H. Jin, S. Dong, and D. Wang, "Measurement of dielectric constant of thin film materials at microwave frequencies," Journal of Electromagnetic Waves and Applications, 23, vol. 5, pp. 809-817, 2009.

[4] A. Vyas, V. Rana, D. Gadani, and A. Prajapati, "Cavity perturbation technique for complex permittivity measurement of dielectric materials at X-band microwave frequency," 2008, pp. 836-838.

[5] M. N. Afsar, "Dielectric measurements of millimeter-wave materials," Microwave Theory and Techniques, IEEE Transactions on, vol. 32, pp. 1598-1609, 1984.

[6] A. Sharma, "Design of Broadband Microwave Absorbers for Application in Wideband Antennas," Master of Science, Electrical and Computer Engineering, TUFTS UNIVERSITY, 2011.
[7] B. Chung, "A convenient method for complex permittivity measurement of thin materials at microwave frequencies," Journal of Physics D: Applied Physics, vol. 39, p. 1926, 2006.
[8] R. Coccioli, G. Pelosi, and S. Selleri, "Characterization of dielectric materials with the finite-element method," Microwave Theory and Techniques, IEEE Transactions on, vol. 47, pp. 1106-1112, 1999.
[9] Y. Xu and R. G. Bosisio, "Analysis of different coaxial discontinuities for microwave permittivity measurements," Instrumentation and Measurement, IEEE Transactions on, vol. 42, pp. 538-543, 1993.
[10] B. Viswanathan, R. Raman, N. Raman, and V. Murthy, "Microwave power loss and XPS measurements on high TcNd---Ba---Cu---Oxide superconducting system," Solid state communications, vol. 66, pp. 409-411, 1988.
[11] V. Murthy and R. Raman, "A method for the evaluation of microwave dielectric and magnetic parameters using rectangular cavity perturbation technique," Solid state communications, vol. 70, pp. 847-850, 1989.
[12] B. Meng, J. Booske, and R. Cooper, "Extended Cavity Perturbation Technique to Determine the Complex permittivity of dielectric materials," Microwave Theory and Techniques, IEEE Transactions on, vol. 43, pp. 2633-2636, 1995.
[13] R. F. Harrington, "Time-Harmonic Fields," New York: McGraw-Hili, 1961.

IOP Conf. Series: Materials Science and Engineering **284** (2017) 012031 doi:10.1088/1757-899X/284/1/012031

Low-cost growth of magnesium doped gallium nitride thin films by sol-gel spin coating method

N Mohd Amin and S S Ng*

Institute of Nano Optoelectronics Research and Technology (INOR), Universiti Sains Malaysia, 11800 USM, Penang, Malaysia
*E-mail: shashiong@usm.my

Abstract. Low-cost sol-gel spin coating growth of magnesium (Mg) doped gallium nitride (GaN) thin films with different concentrations of Mg was reported. The effects of the Mg concentration on the structural, surface morphology, elemental compositions, lattice vibrational, and electrical properties of the deposited films were investigated. X-ray diffraction results show that the Mg-doped samples have wurtzite structure with preferred orientation of GaN(002). The crystallite size decreases and the surface of the films with pits/pores were formed, while the crystalline quality of the films degraded as the Mg concentration increases from 2% to 6. %. All the Raman active phonon modes of the wurtzite GaN were observed while a broad peak attributed to the Mg-related lattice vibrational mode was detected at 669 cm^{-1}. Hall effect results show that the resistivity of the thin films decreases while the hole concentration and hall mobility of thin films increases as the concentration of the Mg increases.

1. Introduction

Wide band gap gallium nitride (GaN) semiconductor has received significant attention as a promising material for the applications of the optoelectronics and high-temperature/high-power devices such light emitting diode (LEDs) and laser diode (LD). The GaN have unique properties such as wide direct band gap (3.4 eV) [1], high breakdown field and high thermal conductivity [2]. This innovation brought about ever increasing research efforts towards more high-performance nitride-based LEDs and LDs with p-type GaN. Until now, magnesium (Mg) doped GaN is the most successful in the development of GaN based optoelectronic devices. Effective research of p-type GaN:Mg layer has been one of the dynamic encounters for realizing blue and green LEDs. The main reason uses of the p-type GaN in the development optoelectronic devices is to supply sufficient concentration of holes so that, when embedded into the active layer, they will recombine with electrons and emit light effectively. Therefore, a higher hole concentration in the p-GaN layer is advantageous to increase the light intensity of LEDs. However, the dopant concentration in the Mg doped GaN layer must be optimized to obtain ohmic metal contacts [3], which is important to decrease the operating voltage of the devices [4]. Despite the steady progress in Mg-doped GaN thin films, several issues remain to be solved. For instance, there are still difficult to produce highly p-type conductivity and difficult to activate the Mg dopant in GaN thin films. Up until now, highly advanced and expensive growth techniques such as metalorganic chemical vapour deposition [5] and molecular beam epitaxy [6] were used to produce p-GaN. While few complex approaches were used to activate the Mg-dopant in GaN thin films, i.e., thermal annealing [7], low-energy electron beam irradiation [8] and microwave treatment [9].

In order to produce more economical p-GaN films, chemical synthesis process using a sol–gel spin coating method was purposed. This method has received considerable attention due to its many benefits such as simple deposition procedure, easier composition control, low equipment cost, and large area coating capability [10-11]. At present, sol-gel spin coating growth of GaN thin films has been successfully demonstrated by Fong et. al [12-14]. Nevertheless, this method has not been applied to grow the p-GaN thin films. For this reason, sol-gel spin coating growth of Mg-doped GaN thin films with different concentrations of Mg was carried out. In this work, the influences of the Mg concentrations on the structural, surface morphology, elemental composition, lattice vibrational and electrical properties of deposited films were investigated systematically using various characterization techniques.

2. Experimental

In this experiment, Mg doped GaN thin films were deposited on commercial AlN template by sol-gel spin coating method. The AlN template was obtained from the Kyma Technologies, Inc. It consists of a thin crystalline AlN layer (thickness around 25 nm) grown epitaxially on silicon (111) substrate.
The sol-gel precursor was prepared using gallium nitrate hydrate ($Ga(NO_3)_3.xH_2O$), magnesium nitrate hydrate ($Mg(NO_3)_2.6H_2O$), diethanolmine (DEA), ethanol (purity, 99.7 %) and distilled water. The function of each material was as Ga source, Mg dopant, stabilizer and solvent, respectively. The precursor solutions with different Mg concentrations (i.e., 0, 2, 3 and 6 %) were prepared by dissolving an appropriate amount of $Ga(NO_3)_3.xH_2O$ and $Mg(NO_3)_2.6H_2O$ into ethanol. Subsequently, DEA and distilled water were slowly added. The resulting mixture was then stirred in ultrasonic bath to form a clear and transparent homogeneous mixture. The precursor was deposited on AlN/Si substrate by the spin-coating method at room temperature with a spinning rate of 3000 rpm for 30 s. The coating process was repeated for few times to obtain films with desired thickness (i.e., approximately 0.5 – 0.6 μm). Next, the coated films were annealed in furnace at 950 ℃ for 15 minutes. Finally, the coated films were nitrided under ammonia gas ambient at 950 ℃ for 75 minutes.
Structural properties of the Mg doped GaN thin films were evaluated by High-resolution X-ray diffraction (XRD, PANalytical X'Pert Pro MRD) with a Cu-$k_{\alpha 1}$ radiation source, (λ= 1.5406 Å). Field emission scanning electron microscopy (FESEM, NOVA NANOSEM 450) and the energy dispersive X-ray spectroscopy (EDX, attached to the FESEM system) were used to examine the surface morphology and elemental compositions of deposited films, respectively. Raman spectroscopy (Horiba Jobin Yvon HR800UV) with a 514.5 nm argon (Ar) ion laser as the excitation light source was used to investigate the lattice vibrational properties of films. Lastly, Hall effect system (Accent HL5500PC) was used to measure the resistivity and the hole concentration of the deposited films.

3. Results and discussion

Figure 1 shows the XRD patterns of Mg-doped GaN thin films with different Mg concentrations grown on AlN/Si substrate. A sharp and intense diffraction peak located at 34.58° and a weak diffraction peak found at 73.03° are correspond to the wurtzite structure GaN(002) and GaN(004) (JCPDS file No. 05-0792), respectively. Besides that, two diffraction peaks originated from the Si substrate and the AlN layer can be observed at 28.47° and 35.73°, respectively. From Figure 1, it can be seen that the intensity of GaN(002) diffraction peak increases with increasing Mg concentration. This is an indication of the incorporation of the Mg atoms into Ga sites. Further increase the Mg concentration to 6% has caused the degradation of the GaN(002) diffraction peak intensity. In other words, the crystallinity of the 6% was deteriorated. Generally, the incorporation of the Mg atoms into the Ga sites will induce disorder arrangement of the Ga and N atoms due to the difference in the ionic radii of magnesium and gallium [3]. While the degree of the disorder is depends on the Mg concentrations. It worthy to mention here that the degradation of the GaN(002) diffraction peak intensity of sample with 6% Mg concentration may most probably due to its porous surface morphology, as will be shown later.
Further investigation on the structural properties of the deposited was conducted. The crystallite size of deposited GaN thin films were calculated using Scherrer's formula [15]:

$$D = \frac{k\lambda}{\beta \cos\theta} \quad (1)$$

Here the k is a dimensionless shape factor (k = 0.94), the λ is the wavelength of the X-ray (λ = 1.5407Å), the β is the full width at half maximum (FWHM) of the diffraction peak, and the θ is the diffracting angle in radian. The in-plane stress of Mg doped GaN thin films were estimated by [14]:

$$\sigma_{xx} = \frac{\varepsilon_{zz} B}{\nu} \tag{2}$$

where B is the bulk modulus (B = 200 GPa) and ν is the Poisson ratio (ν = 0.23) [16]. The ε_{zz} is the relative strain of the c-axis in the sample and can be expressed as:

$$\varepsilon_{zz} = \frac{c - c_0}{c_0} \tag{3}$$

Here, c is the calculated lattice parameter and c_0 is unstrained lattice parameter for wurtzite GaN (c_0 = 5.185 Å) [16]. In this work, the GaN(002) diffraction peak information was used for the calculation of the crystallite size and the strain because it has the highest intensity.

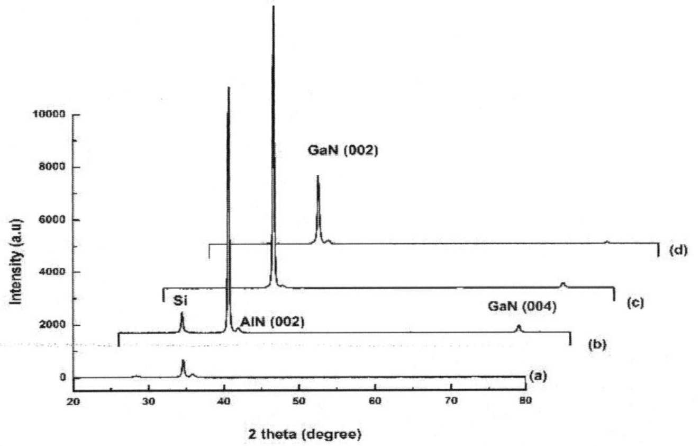

Figure 1. XRD patterns of sol-gel spin coated Mg-doped GaN thin films with different Mg concentrations grown on AlN/ Si: (a) as-grown (0%), (b) 2%, (c) 4 % and (d) 6%.

Table 1 summarizes all the structural parameters extracted from the XRD patterns of the sol-gel spin coated Mg-doped GaN thin films with various Mg dopant concentrations. From Table 1, it was found that films deposited with 2% Mg concentration have the smallest FWHM and biggest crystalline size, however, it has the largest magnitude of the in-plane stress. Note that the positive and negative signs of the in-plane stress indicate the type of tensile and compressive stress, respectively.

TABLE 1. Structural parameters of the sol-gel spin coated Mg-doped GaN thin films with different Mg concentrations extracted from the XRD and Raman results.

Sample	XRD					Raman
	FWHM	Crystallite size (nm)	Relative Strain ($\times 10^{-4}$)	Lattice constant, c ($\times 10^{-10}$ m)	In-plane Strain (GPa)	Stress (GPa)
0%	0.31	46.72	3.857	5.187	0.34	0.24
2%	0.26	55.70	-9.643	5.180	-0.84	-0.73
4%	0.27	53.64	-6.557	5.182	-0.57	-0.49
6%	0.29	49.94	1.929	5.186	0.17	0.24

Figure 2 shows the FESEM microstructure images of the Mg-doped GaN thin films with different Mg concentrations (0, 2, 4 and 6%). It can be seen that all the deposited films exhibit smooth surface morphology and fine granules were well distributed over the surface of the films. As the Mg concentration increases, it was can be observed that the granules size gradually increases and more pits or pores (i.e., black colour dots) were gradually formed on the surface. From Figure 2, it was found that the Mg-doped GaN sample with 2% Mg concentration exhibits the densest and compact surface morphology; while the sample with 6% Mg concentration exhibits porous surface morphology with largest pits/pores.
Table 2 summarizes the elemental compositions of Mg-doped GaN thin films with Mg different concentrations. It can be clearly seen that the Ga and the Mg compositions increase with increasing of Mg doping concentrations. The oxygen is the most commonly found impurity in GaN films grown by sol-gel spin coating approach. Under atmospheric conditions, it was found that the oxidation reaction is more favor as compared to that of the nitridation reaction. Consequently, sol-gel spin coating growth of GaN thin films without oxygen contamination is most challenging task.

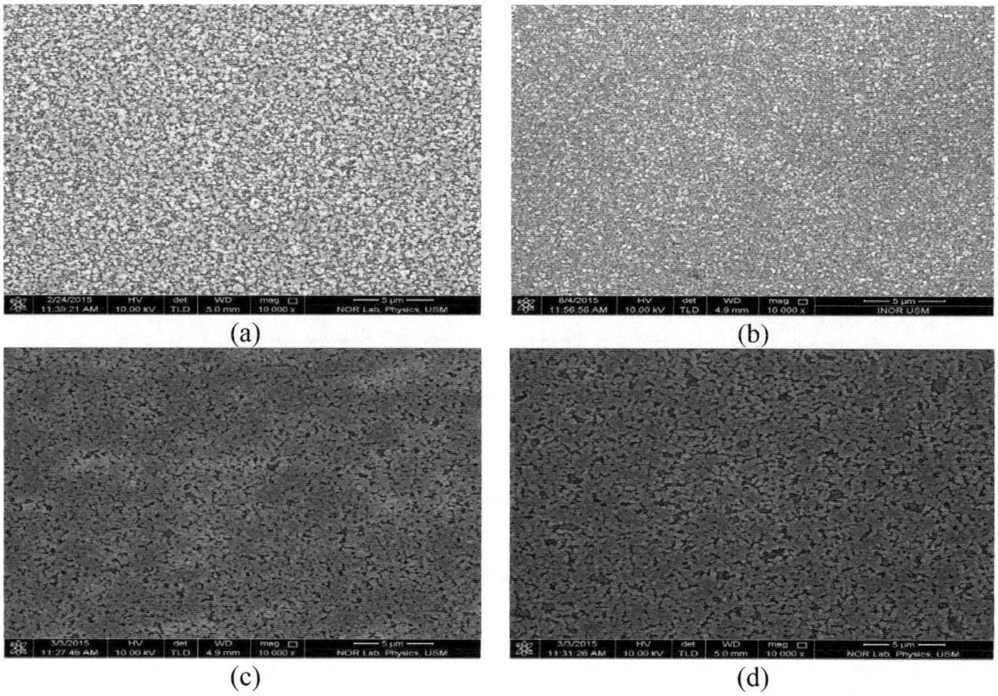

Figure 2. FESEM microstructure images of the Mg-doped GaN thin films with different Mg concentrations: (a) as-grown (0%), (b) 2 %, (c) 4 % and (d) 6%.

TABLE 2. Atomic percentages of Mg-doped GaN thin films grown on AlN/Si substrate.

Sample	N	O	Mg	Al	Si	Ga
As grown	30.08	6.87	-	2.26	38.54	22.25
2%	34.51	2.89	0.35	1.62	37.01	23.62
4%	33.06	2.80	0.63	1.37	27.08	35.06
6%	30.77	4.23	0.92	2.35	41.08	20.66

Figure 3 shows the Raman spectra of Mg-doped GaN thin films with different Mg concentrations deposited on AlN/Si substrate. Under Raman configuration of $z(x,-)\bar{z}$, the allowed Raman active modes for wurtzite structure GaN are the E_2(high) and A_1(LO) phonon modes [17]. From Figure 3, it can be seen that these allowed Raman modes can be detected from all the samples. This result implies that all the deposited GaN thin films have wurtzite structure, i.e., in agreement with the XRD results as discussed previous. For the as-grown sample (Mg concentration = 0%), the E_2(high) and the A_1(LO) phonon modes were observed at 569 cm^{-1} and 730 cm^{-1}, respectively. As the Mg concentration increases, it was found that the E_2(high) phonon mode was shifted towards lower wavenumber. However, for the Mg concentration of 6%, the E_2(high) phonon mode remains unchanged and the underlying cause still remains unknown.

The A_1(LO) phonon mode for all the samples was shifted towards higher wavenumber with increasing Mg concentration. Besides that, the feature of this A_1(LO) phonon mode became broader and weaker as the Mg concentration increases. The behavior of the A_1(LO) phonon mode as a function of the Mg concentration can be explained by the coupling effects of the longitudinal optical phonon and plasmon. Apart from the Raman active modes, all the Mg-doped samples also exhibit a broad peak centered at 669 cm^{-1}. This broad peak is corresponds to the local vibrational mode (LVM) induced by the Mg dopant [18]. The presence of this LVM is a good indication of the incorporation of the Mg atoms into the Ga sites. Note that the strong Raman peak located at 522 cm^{-1} is originated from the Si substrates. It is known that the shift of Raman E_2(high) peak can be directly related to the residual stress in the deposited films. Typically, the E_2(high) peak of an unstressed GaN crystal is located at 568 cm^{-1} [19]. It is reported that the presence of 1 GPa of compressive stress will induce a shift of 4.12 cm^{-1} on the E_2(high) peak [20]. Based on these information, the residual stress in the Mg-doped GaN thin films were estimated and the obtained results were tabulated in Table 1. Overall, it was found that the residual stresses estimated Raman results were comparable to that extracted from the XRD patterns.

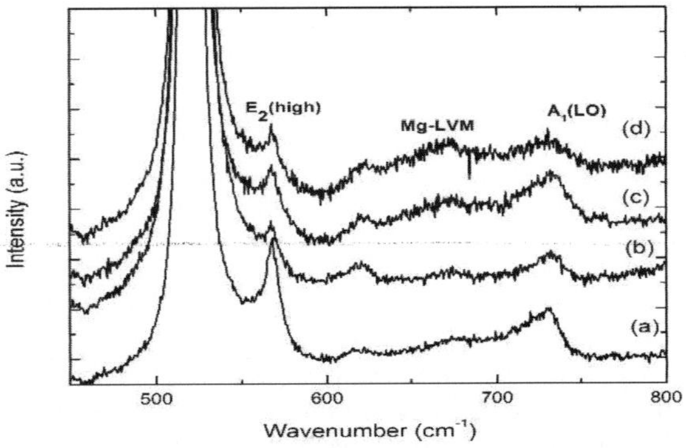

Figure 3. Raman spectra of the Mg-doped GaN thin films deposited on AlN/Si substrate with different Mg concentrations: (a) as-grown (0%), (b) 2%, (c) 4% and (d) 6%.

Figures 4 (a), (b) and (c) show the resistivity, hole concentration and Hall mobility as a function of the Mg concentration of Mg doped GaN thin films on AlN/Si substrate, respectively. As the Mg concentration increases, it was found that the resistivity of the thin films decreases; while the hole concentration and hall mobility of thin films increases. The decrease resistivity with increase in Mg concentration can be attributed due to increase in carrier concentration and also due to increase in mobility, which is seen in the figure. The results indicate that the number of the Mg atoms incorporated into the GaN sites increases with increasing Mg concentration [21].

4. Conclusions

In summary, Mg-doped GaN thin films with different Mg concentrations were successfully grown by low-cost and simple sol-gel spin coating method. The effects of the Mg concentration on characteristics of the deposited films were investigated. All the deposited films exhibit smooth surface morphology and crystalline in wurtzite structure with GaN(002) preferred orientation. The XRD results reveal that the Mg-doped samples have better crystallinity than that the as-grown (undoped) sample. As the Mg concentration increases from 2% to 6%, the crystallite size decreases and the crystalline quality of the films degraded. In addition, pits/pores were gradually formed on the surface of the films with increasing Mg concentration. The crystalline structure of the deposited films was again confirmed by the Raman measurements where all the allowed Raman active phonon modes of the wurtzite GaN were detected. Apart from that, a feature attributed to LVM was observed at 669 cm^{-1}. The resistivity and hole concentration of the deposited films were determined by Hall effect measurements. The results show that the resistivity of the thin films decreases while the hole concentration and hall mobility of thin films increases with increasing concentration of the Mg. For sample with 6% Mg concentration, the hole concentration around 1.36×10^{19} cm^{-3} was obtained.

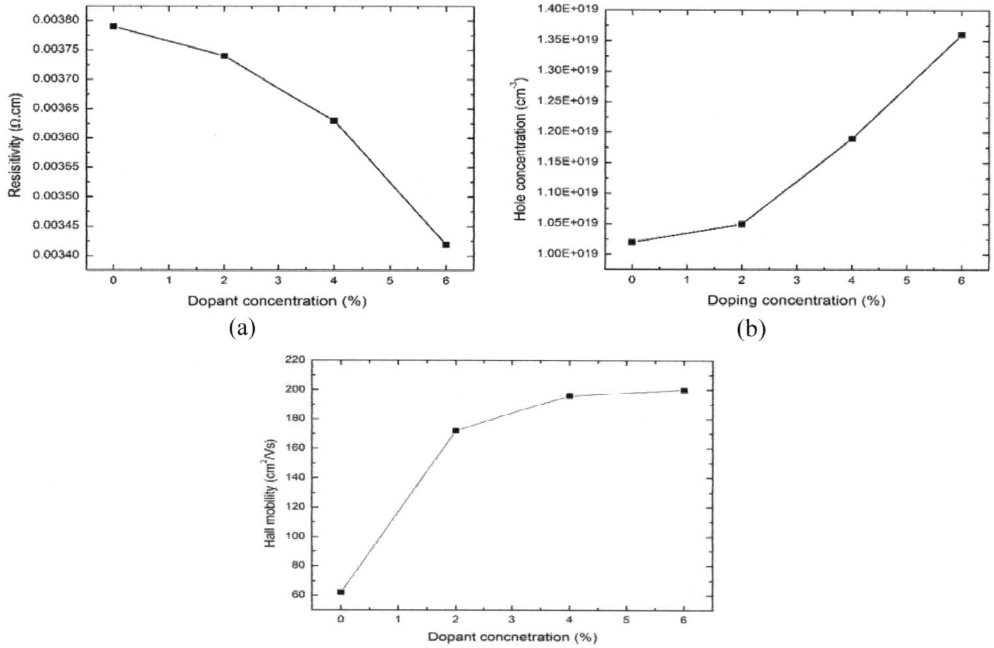

(c)

Figure 4. (a) Resistivity versus dopant concentration, (b) Hole concentration versus dopant concentration, and (c) Hall mobility versus dopant concentration of Mg doped GaN thin films on AlN/Si substrate.

5. References

[1] El-Naggar A M 2012 *J. Mater. Sci: Mater. Electron.* **23** 972
[2] Park C I, Kang J H , Kim K C, Nahm K S, . Suh E -K, Lim K Y 2001 *Thin Solid Films* **401** 60
[3] Ke W C, Jian S R, Chen I C, Jang J S C, Chen W K and Juang J Y 2012 *Mater. Chem. Phys.* **13** 796
[4] Kwon M K, Park I K, Kim J Y, Kim J O, Kim B. and Park S J 2007 *IEEE Photonic. Tech. L.* **19** 1880
[5] Saxler A, Perrin R, Elhamri S and Mitchel W C 2000 *J. Appl. Phys.* **87** 1832
[6] Chuah L S, Hassan Z, Ng S S and Abu Hassan H 2007 *J. Mater. Res.* **22** 2623
[7] Nakamura S,Mukai T, Senoh M and Iwasa N 1992 *Jpn. J. Appl. Phys.* **31** 139
[8] Amano H, Kito M, Hiramatsu K and Akasaki I 1989 *Jpn. J. Appl. Phys.* **28** 2112
[9] Chang S J and Su Y K 2001 *Appl. Phys. Lett.* **78** 312
[10] Shankar S, Saroja M, Venkatachalam M, Balachander M and Kuma V 2014 *J. Nanosci. Nanotechnol.* **2** 769
[11] Nayak P K,Yang J H, Kim J W, Chung S J, Jeong J W,Lee C H and Hong Y T 2009 *J. Phys. D: Appl. Phys.* **42** 035102
[12] Fong C Y, Ng S S, Yam F K, Abu Hassan H and Hassan Z 2013 *J. Sol-Gel Sci. Technol.* **68** 95
[13] Fong C Y, Ng S S, Yam F K, Abu Hassan H and Hassan Z 2015 *Acta Metall. Sin. English Lett.* **28** 362

[14] Fong C Y, Ng S S, Yam F K, Abu Hassan H and Hassan Z 2014 *Mater. Sci. Semicond. Process.* **17**, 63
[15] Kawwama M and Lebbou K 2014 *Appl. Surf. Sci.* **292** 906
[16] Fong C Y, Ng S S, Yam F K, Abu Hassan H and Hassan Z 2015 *Vacuum* **119** 120
[17] Hushur A, Manghnani M H, and Narayan J 2009 *J. Appl. Phys.* **106** 054317
[18] Kirste R, Hoffmann M P, Tweedie J, Bryan Z, Callsen G, Kure T, Nenstiel C, Wagner M R, Collazo R, Hoffmann A, and Sitar Z 2013 *J. Appl. Phys.* **113** 103504
[19] Perlin P, Jauberthie-carillon C, Itic J P, Miguel A S, and Polian A 1992 *Phys. Rev. B* **45** 83
[20] Vajpeyi A P, Chua S J, Tripathy S, and Fitzgerald E, *Electrochem. Solid-State Lett.* **8** 2
[21] Sui Y P and Yu G H 2011 *Chin. Phys. Lett.* **28** 067807 1

Acknowledgments
This project was supported by the Ministry of Science, Technology and Innovation (MOSTI) Malaysia through the Science Fund (Project No.: 03-01-05-SF0750) and the Nippon Sheet Glass Foundation for Materials Science and Engineering, Japan. We would also like to thank USM for support and encouragement.